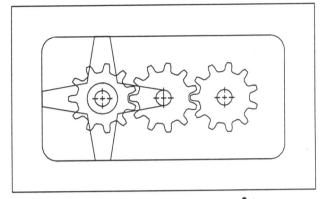

The AutoCAD® Tutor for Engineering Graphics

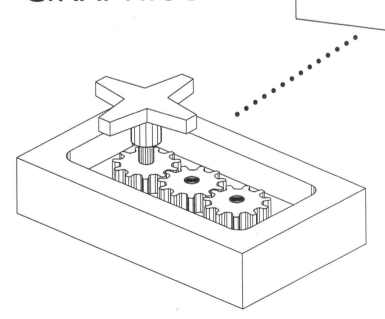

Alan J. Kalameja

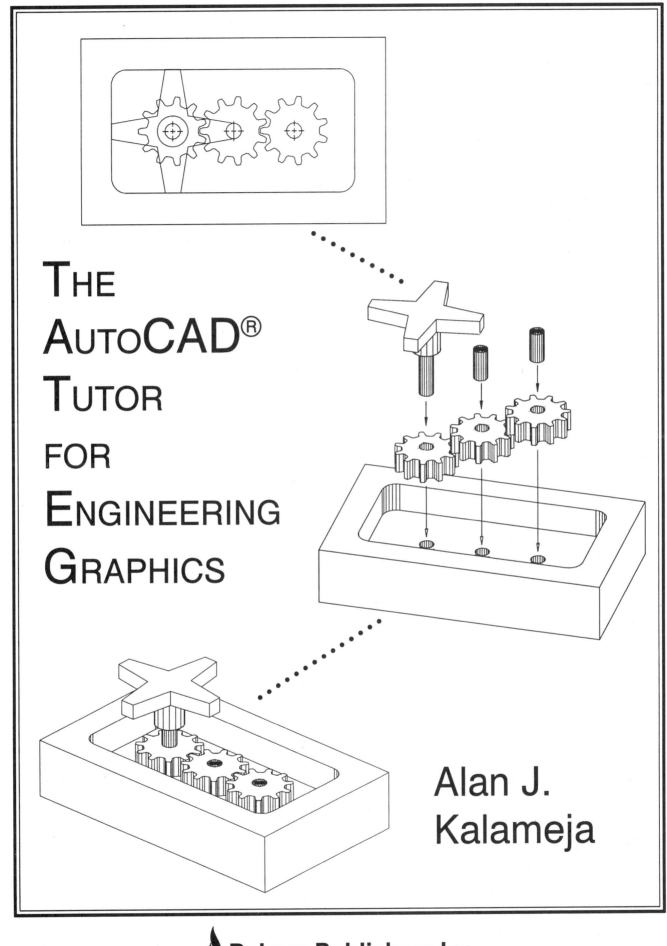

The AutoCAD® Tutor for Engineering Graphics

Alan J. Kalameja

 Delmar Publishers Inc.

NOTICE TO THE READER

Several figures used in the problems at the end of the units are from TECHNICAL DRAWING by Goetsch, Nelson and Chalk, Second Edition, © 1989 by Delmar Publishers Inc.

Cover design by Nancy Gworek.
Cover illustration by Ron Young.

Delmar Staff:
Associate Editor: Kevin Johnson
Developmental Editor: Mary Beth Ray
Project Editor: Judith Boyd Nelson
Production Supervisor: Wendy Troeger
Design Supervisor: Susan Mathews

For information, address Delmar Publishers Inc.
3 Columbia Circle, PO Box 15015
Albany, New York 12212–5015

Printed in the United States of America
Published simultaneously in Canada
by Nelson Canada,
a division of The Thomson Corporation

10 9 8 7 6 5 4 3

Library of Congress Cataloging-in-Publication Data

Kalameja, Alan J.
     The AutoCAD tutor for engineering graphics / Alan J. Kalameja.
          p. cm.
     Includes index.
     ISBN: 0-8273-5081-3
     1. AutoCAD (Computer program) 2. Engineering graphics--Computer
programs. I. Title
T385.K345 1992
620'.0042'02855369--dc20   91-46744
                                                                           CIP

Contents

Preface

Engineering graphics has been around for a long time as a means of defining an object graphically before it is constructed and used by consumers. Previously, the process for producing the drawing involved drawing aids such as pencils, ink pens, triangles, t-squares, etc. to place the idea on paper before making changes and producing blue-line prints for distribution. The ability to produce these drawings on a computer is quite new; however the principles and basics of engineering drawing remain the same.

This text uses engineering drawing basics to produce drawings using AutoCAD and a series of tutorial problems that follow each unit. In some cases, the tutorials may be performed using AutoCAD Release 10; in other cases, AutoCAD Release 11. This is spelled out at the beginning of each tutorial. Following the tutorials, extra problems are provided to add to your skills. A brief description of each unit follows:

Unit 1

This unit provides a brief review on such topics as absolute, relative, and polar coordinates. The use of the Array command is also covered in addition to productive uses of the Offset command especially when used with polylines. A series of tutorials follow used to complement the topics covered in the unit. Additional problems related to the tutorial exercises follow.

Unit 2

This unit discusses how AutoCAD commands and command options may be used for geometric constructions. Two tutorial exercises follow along with additional problems at the end of the chapter.

Unit 3

Shape description and multi-view drawing using AutoCAD is the focus of this unit. The basics of shape description is discussed along with proper use of linetypes, fillets and rounds, chamfers, and runouts. One tutorial follows outlining the steps used for creating a multi-view drawing using AutoCAD. Another tutorial approaches multi-view drawing through the use of .XYZ filters. Additional problems are provided at the end of the unit.

Unit 4

Dimensioning techniques using AutoCAD is the topic of this unit. Basic dimensioning rules are discussed before concentrating on all AutoCAD dimensioning options and dimension variables. Three tutorials follow along with additional dimensioning problems.

Unit 5

Section views are described in this unit including full, half, assembly, aligned, offset, broken, revolved, removed, and isometric sections. Hatching techniques in AutoCAD are also discussed. Two tutorials follow, along with additional problems.

Unit 6

Producing auxiliary views using AutoCAD is discussed in this unit. Tutorials and extra problems follow a discussion on auxiliary view basics.

Unit 7

This unit discusses constructing isometric drawings with particular emphasis on using the Snap-Style option of AutoCAD. In addition to isometric basics, creating circles and angles in isometric are also discussed. Tutorial exercises follow, along with additional problems.

Unit 8

This unit begins the study of three-dimensional modeling with the creation of wireframe and surfaced drawings. A brief comparison on the need for three-dimensional drawings is discussed along with the 3dface, ruled surface, tabulated surface, edge surface, and revolved surface commands. A series of tutorials follow along with additional problems to be drawn as three-dimensionsal wireframe and surfaced drawings.

Unit 9

This unit focuses on using a wireframe drawing to project a front, top, and right side view using the Project.Lsp AutoLISP routine. An isometric view will also be projected. A tutorial follows a discussion of this process in addition to other problems at the end of the unit.

Unit 10

This unit begins the topic of solids modeling and how it compares to wireframe modeling. The Autodesk Advanced Modeling Extension, (AME), will be discussed along with basic creation of solid primitives. Subtraction, union, and intersection operations will be demonstrated for further modeling creation. Methods of displaying solid models will also be discussed. Tutorial exercises follow in addition to other problems at the end of the unit.

A disk is provided with this text to supplement some of the tutorials. Also, all wireframe problems at the end of Unit 9 are complete on disk. Follow the steps to project the views from the wireframe. Each tutorial will alert you to either begin a new drawing or call up an existing drawing provided on the disk and proceed with the remainder of tutorial.

Acknowledgments

I wish to thank the staff at Delmar for their assistance with this document, especially Michael McDermott, Kevin Johnson, and Judith Boyd Nelson. Special thanks go out to Barbara Savins who assisted with the problems at the end of the Shape Description/Multi-View Projection Chapter. I also wish to thank Roy Baker for testing a majority of the 3D tutorial exercises. Finally, I wish to thank my family for their patience during the development of this project.

Alan J. Kalameja

Conventions

All tutorials in this publication use the following conventions in the instructions:

Whenever you are told to enter text, the text appears in **boldface** type. This may take the form of entering an AutoCAD command or entering such information as absolute, relative or polar coordinates. You must follow these and all text inputs by striking the Return or Enter key to execute the input. For example, to draw a line using the Line command from coordinate value (3,1) to coordinate value (8,2), the sequence would look line the following:

Command: **Line**
From point: **3,1**
To point: **8,2**
To point: *(Strike Enter to exit this command)*

Instructions in this tutorial are designed to enter all commands, options, coordinates, etc., from the keyboard. You may enter the same commands by selecting them from the screen menu, digitizing tablet, or pull-down menu area.

Instructions for selecting objects are in italic type. When instructed to select an object, move the pickbox on the object to be selected and press the pick button on the mouse or digitizing puck.

If you enter the wrong command for a particular step, you may cancel the command by holding down the "Ctrl" key and typing the letter "C". The translation for this procedure is commonly called "Control-C" and stands for "Cancel". Use this procedure to cancel any command in operation. Cancels can also be found in the pull-down menu area under Assist, on the digitizing tablet under the selection set modes, or on one button if using a 4-button puck.

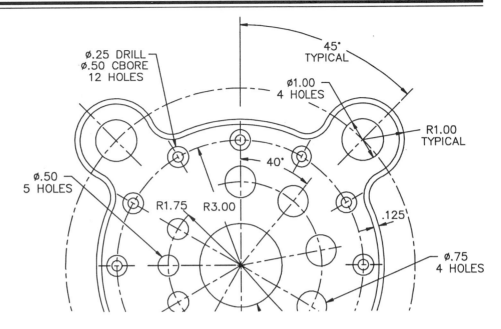

UNIT 1

One-View Drawings

Contents

This unit begins the study of one-view drawings and how they might be constructed. One-view drawings are usually less complicated than multiple-view drawings or section views. Typical examples of these drawings include such items as automotive gaskets and thin sheet metal parts where the thickness of these items are too thin to represent in a drawing form. To assist you in the creation of these and other drawings, a brief review of absolute, relative, and polar coordinates will be given to emphasize the importance of their use. Tutorial exercises follow the brief explanations. Each tutorial explains system preparation, suggested command usage, and plotting information before proceding with the main body of the tutorial. At the end of the unit are problems related to all tutorial problems. It is the hope that the tutorials will guide you to the successful completion of these problems.

Cartesian Coordinates

Before drawing precision geometry such as lines and circles, an understanding of coordinate systems must first be made. The Cartesian, or rectangular, coordinate system is used to place geometry at exact distances through a series of coordinates. A coordinate is made up of an ordered pair of numbers usually identified as X and Y. The coordinates are then plotted on a type of graph or chart. The graph, illustrated at the right, is made up of two perpendicular number lines called coordinate axes. The horizontal axis is called the X axis. The vertical axis is called the Y axis.

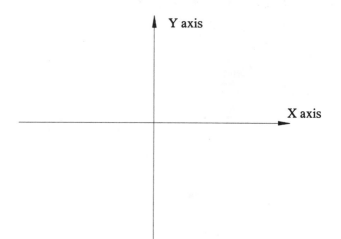

From the illustration at the right, the intersection of the two coordinate axes forms a point called the origin. Coordinates used to describe the origin are 0,0. From the origin, all positive directions move up and to the right. All negative directions move down and to the left.

The coordinate axes are divided into four quadrants that are labeled I, II, III, and IV. In quadrant I, all X and Y values are positive. Quadrant II has a negative X value and positive Y value. Quadrant III has negative values for X and Y. Quadrant IV has a positive X value and a negative Y value.

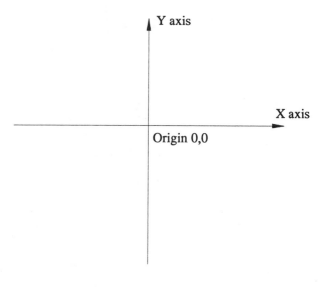

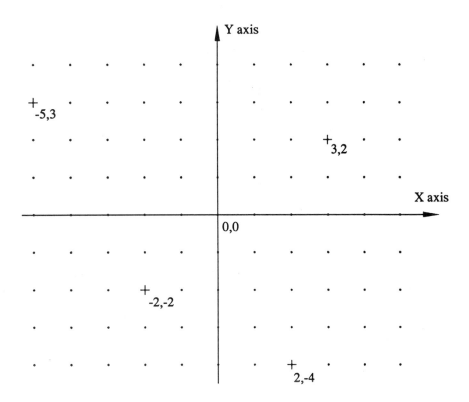

For each ordered pair of (x,y) coordinates, X means to move from the origin to the right if positive and to the left if negative. Y means to move from the origin up if positive and down if negative. The illustration above displays a series of coordinates plotted on the number lines. One coordinate is identified in each quadrant to show the positive and negative values. As an example, coordinate 3,2 located in the quadrant I means to move 3 units to the right of the origin and up 2 units. The coordinate -5,3 located in quadrant II means to move 5 units to the left of the origin and up 3 units. Coordinate -2,-2 located in quadrant III means to move 2 units to the left of the origin and down 2. Lastly, coordinate 2,-4 located in quadrant IV means to move 2 units to the right of the origin and down 4.

When beginning a drawing in AutoCAD, the screen display reflects quadrant I of the Cartesian coordinate system. The origin 0,0 is located in the lower left corner of the drawing screen. The current screen size is measured by the upper right coordinate of the screen which is, by default, 12,9. This value can be changed using the Limits command to accommodate any drawing including architectural and civil engineering.

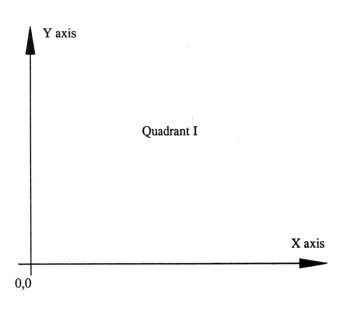

Absolute Coordinates

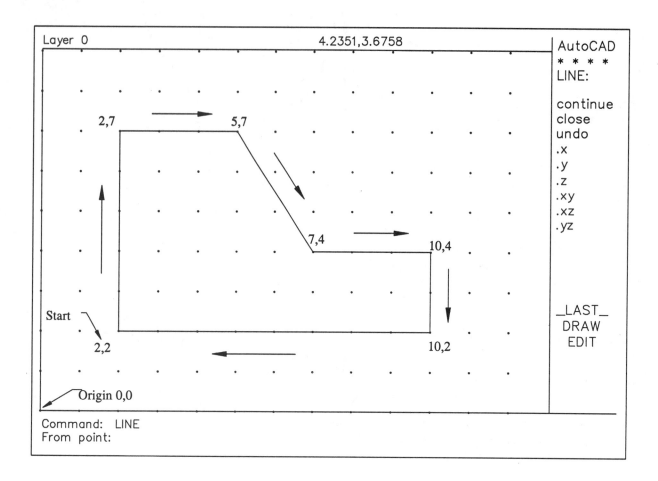

When drawing geometry such as lines, a method of entering precise distances must be used especially when accuracy is important. This is the main purpose of using coordinates. The simplest and most elementary form of coordinate values are absolute coordinates. Absolute coordinates conform to the following format:

$$x,y$$

The one problem with using absolute coordinates is that all coordinate values refer back to the origin 0,0. This origin on the AutoCAD screen is usually located in the lower left corner of a new drawing. The origin will remain in this corner unless it is altered using the Limits command.

Study the Line command prompts below along with the illustration above:

Command: **Line**
From point: **2,2**
To point: **2,7**
To point: **5,7**
To point: **7,4**
To point: **10,4**
To point: **10,2**
To point: **C**

As you can see, all points on the object make reference to the origin at 0,0. Even though absolute coordinates are useful in starting lines, there are more efficient ways to continue lines and draw objects.

Relative Coordinates

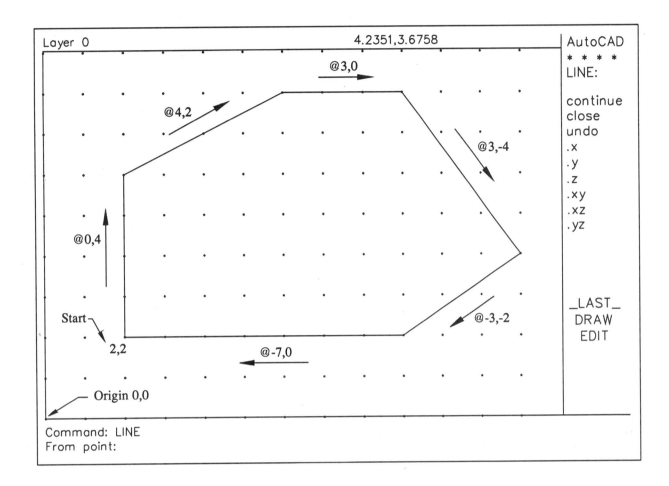

In absolute coordinates, the origin at 0,0 must be kept track of at all times in order to enter the correct coordinate. With complicated objects this is sometimes difficult to accomplish and, as a result, the wrong coordinate is entered. It is possible to reset the last coordinate to become a new origin or 0,0 point. The new point would be relative to the previous point and for this reason this point is called a relative coordinate. The format is as follows:

$$@x,y$$

In the format above, we use the save X and Y values with one exception; the "at" symbol or @ resets the previous point to 0,0 and makes entering coordinates less confusing.

Study the Line command prompts below along with the illustration above:

        Command: **Line**
        From point: **2,2**
        To point: **@0,4**
        To point: **@4,2**
        To point: **@3,0**
        To point: **@3,-4**
        To point: **@-3,-2**
        To point: **@-7,0**
        To point: *(Strike Enter to exit this command)*

In each example above, the @ option resets the previous point to 0,0.

Polar Coordinates

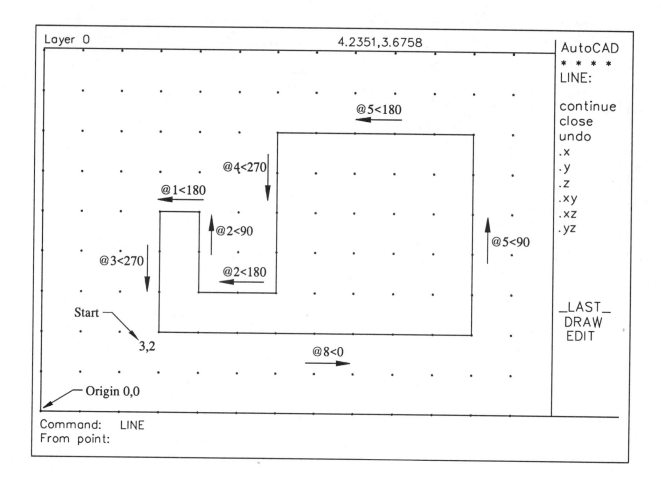

Another popular method of entering coordinates is by polar coordinates. The format is as follows:

@Distance<Direction

As the format above implies, the polar coordinate mode requires a known distance and a direction. The @ symbol resets the previous point to 0,0. The direction is preceded by the < symbol which reads the next number as a polar direction. Above is an illustration describing the directions supported by the polar mode.

Study the Line command prompts below along with the illustration above for the polar coordinate mode:

Command: **Line**
From point: **3,2**
To point: **@8<0**
To point: **@5<90**
To point: **@5<180**
To point: **@4<270**
To point: **@2<180**
To point: **@2<90**
To point: **@1<180**
To point: **@3<270**
To point: *(Strike Enter to exit this command)*

Combining Coordinates

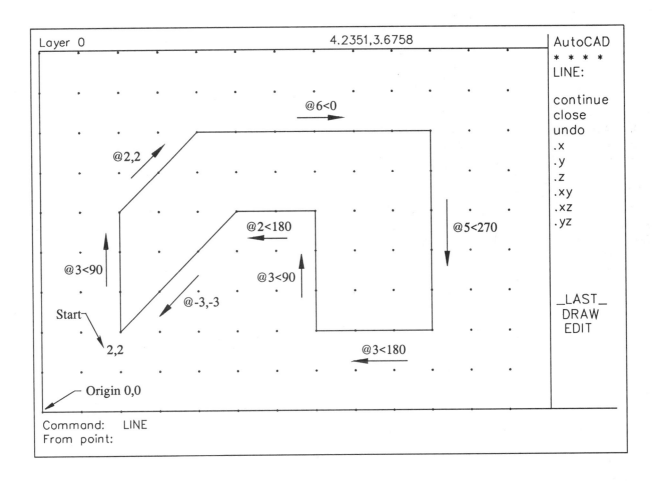

So far, the past three pages concentrated on using each example of coordinate modes, (absolute, relative, and polar), to create geometry. At this point, we do not want to give the impression that once you start with a particular coordinate mode that you must stay with the mode. Rather, drawings are created using one, two, or three coordinate modes in combination with each other. In the illustration above, the drawing starts with an absolute coordinate, changes to a polar coordinate, and changes again to a relative coordinate. It is the responsibility of the CAD operator to choose the most efficient coordinate mode to fit the drawing.

Study the Line command prompts below along with the illustration above:

Command: **Line**
From point: **2,2**
To point: **@3<90**
To point: **@2,2**
To point: **@6<0**
To point: **@5<270**
To point: **@3<180**
To point: **@3<90**
To point: **@2<180**
To point: **@-3,-3**
To point: (*Strike Enter to exit this command*)

Forming Bolt Holes - Method #1

Multiple copies of entities such as bolt holes are easily duplicated in circular patterns. Use the Array command to make multiple arrayed copies of the small circles in the illustration at the right, along with the following prompts:

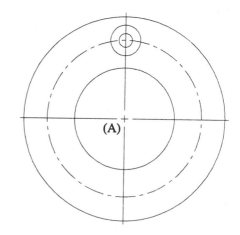

Command: **Array**
Select objects: *(Select both small circles)*
Select objects: *(Strike Enter to continue)*
Rectangular or Polar array (R/P): **P**
Center point of array: **Int**
of *(Select the intersection at "A")*
Number of items: **8**
Angle to fill (+=ccw, -=cw)<360>: *(Strike Enter for default)*
Rotate objects as they are copied? <Y>: *(Strike Enter for default)*

As the circles are copied in the circular pattern, center lines need to be updated to the new bolt hole positions. Again, the Array command is used to copy and duplicate one center line over a 45-degree angle to fill in the counter-clockwise direction as illustrated at the right.

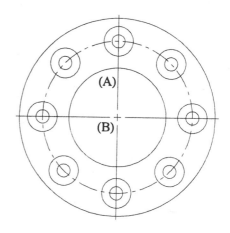

Command: **Array**
Select objects: *(Select the vertical center line at "A")*
Select objects: *(Strike Enter to continue)*
Rectangular or Polar array (R/P): **P**
Center point of array: **Int**
of *(Select the intersection at "B")*
Number of items: **2**
Angle to fill (+=ccw, -=cw)<360>: **45**
Rotate objects as they are copied? <Y>: *(Strike Enter for default)*

Now use the Array command to copy and duplicate the last center line and mark the remaining bolt hole circles as shown in the illustrations at the right.

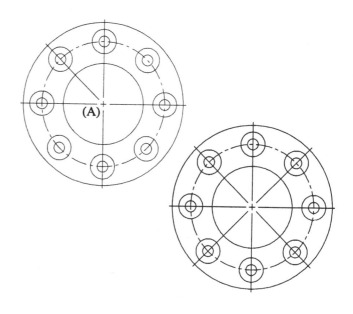

Command: **Array**
Select objects: **L** *(This should select the last line)*
Select objects: *(Strike Enter to continue)*
Rectangular or Polar array (R/P): **P**
Center point of array: **Int**
of *(Select the intersection at "A")*
Number of items: **4**
Angle to fill (+=ccw, -=cw)<360>: *(Strike Enter for default)*
Rotate objects as they are copied? <Y>: *(Strike Enter for default)*

Forming Bolt Holes - Method #2

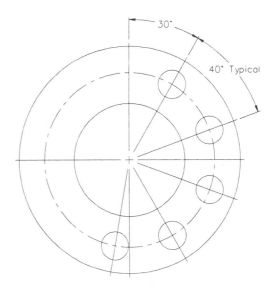

From the previous example, we have seen how easy the Array command can be used for making multiple copies of entities equally spaced around an entire circle. What if the entities are copied only partially around a circle such as the bolt holes in the illustration at the right? The Array command is used here to copy the bolt holes in 40-degree increments. Follow the series of prompts below to perform this operation.

First, use the Array command to rotate and copy the vertical center line at a -30-degree angle to fill. The -30 degrees will copy and rotate the center line in the clockwise direction. Next, place the circle at the intersection of the center lines using the Circle command. Use the Array command, select the circle and centerline, and copy the selected entities at 40-degree increments using the prompt sequence below:

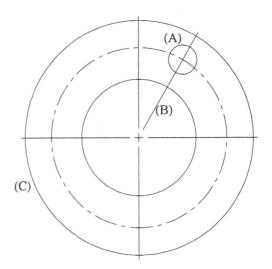

Command: **Array**
Select objects: *(Select the small circle at "A" and line at "B")*
Select objects: *(Strike Enter to continue)*
Rectangular or Polar array (R/P): **P**
Center point of array: **Cen**
of *(Select the large circle anywhere near "C")*
Number of items: *(Strike Enter to continue)*
Angle to fill (+=ccw, -=cw): **-160**
Angle between items: **-40** *(To copy 40 degrees clockwise)*
Rotate objects as they are copied? <Y>: *(Strike Enter for default)*

The result is illustrated at the right. This method of identifying bolt holes shows how the number can be controlled by specifying the total angle to fill and the angle between items. In both angle specifications, a negative value is entered to force the array to be performed in the clockwise direction.

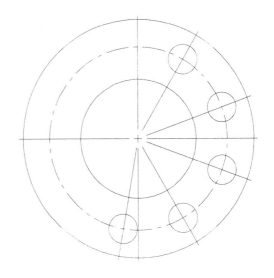

Forming Bolt Holes - Method #3

The object illustrated at the right is similar to the previous example. A new series of holes are to be placed 20 degrees away from each other using the Array command.

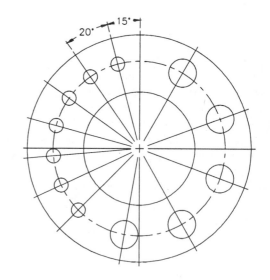

Begin placing the holes by laying out one center line using the Array command with an angle of 15 degrees to fill. Add one circle using the Circle command at the intersection of the center lines. Use the prompts below to add the remaining holes:

Command: **Array**
Select objects: *(Select the small circle at "A" and line at "B")*
Select objects: *(Strike Enter to continue)*
Rectangular or Polar array (R/P): **P**
Center point of array: **Cen**
of *(Select the large circle anywhere near "C")*
Number of items: *(Strike Enter to continue with this command)*
Angle to fill (+=ccw, -=cw)<360>: **120**
Angle between items: **20**
Rotate objects as they are copied? <Y>: *(Strike Enter for default)*

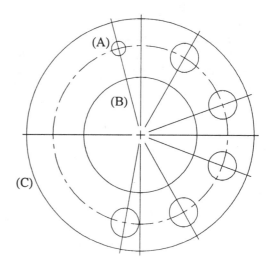

Since the direction of rotation for the array is in the counter-clockwise direction, all angles are specified in positive values.

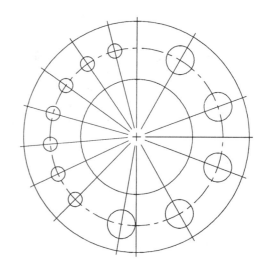

Offsetting Entities - Method #1

Sometimes entities need to be duplicated at a set distance away from existing geometry. The Copy command could be used for this operation, however, a better command would be Offset. This allows the user to specify a distance and a side for the offset to occur. The result is an entity parallel to the original entity at a specified distance. All entities in the illustration at the right need to be offset 0.50 toward the inside of the original object. Follow the prompt sequences below:

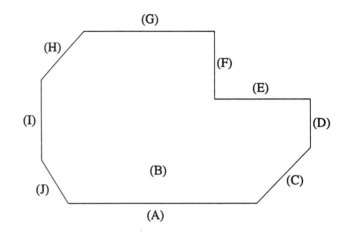

Command: **Offset**
Offset distance or Through <Through>: **0.50**
Select object to offset: *(Select the horizontal line at "A")*
Side to offset? *(Mark a point anywhere on the inside near "B")*

Repeat the above procedure for lines *"C"* through *"J"*.

Notice that when all lines were offset, as in the illustration at the right, the entire original lengths of all line segments were maintained. Since all offsetting occurred inside, this resulted in segments that overlap at their intersection points. In one case, at "A" and "B," the lines did not even meet at all. The Chamfer command can be used to edit all lines to form a sharp corner. This is accomplished by assigning a value of 0.00 for the chamfer distances; this is the default value for this command.

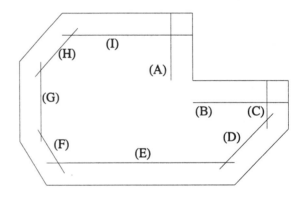

Command: **Chamfer**
Polyline/Distances/<Select first line>: *(Select the line at "A")*
Select second line: *(Select the line at "B")*

Repeat the above procedure for lines *"A"* through *"I"*.

Using the Offset command along with the Chamfer command produces the result illustrated at the right. The Chamfer command must be set to a value of 0 for this special effect. The Fillet command performs the same result when set to a radius value of 0.

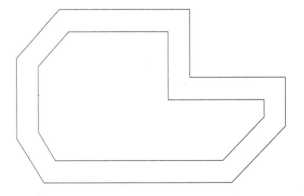

Offsetting Entities - Method #2

The object illustrated at the right is identical to the previous example. Also in the previous example, each individual line had to be offset to copy the lines parallel at a specified distance. Then the Chamfer command had to be used to clean up the corners. There is an easier way to perform this operation. First, convert all individual line segments into one polyline using the Pedit command.

Command: **Pedit**
Select polyline: *(Select the line at "A")*
Entity selected is not a polyline.
Do you want it to turn into one? **Yes**
Close/Join/Width/Edit vertex/Fit curve/Spline curve/
    Decurve/Undo/eXit <X>: **J**
Select objects: *(Select lines "B" through "I")*
8 lines added to polyline.
Close/Join/Width/Edit vertex/Fit curve/Spline curve/
    Decurve/Undo/eXit <X>:   *(Strike Enter to exit this command)*

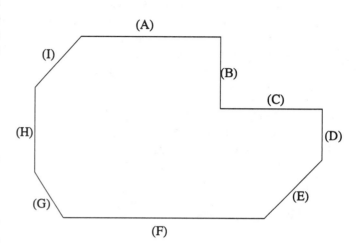

The Offset command is used to copy the shape 0.50 units on the inside. Since the object was converted into a polyline, all entities are offset at the same time. This procedure bypasses the need to use the Chamfer or Fillet commands to corner all intersections.

Command: **Offset**
Offset distance or Through <Through>: **0.50**
Select object to offset: *(Select the polyline at "A")*
Side to offset? *(Select a point anywhere near "B")*
Select object to offset: *(Strike Enter to exit this command)*

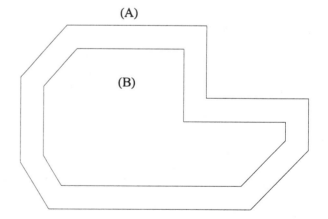

Tutorial Exercise #1
Template.Dwg

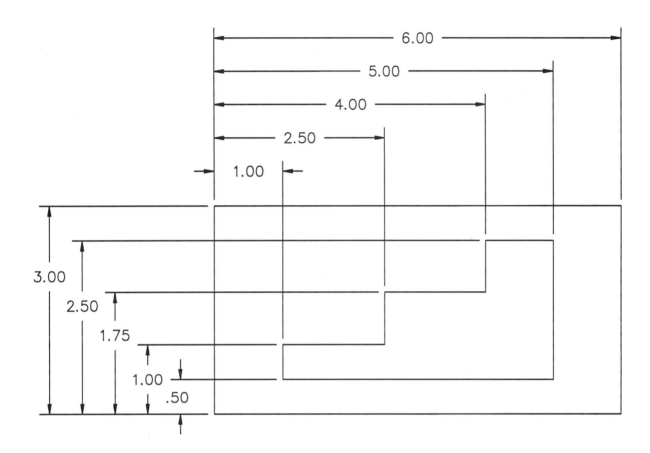

PURPOSE:
This tutorial is designed to allow the user to construct a one view drawing of the Template using the absolute, relative, and polar coordinate modes.

SYSTEM SETTINGS:
Begin a new drawing called "Template." Use the Units command to change the number of decimal places past the zero from 4 to 2. Keep the remaining default unit values. Using the Limits command, keep 0,0 for the lower left corner and change the upper right corner from 12,9 to 10.50,8.00. Use the Grid command to change the grid spacing from 1.00 to 0.25 units. Do not turn the Snap or Ortho on.

LAYERS:
Create the following layers with the format:

Name-Color-Linetype
Object - White - Continuous

SUGGESTED COMMANDS:
The Line command will be used entirely for this tutorial in addition to a combination of coordinate systems. The Erase command could be used although a more elaborate method of correcting a mistake would be to use the Line-undo command to erase a previously drawn line and still stay in the Line command.

DIMENSIONING:
Dimensions can be added to this problem at a later time. Consult your instructor.

PLOTTING:
This tutorial exercise may be plotted on "A"-size paper (8.5" x 11"). Use a plotting scale of 1=1 to produce a full size plot.

VERSION OF AUTOCAD:
This tutorial exercise may be completed using either AutoCAD Release 10 or Release 11.

Step #1

Begin this tutorial exercise with the Line command and draw the outer perimeter of the box. Use an absolute coordinate point followed by polar coordinates as illustrated at the right.

Command: **Line**
From point: **2.00,2.00**
To point: **@6.00<0**
To point: **@3.00<90**
To point: **@6.00<180**
To point: **@3.00<270**
To point: *(Strike Enter to exit this command)*

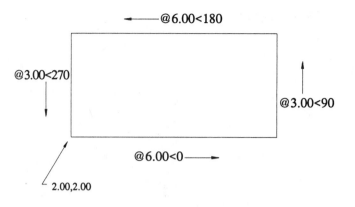

Step #2

The next step will be to draw the stair step outline of the template using the Line command again. However, we first need to identify the starting point of the template. Absolute coordinates could be calculated but in more complex objects this would be difficult. A more efficient method would be to use the Line command again and at the "From point" prompt, enter a relative coordinate. This would position the new point at a specified distance from the previously used point as illustrated at the right.

Command: **Line**
From point: **@1.00,0.50**

This relative coordinate begins a new line a distance of 1 unit in the X direction and 0.50 units in the Y direction from the last point.

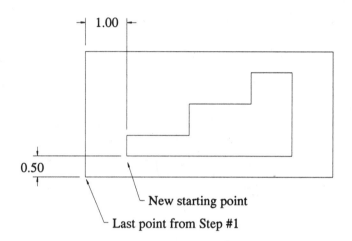

Step #3

Once the new starting point has been identified by the relative coordinate @1.00,0.50, continue with the Line command to complete the inner part of the template by using the polar coordinate mode as illustrated at the right.

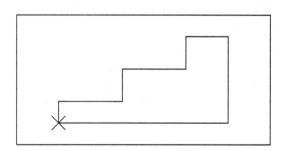

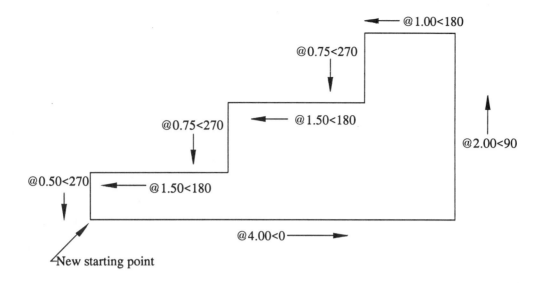

Step #4

Continue using the Line command and polar coordinate mode to complete the inner part of the template as illustrated above.

Command: **Line**
From point: **@1.00,0.50**
To point: **@4.00<0**
To point: **@2.00<90**
To point: **@1.00<180**
To point: **@0.75<270**
To point: **@1.50<180**
To point: **@0.75<270**
To point **@1.50<180**
To point **@0.50<270**
To point: *(Strike Enter to exit this command)*

Step #5

The completed problem is illustrated at the right. Dimensions may be added at a later time upon the request of your instructor.

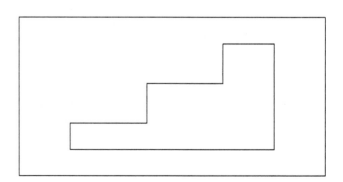

Tutorial Exercise #2
Tile.Dwg

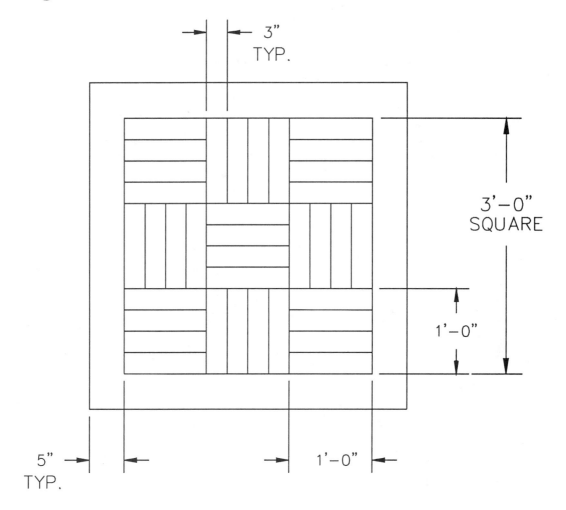

PURPOSE:
This tutorial is designed to use the Offset and Trim commands to complete the drawing of the Floor Tile.

SYSTEM SETTINGS:
Begin a new drawing called "Tile." Use the Units command to change the units of measure from decimal to architectural. Keep the remaining default unit values. Using the Limits command, keep 0,0 for the lower left corner and change the upper right corner to 15'6,9'6. Use the Snap command and change the value from 1" to 4" (with the Grid set to zero by default, the snap setting of 4" will also change the grid spacing to 4"). Turn the snap Off.

LAYERS:
Create the following layers with the format:

Name-Color-Linetype
Object - White - Continuous

SUGGESTED COMMANDS:
The Line command will be used to begin the Tile. The Offset command is used to copy selected line segments at a specified distance. The Trim command is then used to clean up intersecting corners. The Erase command can be used to delete entities from the drawing. (Remember to use the Oops command to bring back previously erased entities deleted by mistake.)

DIMENSIONING:
Dimensions may be added to this problem at a later time. Consult your instructor.

PLOTTING:
This tutorial exercise may be plotted on "B"-size paper (11" x 17"). Use a plotting scale of 1=12 to produce a scaled plot.

VERSION OF AUTOCAD:
This tutorial exercise may be completed using either AutoCAD Release 10 or Release 11.

Step #1

Begin this exercise by using the Line command and polar coordinate mode to draw a 3′-0″ square as illustrated at the right.

Command: **Line**
From point: **5′,3′**
To point: **@3′<0**
To point: **@3′<90**
To point: **@3′<180**
To point: **@3′<270**
To point: *(Strike Enter to exit this command)*

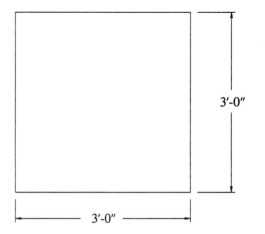

3′-0″

3′-0″

Step #2

Use the Array command to copy the top line in a rectangular pattern and have all lines spaced 3 units away from each other as illustrated at the right. Select the top line as the entity to array and perform a rectangular array consisting of 12 rows and 1 column. Since the top line selected will be copied straight down, a negative distance must be entered to perform this operation. Another popular command that could be used here is Offset. However, since each line must be offset separately, the Array command is the more efficient command to be used.

Command: **Array**
Select objects: *(Select the top horizontal line at "A")*
Select objects: *(Strike Enter to continue)*
Rectangular or Polar array (R/P): **R**
Number of rows (---) <1>: **12**
Number of columns (||||) <1>: **1**
Unit cell or distance between rows (---): **-3**

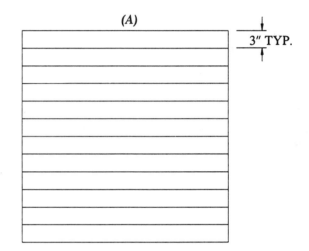

(A)

3″ TYP.

Step #3

Again, use the Array command but this time copy the right vertical line in a rectangular pattern and have all lines spaced 3 units away from each other as illustrated at the right. To assist in selecting this line, be sure the snap has been turned off by using the F9 key. Select the right line as the entity to array and perform a rectangular array consisting of 1 row and 12 columns. Since the right line selected will be copied to the left, a negative distance must be entered to perform this operation.

Command: **Array**
Select objects: *(Select the right vertical line at "A")*
Select objects: *(Strike Enter to continue)*
Rectangular or Polar array (R/P): **R**
Number of rows (---) <1>: **1**
Number of columns (||||) <1>: **12**
Distance between columns (||||): **-3**

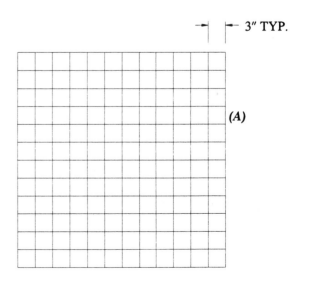

3″ TYP.

(A)

Step #4

Use the Trim command, select the two vertical dashed lines, shown at the right, as cutting edges, and use figure in Step #5 as a guide for which lines to trim out. Again, be sure the snap is turned off to assist you in selecting the line more easily.

Command: **Trim**
Select cutting edges...
Select objects: *(Select the two dashed lines, "A" and "B")*
Select objects: *(Strike Enter to continue)*
Select objects to trim: *(Select "A", "B", "C", and "D" in the figure illustrated in Step #5)*
Select objects to trim: *(Strike Enter to exit this command)*

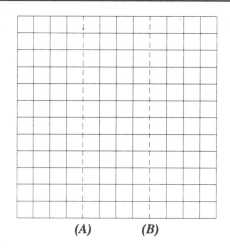

(A)   (B)

Step #5

For the last prompt of the Trim command in Step #4, select all horizontal lines in the areas marked *"A"*, *"B"*, *"C"*, and *"D"* at the right. When finished selecting the entities to trim, strike the "Enter" key to exit the command.

Select objects to trim: *(Select all horizontal lines in areas "A", "B", "C", and "D")*
Select objects to trim: *(Strike Enter to exit this command)*

The above prompts for the Trim command illustrate the Release 10 version of AutoCAD. Release 11 users have the following prompt sequence for Trim:

<Select objects to trim>Undo:

A built-in undo is provided in Release 11 if a mistake is made by trimming the wrong entity.

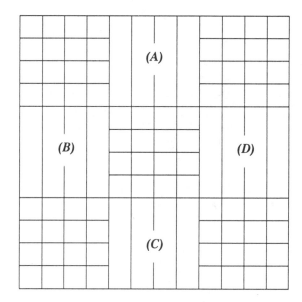

Step #6

Use the Trim command, select the two horizontal dashed lines, shown at the right, as cutting edges, and use the illustration in Step 7 as a guide for which lines to trim out.

Command: **Trim**
Select cutting edges...
Select objects: *(Select the two dashed lines, "A" and "B")*
Select objects: *(Strike Enter to continue)*
Select objects to trim: *(Select "A", "B", "C", "D", and "E" in the figure illustrated in Step #7)*

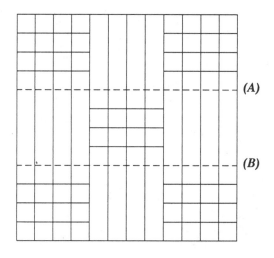

(A)

(B)

Step #7

For the last step of the Trim command in Step #6, select all vertical lines in the areas marked *"A"*, *"B"*, *"C"*, *"D"*, and *"E"*. When finished selecting the entities to trim, strike the Enter key to exit the command as illustrated at the right.

Select objects to trim: *(Select all vertical lines in areas "A", "B", "C", "D", and "E")*
Select objects to trim: *(Strike Enter to exit this command)*

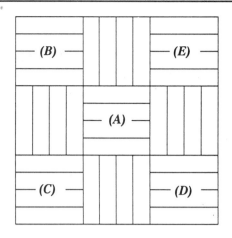

Step #8

Use the Offset command again to offset the lines *"A"*, *"B"*, *"C"*, and *"D"* 5 units in the directions illustrated at the right.

Command: **Offset**
Offset distance or Through<Through>: **5** *(understood as inches)*
Select object to offset: *(Select the line at "A")*
Side to offset? *(Pick a point above the line)*
Select object to offset: *(Select the line at "B")*
Side to offset? *(Pick a point right of the line)*
Select object to offset: *(Select the line at "C")*
Side to offset? *(Pick a point below the line)*
Select object to offset: *(Select the line at "D")*
Side to offset? *(Pick a point left of the line)*
Select object to offset: *(Strike Enter to exit this command)*

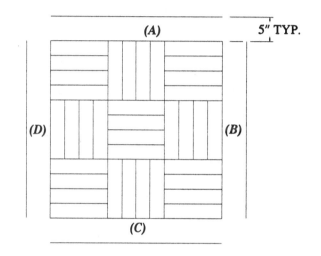

Step #9

Use the Fillet command to set a radius of 0 and place a corner at the intersection of lines *"A"* and *"B"*. The radius should already be set to 0 by default so simply pick the two lines and the corner is formed. Repeat this procedure for the other corners as illustrated at the right.

Command: **Fillet**
Polyline/Radius/<Select two objects>: *(Select lines "A" and "B")*

The above prompt sequence is for Release 10 AutoCAD users. Release 11 users will use the following sequence for the Fillet command:

Command: **Fillet**
Polyline/Radius/<Select first object>: *(Select line "A")*
Select second object: *(Select line "B")*

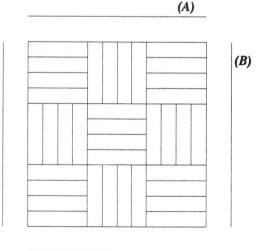

Step #10

The completed tile drawing is illustrated at the right. Follow the next step to add more tiles in a rectangular pattern using the Array command.

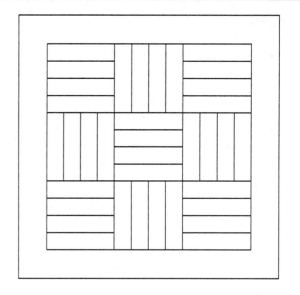

Step #11

As an alternate step, use the Array command to copy the initial design in a rectangular pattern by row and column as illustrated at the right. First, change the upper right corner to 33',21'. Follow the prompts below to perform this operation:

Command: **Array**
Select objects: **W**
First corner: **4',2'**
Other corner: **9',7'**
Select objects: *(Strike Enter to continue)*
Rectangular or Polar array (R/P): **R**
Number of rows(---) <1>: **3**
Number of columns (|||) <1>: **2**
Unit cell or distance between rows (---): **3'10**
Distance between columns (|||): **3'10**

Plot this drawing at a scale of 1=12 on a "D" size drawing sheet.

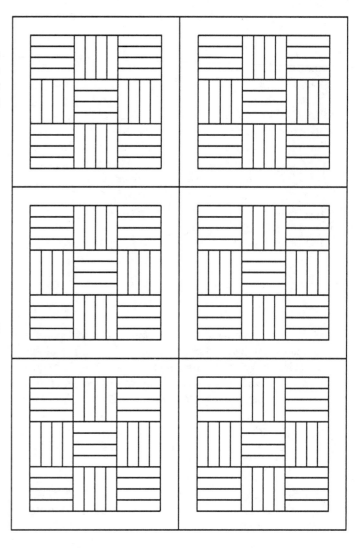

Tutorial Exercise #3
Inlay.Dwg

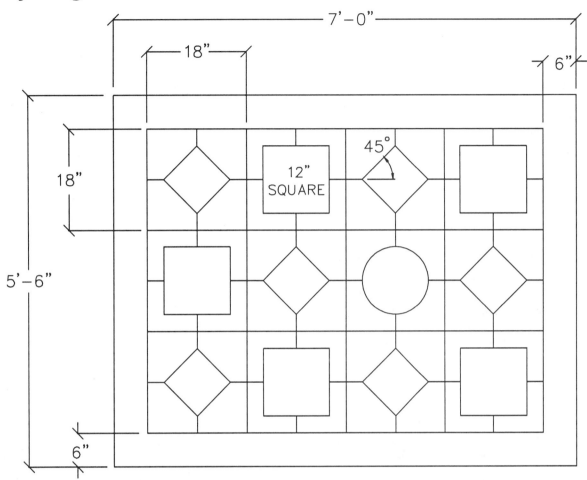

PURPOSE:
This tutorial is designed to allow the user to construct a drawing of the Inlay using the Copy command and the Multiple option.

SYSTEM SETTINGS:
Begin a new drawing called "Inlay." Use the Units command to change the units of measure from decimal to architectural. Keep the remaining default unit values. Use the Limits command, keep 0,0 for the lower left corner and change the upper right corner to 15'6,9'6. Use the Snap command and change the value from 1" to 3". (With the Grid set to zero by default, the snap setting of 3" will also change the grid spacing to 3".) Turn the snap Off.

LAYERS:
Create the following layers with the format:

Name-Color-Linetype
Object - White - Continuous

SUGGESTED COMMANDS:
Begin this tutorial by drawing a 6'-0" x 4'-6" rectangle using the Line command. Offset the edges of the rectangle by a distance of 18". Then, using the 3" grid as a guide along with the Snap-On, draw the diamond and square shapes. Use the Copy command along with the Multiple option to copy the diamond and square shapes numerous times at the designated areas.

DIMENSIONING:
Dimensions may be added to this problem at a later time. Consult your instructor.

PLOTTING:
This tutorial exercise may be plotted on "B"-size paper (11" x 17"). Use a plotting scale of 1=12 to produce a full size plot.

VERSION OF AUTOCAD:
This tutorial exercise may be completed using either AutoCAD Release 10 or Release 11.

Step #1

Begin this exercise by using the Line command and polar coordinate mode to draw a rectangle 6'-0" by 4'-6" as illustrated at the right.

Command: **Line**
From point: **4',2'**
To point: **@6'<0**
To point: **@4'6<90**
To point: **@6'<180**
To point: **@4'6<270**
To point: *(Strike Enter to exit this command)*

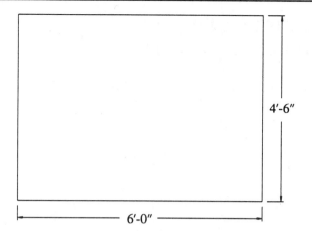

4'-6"

6'-0"

Step #2

Use the Array command to copy the top line in a rectangular pattern and have all lines spaced 18 units away from each other. Select the top horizontal line at "A" as the entity to array. Since this line will be copied straight down, a value of -18 for the spacing between rows will perform this operation. Repeat this command to copy the right vertical line at "B" 3 times to the left at a distance of 18 units. Enter a value of -18 units for the spacing in between columns since the copying is performed in the left direction as illustrated at the right.

Command: **Array**
Select objects: *(Select the top horizontal line at "A")*
Select objects: *(Strike Enter to continue)*
Rectangular or Polar array (R/P): **R**
Number of rows (---) <1>: **3**
Number of columns (|||) <1>: **1**
Unit cell or distance between rows (---): **-18**

(Repeat the Rectangular Array command above for the vertical line at "B". Use 1 for the number of rows, 4 for the number of columns, and −18 as the distance between columns.)

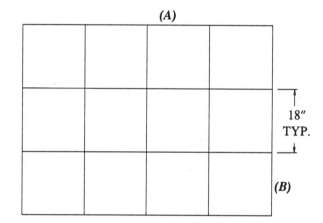

(A)

18"
TYP.

(B)

Step #3

Begin drawing one 12" x 12" diamond figure in the position illustrated at the right. Use the Zoom-Window option to magnify the area around the position of the diamond figure.

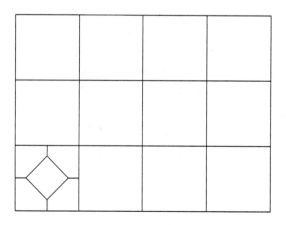

Step #4

Be sure the Grid and Snap values are set to 3″. Use the F9 key to turn the snap On. Then, use the Line command to draw the four lines illustrated at the right. Use "A", "B", "C", and "D" as the starting points for the four lines.

Command: **Line**
From point: *(Pick the grid-snap point at "A")*
To point: **@3 < 270**
To point: *(Strike Enter to exit this command)*

Repeat the above sequence for "B", "C", and "D". Use a polar direction of 180 for "B", 90 for "C", and 0 for "D".

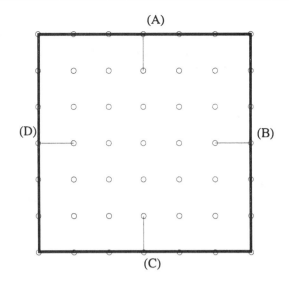

Step #5

Use the Line command to draw the diamond-shaped figure illustrated at the right. Be sure the ortho mode is Off and the snap mode is On.

Command: **Line**
From point: **Endp**
of *(Select the endpoint of the line at "A")*
To point: **Endp**
of *(Select the endpoint of the line at "B")*
To point: **Endp**
of *(Select the endpoint of the line at "C")*
To point: **Endp**
of *(Select the endpoint of the line at "D")*
To point: **C**

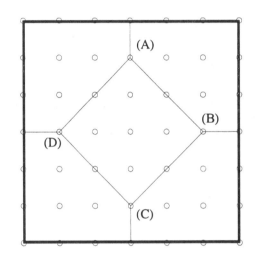

Step #6

Use the Copy command and the Multiple option to repeat the diamond-shaped pattern at the right. Use the Osnap-Intersec option to assist you during this operation. Turn snap Off to assist in selecting the entities to copy.

Command: **Copy**
Select objects: *(Select the dashed lines at the right)*
Select objects: *(Strike Enter to continue)*
<Base point or displacement>/ Multiple: **M**
Base point: **Int**
of *(Select the intersection at "A")*
Second point of displacement: **Int**
of *(Select the intersection at "B")*
Second point of displacement: **Int**
of *(Select the intersection at "C")*
Second point of displacement: **Int**
of *(Select the intersection at "D")*
(Repeat the above procedure for "E" and "F")
(Strike Enter to exit this command)

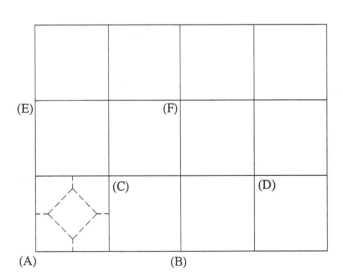

Step #7

The diamond pattern of the inlay floor tile should be similar to the illustration at the right.

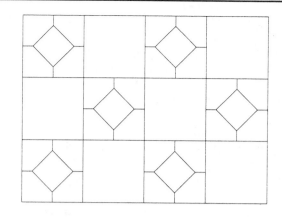

Step #8

Begin drawing one 12″ x 12″ square figure in the position illustrated at the right. Use the Zoom-Window option to magnify the area around the position of the square figure.

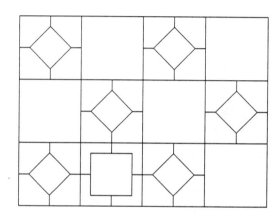

Step #9

Turn the snap On by striking the F9 function key. Then, use the Line command to draw the four lines illustrated at the right. Use "A", "B", "C", and "D" as the starting points for the four lines.

Command: **Line**
From point: *(Pick the grid-snap point at "A")*
To point: **@3 < 270**
To point: *(Strike Enter to exit this command)*

Repeat the above sequence for "B", "C", and "D". Use a polar direction of 180 for "B", 90 for "C", and 0 for "D".

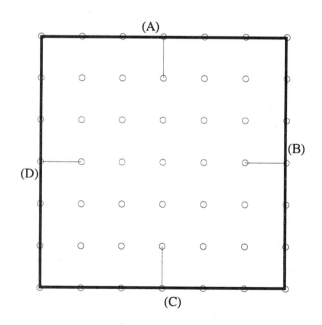

Step #10

Use the Line command to draw the square shaped object illustrated at the right.

Command: **Line**
From point: **Endp**
of *(Select the endpoint of the line at "A")*
To point: **@6<0** *(To "B")*
To point: **@12<270** *(To "C")*
To point: **@12<180** *(To "D")*
To point: **@12<90** *(To "E")*
To point: **C** *(Back to "A")*

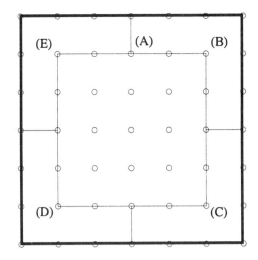

Step #11

Use the Copy command and the Multiple option to repeat the square-shaped pattern at the right. Use the Osnap-Intersec option to assist you during this operation. Be sure the snap mode is Off to assist in selecting the entities to copy.

Command: **Copy**
Select objects: *(Select the dashed lines at the right)*
Select objects: *(Strike Enter to continue)*
<Base point or displacement>/ Multiple: **M**
Base point: **Int**
of *(Select the intersection at "A")*
Second point of displacement: **Int**
of *(Select the intersection at "B")*
Second point of displacement: **Int**
of *(Select the intersection at "C")*
Second point of displacement: **Int**
of *(Select the intersection at "D")*
(Repeat the above procedure for "E" and "F")
(Strike Enter to exit this command)

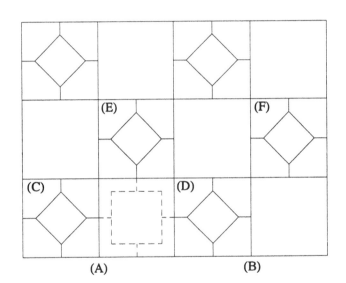

Step #12

The drawing of the inlay should appear similar to the illustration at the right.

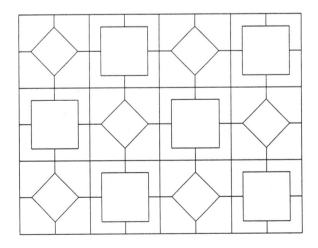

Step #13

Command: Offset
Offset distance or Through<Through>: **6**
Select object to offset: *(Select line "A")*
Side to offset? *(Pick a point above the line)*
Select object to offset: *(Select line "B")*
Side to offset? *(Pick a point left of the line)*
Select object to offset: *(Select line "C")*
Side to offset? *(Pick a point below the line)*
Select object to offset: *(Select line "D")*
Side to offset? *(Pick a point right of the line)*
Select object to offset: *(Strike Enter to exit this command)*

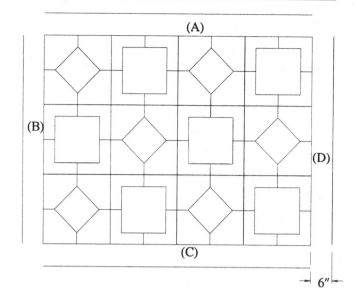

Step #14

Use the Fillet command set to a radius of 0 to place a corner at the intersection of lines "A" and "B". The radius should already be set to 0 by default so simply pick the two lines and the corner is formed. Repeat this procedure for the other corners.

Command: Fillet
Polyline/Radius/<Select two objects>: *(First, select the line at "A"; then select the line at "B")*

The above prompt sequence is for Release 10 AutoCAD users. Release 11 users will use the following sequence for the Fillet command:

Command: Fillet
Polyline/Radius/<Select first object>: *(Select line "A")*
Select second object: *(Select line "B")*

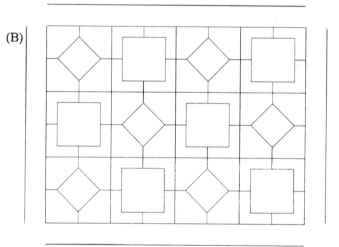

Step #15

The completed problem is illustrated at the right. Dimensions may be added upon the request of your instructor.

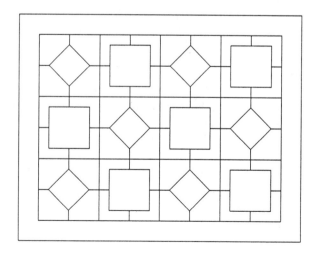

Tutorial Exercise #4
Clutch.Dwg

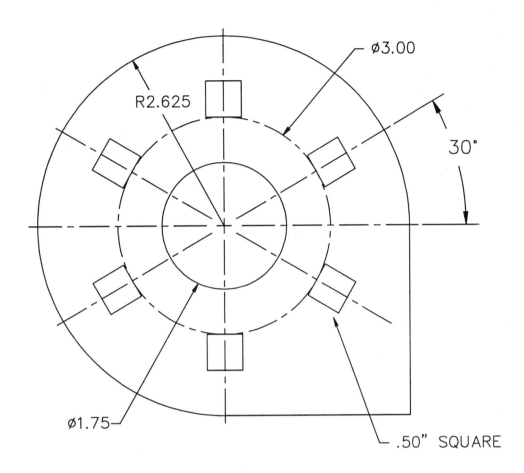

PURPOSE:
This tutorial is designed to allow the user to construct a one-view drawing of the Clutch using coordinates and the Array command.

SYSTEM SETTINGS:
Begin a new drawing called "Clutch." Use the Units command to change the number of decimal places past the zero from 4 to 3. Keep the remaining default unit values. Using the Limits command, keep 0,0 for the lower left corner and change the upper right corner from 12,9 to 10.500,8.000. Use the Grid command and change the grid spacing from 1.0000 to 0.250 units. Do not turn the snap or ortho on.

LAYERS:
Create the following layers with the format:

Name-Color-Linetype
Object - White - Continuous
Center - Yellow - Center

SUGGESTED COMMANDS:
Draw the basic shape of the object using the Line and Circle commands. Layout a center line circle, draw one square shape and use the Array command to create a multiple copy of the square in a circular pattern.

DIMENSIONING:
Dimensions may be added to this problem at a later time. Consult your instructor.

PLOTTING:
This tutorial exercise may be plotted on "A"-size paper (8.5" x 11"). Use a plotting scale of 1=1 to produce a full size plot.

VERSION OF AUTOCAD:
This tutorial exercise may be completed using either AutoCAD Release 10 or Release 11.

Step #1

Begin drawing the clutch by placing a circle with the center at absolute coordinate 5.250,4.250 and radius of 2.625 units. Check to see the current layer is "Object". Next, prepare to place a center mark at the center of the circle by changing the dimension variable, Dimcen from a value of 0.090 to -.090. The negative value will extend the center lines past the extremities of the circle. Then use the Dim-Cen command, identify the circle, and place the center marker. Follow the prompts carefully below.

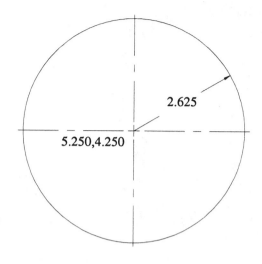

Command: **Circle**
3P/2P/TTR/Center point>: **5.250,4.250**
Diameter/<Radius>: **2.625**

Command: **Dim**
Dim: **Dimcen**
Current value <0.090> New value: **-0.090**

Dim: **Center**
Select arc or circle: *(Select anywhere along the circle)*
Dim: **Exit** *(This returns you to the Command prompt)*

Step #2

Draw two lines at a distance of 2.625 using the polar coordinate mode and using the Osnap-Intersec option. Begin the line at point "A" illustrated at the right.

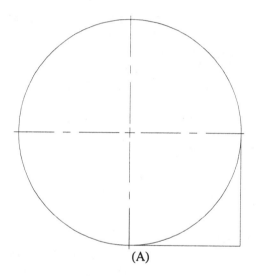

Command: **Line**
From point: **Int**
of *(Select the intersection of the line and circle at "A")*
To point: **@2.625<0**
To point: **@2.625<90**
To point: *(Strike Enter to exit this command)*

(A)

Step #3

Use the Trim command to partially delete one-fourth of the circle. Use the dashed lines illustrated at the right as the cutting edges and select the circle at "A" as the entity to trim.

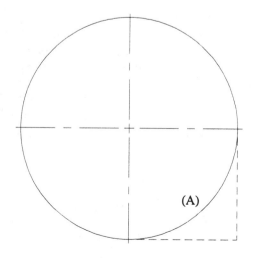

Command: **Trim**
Select Cutting edge(s)...
Select objects: *(Select the two dashed lines at the right)*
Select objects: *(Strike Enter to continue)*
Select object to trim: *(Select the circle at "A")*
Select object to trim: *(Strike Enter to exit this command)*

(A)

Step #4

Use the Erase command to delete the bottom vertical center line at "A". This center line will be placed back in its original position at a later step.

Command: **Erase**
Select objects: *(Select the bottom vertical center line at "A")*
Select objects: *(Strike Enter to execute this command)*

Place a circle of 1.75 diameter from absolute coordinate 5.250,4.250.

Command: **Circle**
3P/2P/TTR/<Center point>: **5.250,4.250**
Diameter/<Radius>: **D**
Diameter: **1.75**

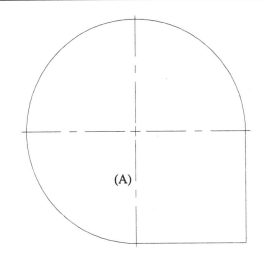

Step #5

Place a circle of 3.000 diameter from absolute coordinate 5.250,4.250. Since this circle needs to be converted from an object line to a center line, use the Chprop command and the LAyer option and change the circle and lines to the Center layer. (Be sure to have the Center layer previously created using the Layer command.)

Command: **Circle**
3P/2P/TTR/<Center point>: **5.250,4.250**
Diameter/<Radius>: **D**
Diameter: **3.00**

Command: **Chprop**
Select objects: *(Select the 3.00-diameter circle and lines "A," "B," and "C")*
Select objects: *(Strike Enter to continue)*
Change what property (Color/LAyer/LType/Thickness): **LA**
New layer <object>: **Center**
Change what property (Color/LAyer/LType/Thickness):
   (Strike Enter to exit this command)

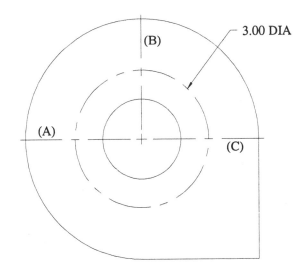

Step #6

Use the Zoom command with the Window option to magnify the upper portion of the clutch for constructing a square in the next step.

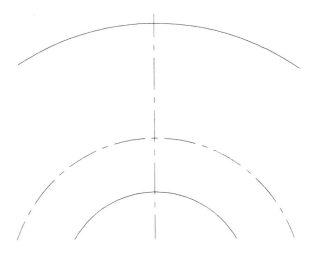

Step #7

Begin constructing the 0.500-unit square illustrated at the right. Use the Line command, begin at the intersection of the vertical and circular center lines, and use polar coordinates to assist in this operation.

Command: **Line**
From point: **Int**
of *(Select the intersection of the line and circle at "A")*
To point: **@0.250<0**
To point: **@0.500<90**
To point: **@0.500<180**
To point: **@0.500< 270**
To point: **C** *(This will close the square and exit Line)*

Use the Zoom-Previous command to return to the previous view of the Clutch before performing the next step.

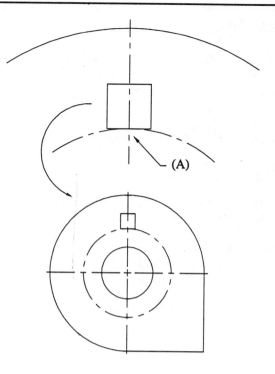

Step #8

Use the Array command to copy the square and vertical center line in a circular pattern six times. Follow the prompts below to perform this operation and complete the clutch.

Command: **Array**
Select objects: **W**
First corner: *(Select a point approximately at "A")*
Other corner: *(Select a point approximately at "B")*
Select objects: *(Select the vertical center line at "C")*
Select objects: *(Strike Enter to continue)*
Rectangular or Polar array (R/P): **P**
Center of array: **Int**
of *(Select the intersection of the center mark at "D")*
Number of items: **6**
Angle to fill (+=ccw, -=cw) <360>: *(Strike Enter to continue)*
Rotate objects as they are copied?<Y> *(Strike Enter to continue)*

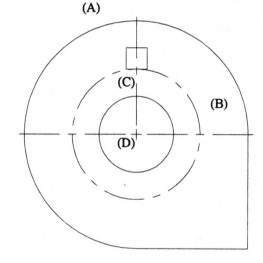

Step #9

The completed problem is illustrated at the right. Dimensions may be added at a later date.

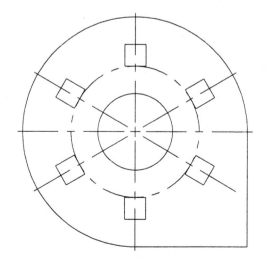

Problems for Unit 1

Directions for Problems 1-1 through 1-7:
Supply the appropriate absolute, relative, and/or polar coordinates for these figures in the matrix below each object.

Problem 1-1

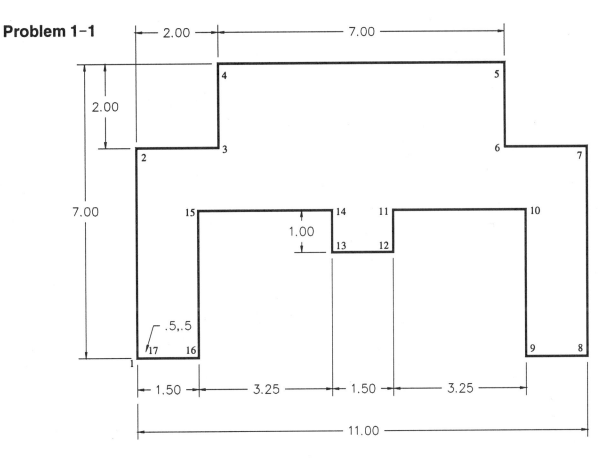

	ABSOLUTE	RELATIVE	POLAR
FROM PT(1)	.5,.5	.5,.5	.5,.5
TO PT(2)			
TO PT(3)			
TO PT(4)			
TO PT(5)			
TO PT(6)			
TO PT(7)			
TO PT(8)			
TO PT(9)			
TO PT(10)			
TO PT(11)			
TO PT(12)			
TO PT(13)			
TO PT(14)			
TO PT(15)			
TO PT(16)			
TO PT(17)			
TO PT	*(ENTER)*	*(ENTER)*	*(ENTER)*

Problem 1-2

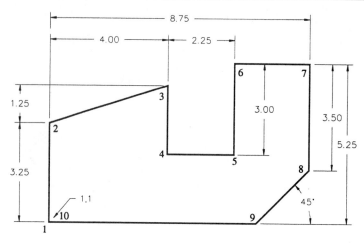

	ABSOLUTE	RELATIVE
FROM PT(1)	1,1	1,1
TO PT(2)		
TO PT(3)		
TO PT(4)		
TO PT(5)		
TO PT(6)		
TO PT(7)		
TO PT(8)		
TO PT(9)		
TO PT(10)		
TO PT	(ENTER)	(ENTER)

Problem 1-3

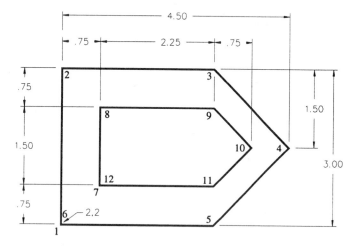

	ABSOLUTE	RELATIVE
FROM PT(1)	2,2	2,2
TO PT(2)		
TO PT(3)		
TO PT(4)		
TO PT(5)		
TO PT(6)		
TO PT	(ENTER)	(ENTER)
FROM PT(7)		
TO PT(8)		
TO PT(9)		
TO PT(10)		
TO PT(11)		
TO PT(12)		
TO PT	(ENTER)	(ENTER)

Problem 1-4

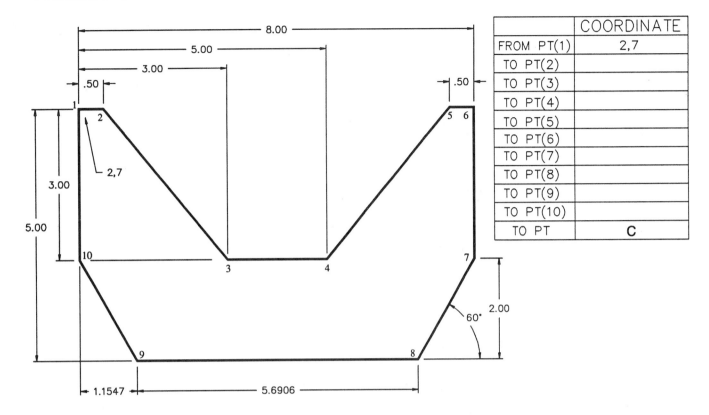

	COORDINATE
FROM PT(1)	2,7
TO PT(2)	
TO PT(3)	
TO PT(4)	
TO PT(5)	
TO PT(6)	
TO PT(7)	
TO PT(8)	
TO PT(9)	
TO PT(10)	
TO PT	C

Problem 1-5

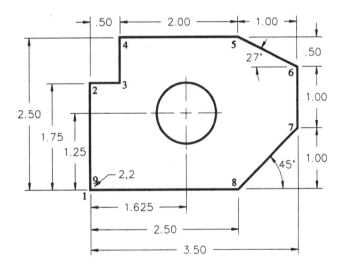

	ABSOLUTE	RELATIVE
FROM PT(1)	2,2	2,2
TO PT(2)		
TO PT(3)		
TO PT(4)		
TO PT(5)		
TO PT(6)		
TO PT(7)		
TO PT(8)		
TO PT(9)		
TO PT	(ENTER)	(ENTER)
CENTER PT(10)		

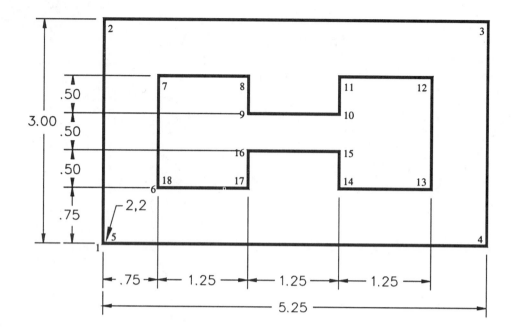

	ABSOLUTE	RELATIVE	POLAR
FROM PT(1)	2,2	2,2	2,2
TO PT(2)			
TO PT(3)			
TO PT(4)			
TO PT(5)			
TO PT	*(ENTER)*	*(ENTER)*	*(ENTER)*
FROM PT(6)			
TO PT(7)			
TO PT(8)			
TO PT(9)			
TO PT(10)			
TO PT(11)			
TO PT(12)			
TO PT(13)			
TO PT(14)			
TO PT(15)			
TO PT(16)			
TO PT(17)			
TO PT(18)			
TO PT	*(ENTER)*	*(ENTER)*	*(ENTER)*

Problem 1-7

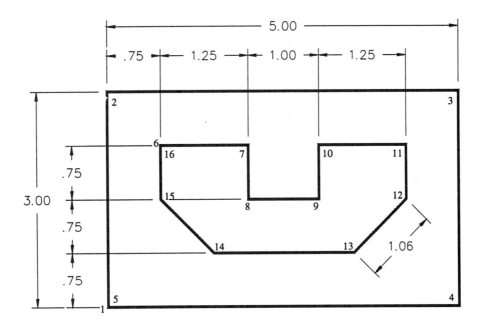

	ABSOLUTE	RELATIVE	POLAR
FROM PT(1)	2,2	2,2	2,2
TO PT(2)			
TO PT(3)			
TO PT(4)			
TO PT(5)			
TO PT	(ENTER)	(ENTER)	(ENTER)
FROM PT(6)			
TO PT(7)			
TO PT(8)			
TO PT(9)			
TO PT(10)			
TO PT(11)			
TO PT(12)			
TO PT(13)			
TO PT(14)			
TO PT(15)			
TO PT(16)			
TO PT	(ENTER)	(ENTER)	(ENTER)

Problem 1-8

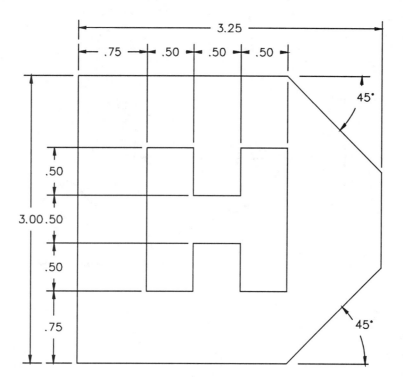

Problem 1-9

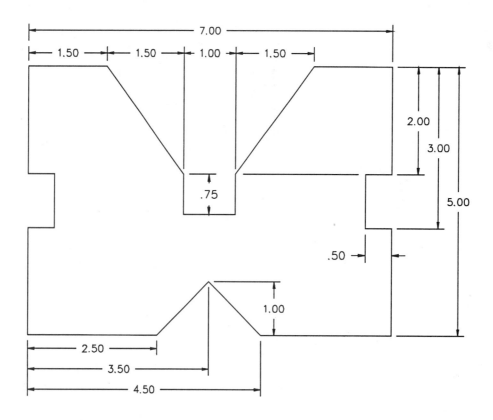

Problem 1-10

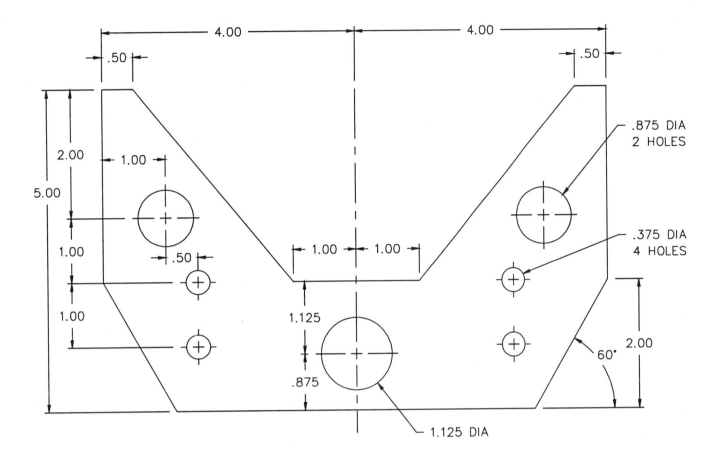

Problem 1-11

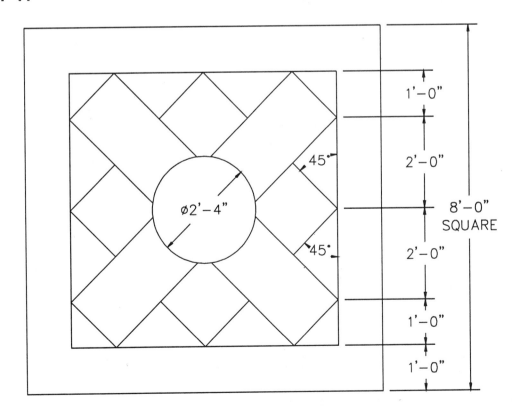

Problem 1-12

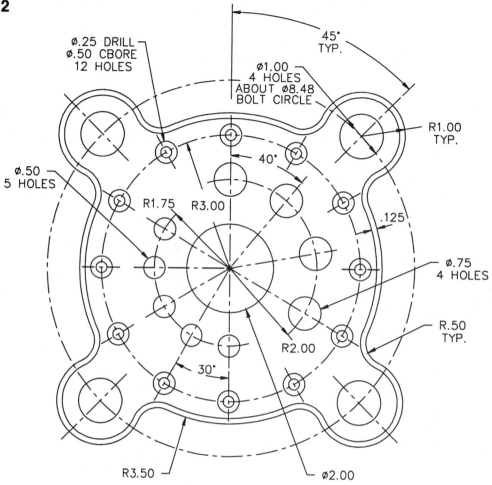

Ø.25 DRILL
Ø.50 CBORE
12 HOLES

Ø1.00
4 HOLES
ABOUT Ø8.48
BOLT CIRCLE

45° TYP.

R1.00 TYP.

Ø.50
5 HOLES

40°

.125

R1.75

R3.00

Ø.75
4 HOLES

R.50 TYP.

R2.00

30°

R3.50

Ø2.00

Problem 1-13

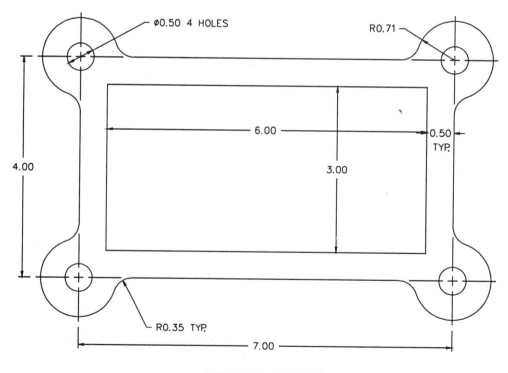

Ø0.50 4 HOLES

R0.71

6.00

0.50 TYP.

4.00

3.00

R0.35 TYP.

7.00

.125 GASKET THICKNESS

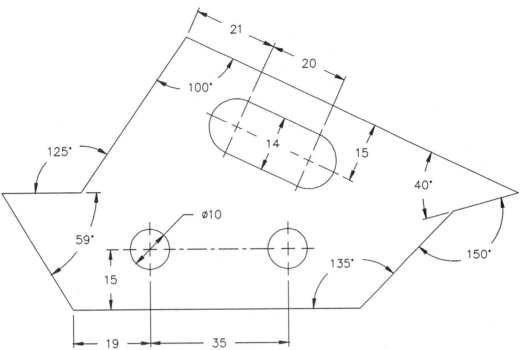

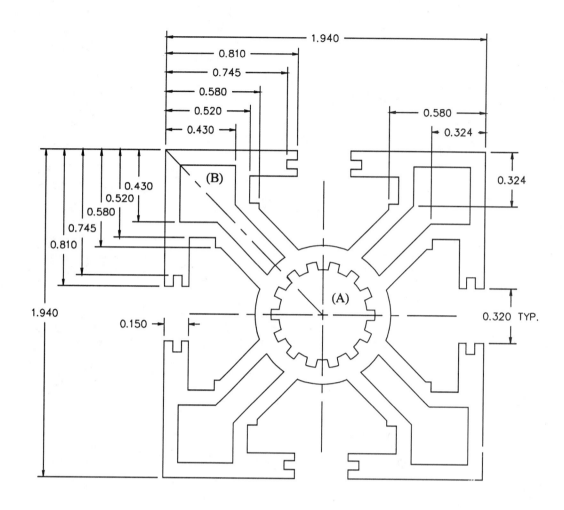

Area "A"

Area "B"

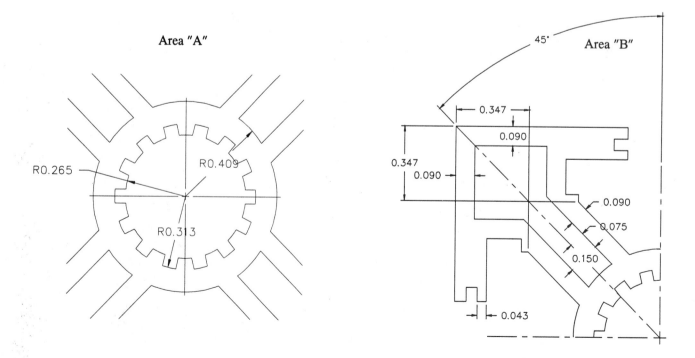

Problem 1-16

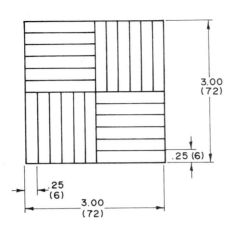

3.00
(72)

.25 (6)

.25
(6)

3.00
(72)

Problem 1-17

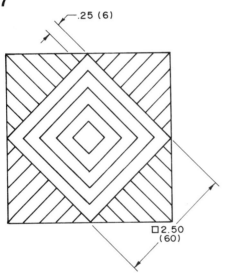

.25 (6)

□2.50
(60)

Problem 1-18

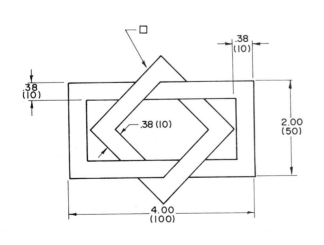

□

.38
(10)

.38
(10)

.38 (10)

2.00
(50)

4.00
(100)

Problem 1-19

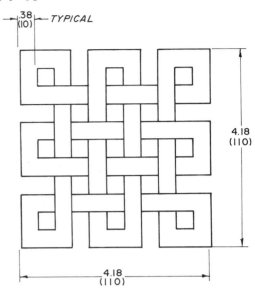

.38
(10)   TYPICAL

4.18
(110)

4.18
(110)

Problem 1-20

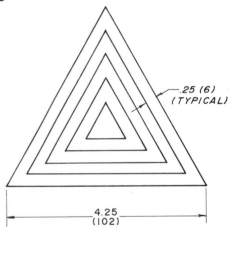

.25 (6)
(TYPICAL)

4.25
(102)

Problem 1-21

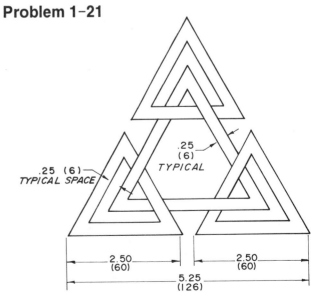

.25
(6)
TYPICAL

.25 (6)
TYPICAL SPACE

2.50
(60)

2.50
(60)

5.25
(126)

Problem 1-22

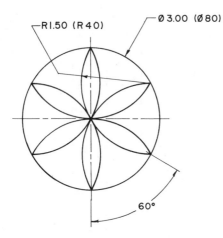

R1.50 (R40)

Ø 3.00 (Ø80)

60°

Problem 1-23

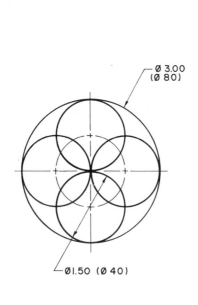

Ø 3.00 (Ø 80)

Ø1.50 (Ø 40)

Problem 1-24

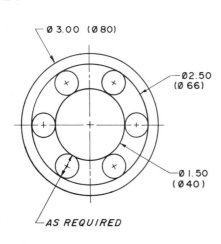

Ø 3.00 (Ø80)

Ø2.50 (Ø 66)

Ø1.50 (Ø40)

AS REQUIRED

Problem 1-25

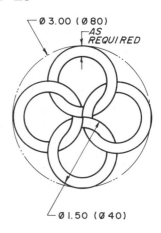

Ø 3.00 (Ø80)

AS REQUIRED

Ø1.50 (Ø 40)

Problem 1-26

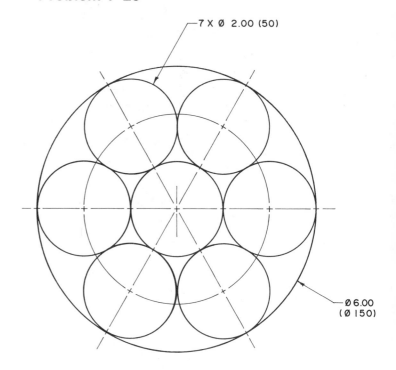

7 X Ø 2.00 (50)

Ø 6.00 (Ø 150)

Problem 1-27

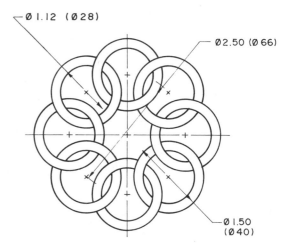

Ø 1.12 (Ø28)

Ø2.50 (Ø 66)

Ø1.50 (Ø40)

Problem 1-28

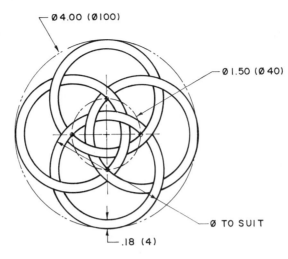

Ø4.00 (Ø100)

Ø1.50 (Ø40)

Ø TO SUIT

.18 (4)

Problem 1-29

Ø5.50 (Ø132)

ALL SPACES = .25 (6)

Problem 1-30

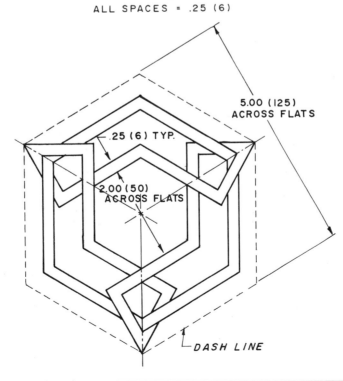

5.00 (125)
ACROSS FLATS

.25 (6) TYP.

2.00 (50)
ACROSS FLATS

DASH LINE

Problem 1-31

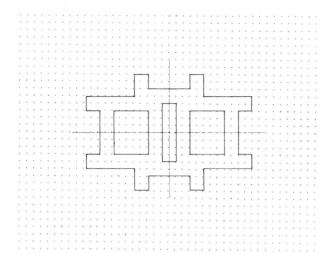

Problem 1-34

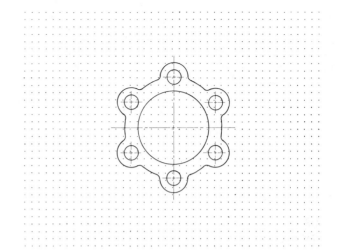

Problem 1-32

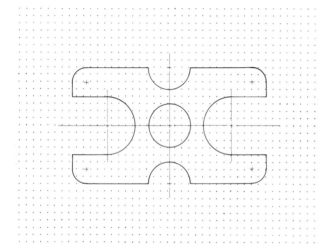

Problem 1-35

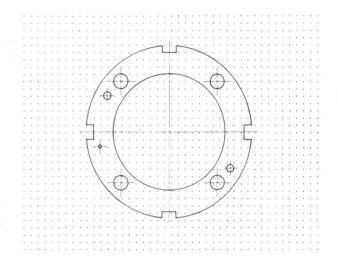

Problem 1-33

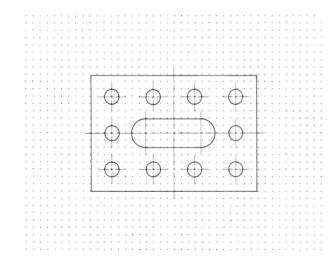

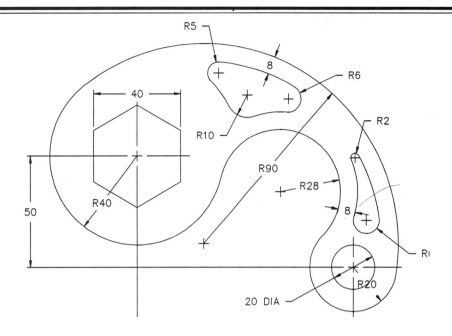

UNIT 2

Geometric Constructions

Contents

Many of the elements that go into the creation of a design revolve around geometric shapes. Applying and manipulating these shapes is the next step toward a successful design. Important manual drafting tools to assist in the construction of geometric shapes range from the T-square and drafting machine for drawing parallel and perpendicular lines to the compass and dividers for drawing circles and arcs and for setting off distances. The computer offers the designer superior control and accuracy when dealing with geometric constructions. Much of the focus of this unit will be on how to manipulate the numerous object snap modes supplied by AutoCAD through construction examples and practical applications in the form of a tutorial. Numerous other problems are supplied at the end of this chapter to challenge the designer for a complete and correct solution.

Bisecting Lines and Arcs

Illustrated at the right is an arc intersected by a line. The purpose of this problem is to locate the midpoint of the line and the arc. The Divide command will be used to accomplish this. However, if the command is used to find the midpoint, it may not be visible. This is because the Divide command places a point depending on the amount of divisions asked for. The appearance of the point is controlled by the system variable PDMODE. There is also a system variable to control the size of the point called Pdsize. Follow the steps below to change the point appearance and size before using the Divide command.

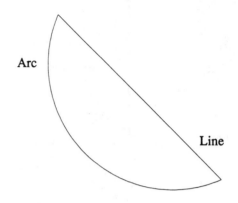

Command: **Setvar**
Variable name or ?: **PDMODE**
New value for pdmode <0>: **2**

Command: **Setvar**
Variable name or ? <PDMODE>: **Pdsize**
New value for pdsize <0>: **.25**

Next, use the Point command along with the Osnap-Midpoint option to locate the midpoint of the line and arc using the following steps:

Command: **Point**
Point: **Mid**
of *(Select the line or arc at any convenient location)*

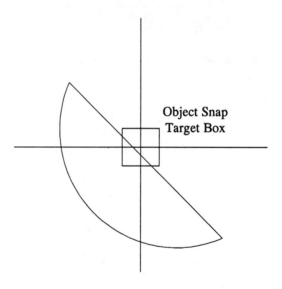

The illustration at the right shows the midpoint locations of the line and arc. The current point is controlled by the system variable PDMODE. The style of point, namely the "plus", reflects the current PDMODE value of 2.

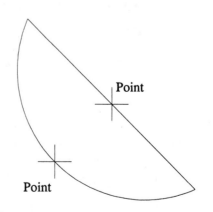

Bisecting Angles

Illustrated at the right are two lines forming an acute angle. (An acute angle measures any value less than 90 degrees.) The purpose of this problem is to bisect or divide the angle equally in two parts. A construction line will be drawn from the endpoints of the angle. Then, a line will be drawn from the vertex of the angle to the midpoint of the construction line using the Line command and the Osnap-Midpoint option.

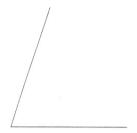

First, draw the construction line.

Command: **Line**
From point: **Endp**
of *(Select the endpoint of the line at "A")*
To point: **Endp**
of *(Select the endpoint of the line at "B")*
To point: *(Strike Enter to exit this command)*

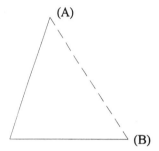

Next, use the Line command to draw a line from the vertex of the angle to the midpoint of the construction line. Use the Osnap-Endpoint and Osnap-Midpoint options to assist you in this operation.

Command: **Line**
From point: **Endp**
of *(Select the endpoint of the line at "A")*
To point: **Mid**
of *(Select the midpoint of the construction line at "B")*
To point: *(Strike Enter to exit this command)*

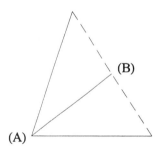

Use the Erase command to delete the construction line, leaving the angle bisected or divided into two equal angles.

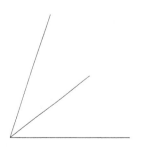

Dividing an Entity into an Equal Number of Parts

Illustrated at the right is an inclined line. The purpose of this problem is to divide the line into an equal number of parts. This proved to be a tedious task using manual drafting methods but thanks to the AutoCAD Divide command, this operation is much easier to perform. The Divide command instructs the user to supply the number of divisions and performs the division by placing a point along the entity to be divided. The size and shape are controlled by the system variables Pdsize and Pdmode, respectively. Be sure the Pdmode variable is set to a value that will produce a visible point. Otherwise, the results of the Divide command will not be obvious.

Command: **Setvar**
Variable name or ?: **PDMODE**
New value for pdmode <0>: **2**

Command: **Setvar**
Variable name or ? <PDMODE>: **Pdsize**
New value for pdsize <0>: **.25**

Next, use the Divide command, select the inclined line as the entity to divide, enter a value for the number of segments, and the command divides the entity by a series of points.

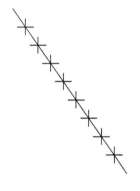

Command: **Divide**
Select object to divide: *(Select the inclined line)*
<Number of segments>/Block: **9**

A practical application of the Divide command may be in the area of screw threads where a number of threads per inch is needed to form the profile of the thread.

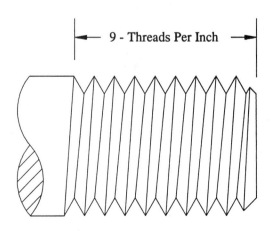

9 - Threads Per Inch

Drawing Entities Parallel to Each Other

Another fundamentally important operation regarding geometric constructions is the ability to construct entities parallel to each other. Illustrated at the right is an inclined line. The purpose of this problem is to create a matching entity parallel to the original line at a set distance. The Offset command is used to accomplish this.

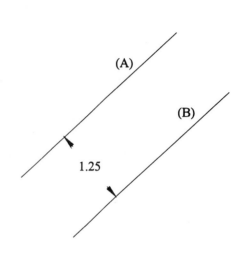

Use the Offset command, set the distance to offset, and select the entity to offset and the side of the offset. The result will be similar to the illustration at the right.

Command: **Offset**
Offset distance or Through<Through>: **1.25**
Select object to offset: *(Select the line at "A")*
Side to offset? *(Select anywhere near "B")*
Select object to offset: *(Strike Enter to exit this command)*

The Offset command will also produce concentric circles or arcs. Illustrated at the right is an arc. The purpose of this problem is to create an additional arc at a distance of 1.25 units away from the original arc.

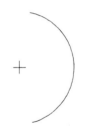

Use the Offset command, set the distance to offset, select the entity (this time the arc) to offset, and the side of the offset. The result will be similar to the illustration at the right.

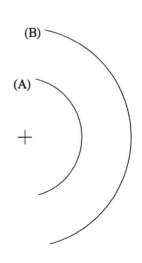

Command: **Offset**
Offset distance or Through<Through>: **1.25**
Select object to offset: *(Select the arc at "A")*
Side to offset? *(Select anywhere near "B")*
Select object to offset: *(Strike Enter to exit this command)*

Constructing Hexagons

The Hexagon is an important geometric shape commonly used for such items as the plan view of a bolt or screw type of fastener. The illustrations at the right show two types of hexagons; one drawn in relation to its flat edges, and the other drawn in relation to its corners. The Polygon command is used for drawing either example. Simply supply the number of the sides and whether the figure is inscribed or circumscribed about a circle and the radius of the circle and a polygon is drawn. One interesting characteristic of polygons is that they are constructed using a series of polylines, making the polygon one entity.

One other note concerning polygons is that the examples on this page illustrate constructing a hexagon. Using the Polygon command, any size figure with up to 1024 sides may be constructed; not just hexagons.

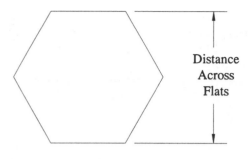

Distance Across Flats

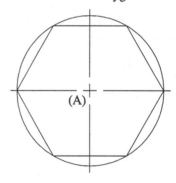

Distance Across Corners

Illustrated at the right is an example of an inscribed hexagon, or a figure constructed inside of a circle. The Polygon command is used to create this type of shape.

Command: **Polygon**
Number of sides: **6**
Edge/<Center of Polygon>: *(Select the center at "A")*
Inscribed in circle/Circumscribed about circle (I/C): **I**
Radius of circle: *(Enter a numerical value)*

Inscribed Polygon

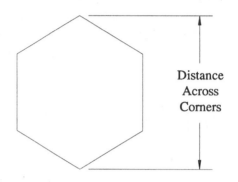

(A)

Yet another use of the Polygon is illustrated at the right when the figure needs to be drawn outside of a circle. The Circumscribed option would be used for this example.

Command: **Polygon**
Number of sides: **6**
Edge/<Center of Polygon>: *(Select the center at "A")*
Inscribed in circle/Circumscribed about circle (I/C): **C**
Radius of circle: *(Enter a numerical value)*

Circumscribed Polygon

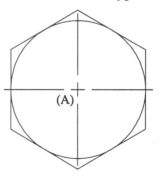

(A)

Constructing an Ellipse

Main parts of an ellipse are the major diameter or axis and minor diameter or axis illustrated at the right. Numerous construction arcs and lines were needed to construct the ellipse using manual methods. The Ellipse command prompts the user for the center of the ellipse, the endpoint of one axis, and the endpoint of the other axis. The ellipse is drawn as a series of polylines representing one entity.

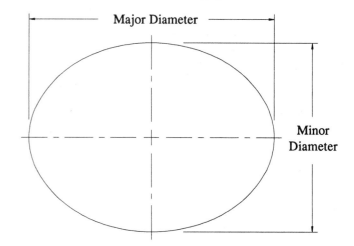

In the example at the right, a polar coordinate can be used to identify both axes of the ellipse following the prompts below.

Command: **Ellipse**
<Axis endpoint 1>/ Center: **C**
Center of ellipse: *(Select a point at "A")*
Axis endpoint: **@4<0** *(Toward a point at "B")*
<Other axis distance>/ Rotation: **@3<0** *(Toward a point at "C")*

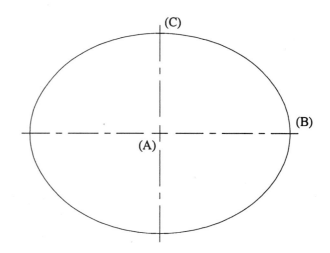

The object at the right is an example of how to outline the view with an ellipse and how to use the Offset command to offset the ellipse in the direction inside of the object.

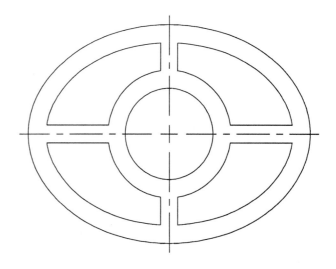

Drawing an Arc Tangent to Two Lines

Illustrated at the right are two inclined lines. The purpose of this problem is to connect an arc tangent to the two lines at a specified radius. The Circle-TTR (Tangent-Tangent-Radius) command will be used here along with the Trim command to clean up the excess geometry.

First, use the Circle-TTR command to construct an arc tangent to both lines.

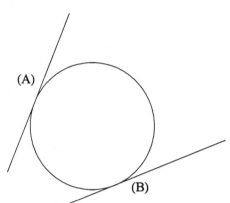

Command: **Circle**
3P/2P/TTR/<Center point>: **TTR**
Enter Tangent spec: *(Select the line at "A")*
Enter second Tangent spec: *(Select the line at "B")*
Radius: *(Enter a desired radius value)*

Use the Trim command to clean up the lines and arc. The completed result is illustrated at the right. It is interesting to note that the Fillet command could have been used for this procedure. Not only will the curve be drawn, but this command will automatically trim the lines.

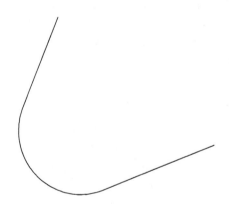

The object at the right is an example of a typical application where this procedure might be used.

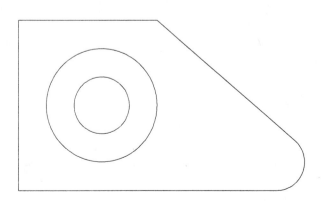

Drawing an Arc Tangent to a Line and Arc

Illustrated at the right is an arc and an inclined line. The purpose of this problem is to connect an additional arc tangent to the original arc and line at a specified radius. The Circle-TTR command will be used here along with the Trim command to clean up the excess geometry.

First, use the Circle-TTR command to construct an arc tangent to the arc and inclined line.

Command: **Circle**
3P/2P/TTR/<Center point>: **TTR**
Enter Tangent spec: *(Select the arc at "A")*
Enter second Tangent spec: *(Select the line at "B")*
Radius: *(Enter a desired radius value)*

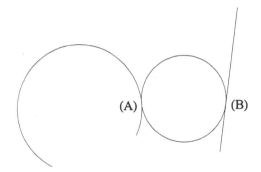

Use the Trim command to clean up the arc and line. The completed result is illustrated at the right.

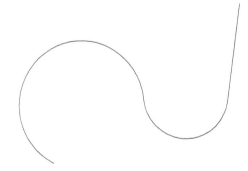

The object at the right is an example of a typical application where this procedure might be used.

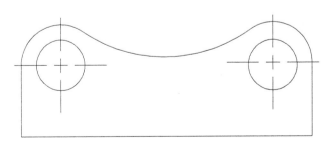

Drawing an Arc Tangent to Two Arcs
Method #1

Illustrated at the right are two arcs. The purpose of this problem is to connect a third arc tangent to the original two at a specified radius. The Circle-TTR command will be used here along with the Trim command to clean up the excess geometry.

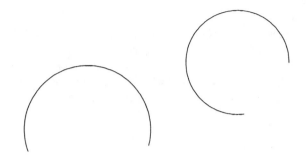

Use the Circle-TTR command to construct an arc tangent to the two original arcs.

Command: **Circle**
3P/2P/TTR/<Center point>: **TTR**
Enter Tangent spec: *(Select the first arc at "A")*
Enter second Tangent spec: *(Select the second arc at "B")*
Radius: *(Enter a desired value)*

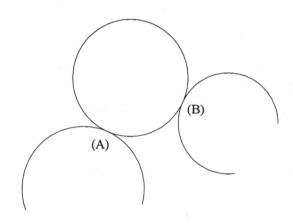

Use the Trim command to clean up the two arcs using the circle as a cutting edge. The completed result is illustrated at the right.

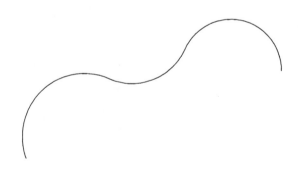

The object at the right is an example of a typical application where this procedure might be used.

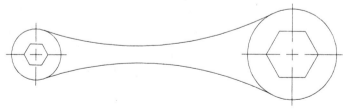

Drawing an Arc Tangent to Two Arcs
Method #2

Illustrated at the right are two arcs. The purpose of this problem is to connect an additional arc tangent to and enclosing both arcs at a specified radius. The Circle-TTR command will be used here along with the Trim command.

First, use the Circle-TTR command to construct an arc tangent to and enclosing both arcs.

Command: **Circle**
3P/2P/TTR/<Center point>: **TTR**
Enter Tangent spec: *(Select the arc at "A")*
Enter second Tangent spec: *(Select the arc at "B")*
Radius: *(Enter a desired radius value)*

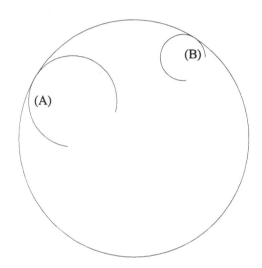

Use the Trim command to clean up all arcs. The completed result is illustrated at the right.

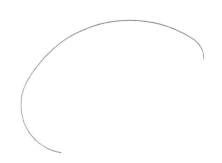

The object at the right is an example of a typical application where this procedure might be used.

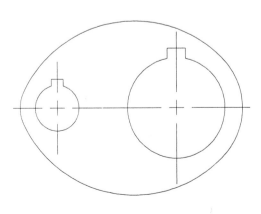

Drawing an Arc Tangent to Two Arcs
Method #3

Illustrated at the right are two arcs. The purpose of this problem is to connect an additional arc tangent to one arc and enclosing the other. The Circle-TTR command will be used here along with the Trim command to clean up unnecessary geometry.

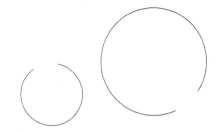

First, use the Circle-TTR command to construct an arc tangent to the two arcs. Study the illustration at the right and the prompts below to understand the proper pick points for this operation.

Command: **Circle**
3P/2P/TTR/<Center point>: **TTR**
Enter Tangent spec: *(Select the arc at "A")*
Enter second Tangent spec: *(Select the line at "B")*
Radius: *(Enter a desired radius value)*

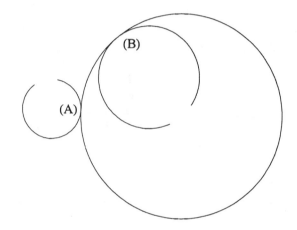

Use the Trim command to clean up the arcs. The completed result is illustrated at the right.

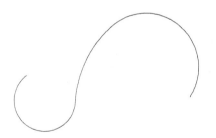

The object at the right is an example of a typical application where this procedure might be used.

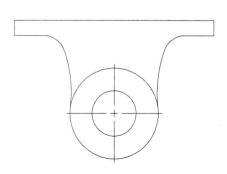

Drawing a Line from a Point Perpendicular to Another Line

Illustrated at the right is an inclined line and a point. The purpose of this problem is to construct a line from a point perpendicular to another line. The Line command will be used for this operation in addition to the Osnap-Node and Osnap-Perpend options.

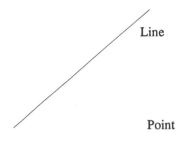

Line

Point

Use the Line command and the Osnap-Node option to snap to the point illustrated at the right.

Command: **Line**
From point: **Node**
of *(Select the point at "A")*

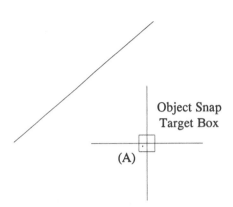

Object Snap
Target Box

(A)

Continue the Line command and respond to the prompt, "To point" by using the Osnap-Perpend option and selecting anywhere along the inclined line.

To point: **Perpend**
of *(Select the line at "A")*
To point: *(Strike Enter to exit this command)*

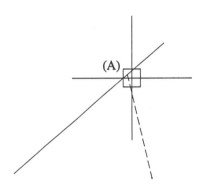

(A)

The completed solution is illustrated at the right.

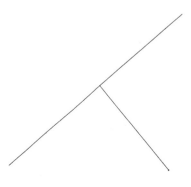

Drawing a Line Tangent to Two Circles or Arcs

Illustrated at the right are two circles. The purpose of this problem is to connect the two circles with two tangent lines. This can be accomplished using the Line command and the Osnap-Tangent option.

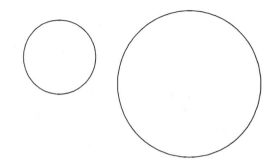

Use the Line command to connect two lines tangent to the circles. The following procedure is used for the first line. Use the same procedure for the second.

Command: **Line**
From point: **Tan**
to *(Select the circle near "A")*
To point: **Tan**
to *(Select the circle near "B")*
To point: *(Strike Enter to exit this command)*

When using the Tangent option, the rubberband cursor is not present when drawing the beginning of the line. This is due to calculations required when identifying the second point.

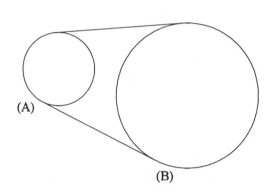

(A)

(B)

Use the Trim command to clean up the circles so the appearance of the object is similar to the illustration at the right.

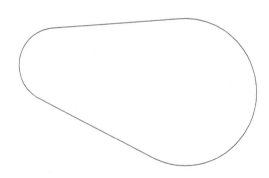

The object at the right is an example of a typical application where drawing lines tangent to circles might be used.

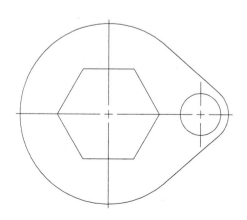

Quadrant vs Tangent Osnap Option

Various examples have been given on previous pages concerning drawing lines tangent to two circles, two arcs, or any combination of the two. The object at the right illustrates the use of the Osnap-Tangent option when used along with the Line command.

Command: **Line**
From point: **Tan**
to *(Select the arc at "A")*
To point: **Tan**
to *(Select the arc at "B")*
To point: *(Strike Enter to exit this command)*

Note that the angle of the line formed by points "A" and "B" is neither horizontal or vertical. The object at the right is a typical example of the capabilities of the Osnap option.

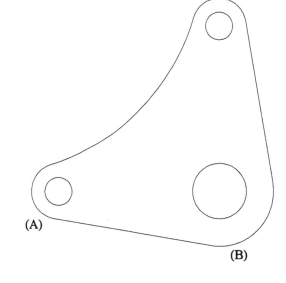

(A)

(B)

The object illustrated at the right is a modification of the drawing above with the inclined tangent lines changing to horizontal and vertical tangent lines. This example is cited to warn the user that two Osnap options are available to perform tangencies, namely Osnap-Tangent and Osnap-Quadrant. However, it is up to the user to evaluate under what conditions the Osnap options are to be used. At the right, the Osnap-Tangent or Osnap-Quadrant option could be used to draw the lines tangent to the arcs. The Quadrant option could be used only since the lines to be drawn are perfectly horizontal or vertical. Usually it is impossible to know this ahead of time, and in this case, the Osnap-Tangent option should be used whenever possible.

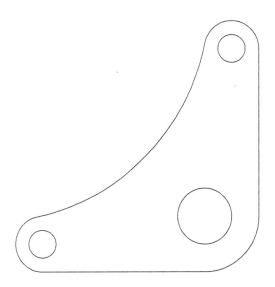

The slot at the right is an excellent example of when to use the Osnap-Quadrant option. Slots have two semi-circles connected together by two horizontal or vertical lines enabling the user to use the Quadrant option. If the slot is positioned at an odd angle, simply construct it and use the Rotate command to position it.

Tutorial Exercise #5
Pattern1.Dwg

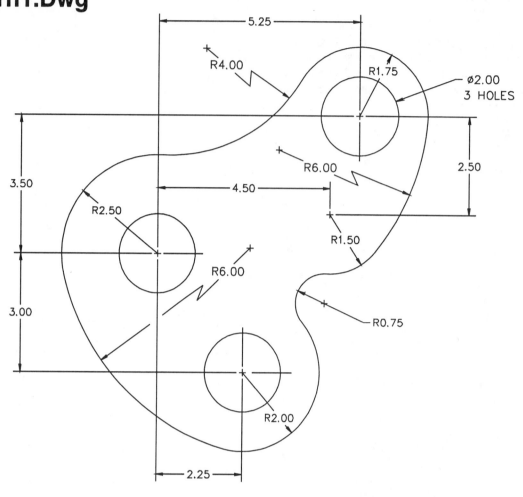

PURPOSE:
This tutorial is designed to use geometric commands to construct a one-view drawing of Pattern1. Refer to the following for special system settings and suggested command sequences.

SYSTEM SETTINGS:
Begin a new drawing called "Pattern1." Use the Units command to change the number of decimal places past the zero from 4 to 2. Keep the remaining default unit values. Using the Limits command, keep 0,0 for the lower left corner and change the upper right corner from 12,9 to 21.00,16.00. Use the Grid command and change the grid spacing from 1.00 to 0.50 units. Do not turn the snap or ortho On.

LAYERS:
Create the following layers with the format:

Name-Color-Linetype
Object - White - Continuous
Center - Yellow - Center
Dim - Yellow - Continuous

SUGGESTED COMMANDS:
Begin constructing this object by first laying out four points which will be used as centers for circles. Use the Circle-TTR command to construct tangent arcs to the circles already drawn. Use the Trim command to clean up and partially delete circles to obtain the outline of the pattern. Then, add the 2.00 diameter holes followed by the center markers using the Dim-Cen command.

DIMENSIONING:
This drawing may be dimensioned at a later date. Consult your instructor before continuing.

PLOTTING:
This tutorial exercise may be plotted on "C"-size paper (18" x 24"). Plot Pattern1 at a scale of full size, or 1=1.

VERSION OF AUTOCAD:
This tutorial exercise may be completed using either AutoCAD Release 10 or Release 11.

Step #1

Use the Setvar command and set the Pdmode system variable to a value of 2. This will form a "plus sign" when using the Point command. Locate one point at absolute coordinate 7.50,7.50. Then, use the Copy command and the dimensions at the right as a guide for duplicating the remaining points.

Command: **Setvar**
Variable name or ?: **Pdmode**
New value for Pdmode <0>: **2**

Command: **Point**
Point: **7.50,7.50** *(Locates the point at "A")*

Command: **Copy**
Select objects: **L** *(This should select the point)*
Select objects: *(Strike Enter to continue)*
<Base point or displacement>/Multiple: **M**
Base point: **Node**
of *(Select the point at "A")*
Second point of displacement: **@2.25,-3.00** *(Locates point "B")*
Second point of displacement: **@4.50,1.00** *(Locates point "C")*
Second point of displacement: **@5.25,3.50** *(Locates point "D")*
Second point of displacement: *(Strike Enter to exit this command)*

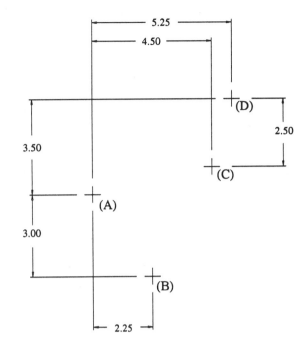

Step #2

Use the Circle command to place four circles of different sizes from points located at "A", "B", "C", and "D". When you have completed drawing the four circles, use the Erase command to erase points "A", "B", "C", and "D".

Command: **Circle**
3P/2P/TTR/<Center point>: **Node**
of *(Select the point at "A")*
Diameter/<Radius>: **2.50**

Command: **Circle**
3P/2P/TTR/<Center point>: **Node**
of *(Select the point at "B")*
Diameter/<Radius>: **2.00**

Command: **Circle**
3P/2P/TTR/<Center point>: **Node**
of *(Select the point at "C")*
Diameter/<Radius>: **1.50**

Command: **Circle**
3P/2P/TTR/<Center point>: **Node**
of *(Select the point at "D")*
Diameter/<Radius>: **1.75**

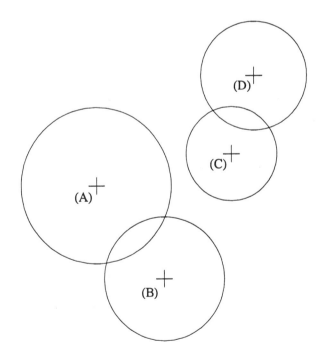

Step #3

Use the Circle-TTR command to construct a 4.00-radius circle tangent to the two dashed circles at the right. Then, use the Trim command to trim away part of circle "C".

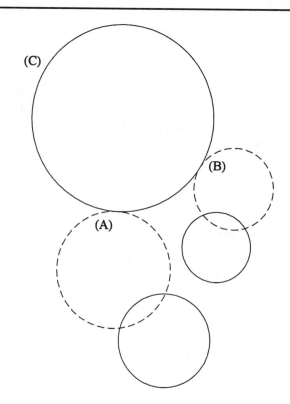

Command: **Circle**
3P/2P/TTR/<Center point>: **TTR**
Enter Tangent spec: *(Select the dashed circle at "A")*
Enter second Tangent spec: *(Select the dashed circle at "B")*
Radius: **4.00**

Command: **Trim**
Select cutting edges...
Select objects: *(Select the two dashed circles at the right)*
Select objects: *(Strike Enter to continue)*
Select object to trim: *(Select the large circle at "C")*
Select object to trim: *(Strike Enter to exit this command)*

The above prompts illustrate the Trim command for Release 10 AutoCAD. Release 11 users have the following prompt sequence:

<Select object to trim>Undo:

A built-in undo is provided if a mistake is made during the trimming process.

Step #4

Use the Circle-TTR command to construct a 6.00-radius circle tangent to the two dashed circles at the right. Then, use the Trim command to trim away part of circle "C".

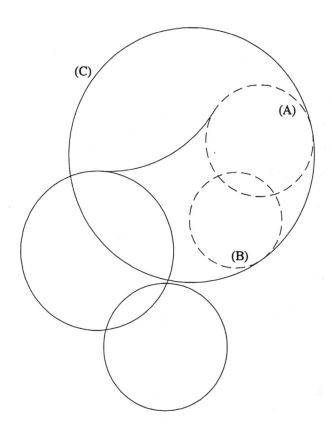

Command: **Circle**
3P/2P/TTR/<Center point>: **TTR**
Enter Tangent spec: *(Select the dashed circle at "A")*
Enter second Tangent spec: *(Select the dashed circle at "B")*
Radius: **6.00**

Command: **Trim**
Select cutting edges...
Select objects: *(Select the two dashed circles at the right)*
Select objects: *(Strike Enter to continue)*
Select object to trim: *(Select the large circle at "C")*
Select object to trim: *(Strike Enter to exit this command)*

Step #5

Use the Circle-TTR command to construct a 6.00-radius circle tangent to the two dashed circles at the right. Then, use the Trim command to trim away part of circle "C".

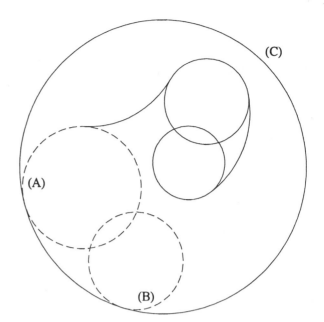

Command: **Circle**
3P/2P/TTR/<Center point>:   **TTR**
Enter Tangent spec: *(Select the dashed circle at "A")*
Enter second Tangent spec: *(Select the dashed circle at "B")*
Radius: **6.00**

Command: **Trim**
Select cutting edges...
Select objects: *(Select the two dashed circles at the right)*
Select objects: *(Strike Enter to continue)*
Select object to trim: *(Select the large circle at "C")*
Select object to trim: *(Strike Enter to exit this command)*

Step #6

Use the Circle-TTR command to construct a 0.75-radius circle tangent to the two dashed circles at the right. Then, use the Trim command to trim away part of circle "C".

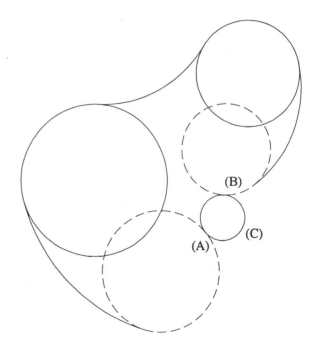

Command: **Circle**
3P/2P/TTR/<Center point>:   **TTR**
Enter Tangent spec: *(Select the dashed circle at "A")*
Enter second Tangent spec: *(Select the dashed circle at "B")*
Radius: **0.75**

Command: **Trim**
Select cutting edges...
Select objects: *(Select the two dashed circles at the right)*
Select objects: *(Strike Enter to continue)*
Select object to trim: *(Select the circle at "C")*
Select object to trim: *(Strike Enter to exit this command)*

Step #7

Use the Trim command, select all dashed arcs at the right as cutting edges, and trim away the circular segments to form the outline of the pattern1 drawing.

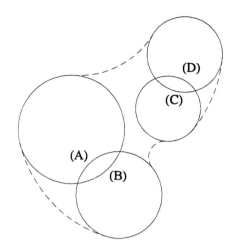

Command: **Trim**
Select cutting edges...
Select objects: *(Select the four dashed arcs at the right)*
Select objects: *(Strike Enter to continue)*
Select object to trim: *(Select the circle at "A")*
Select object to trim: *(Select the circle at "B")*
Select object to trim: *(Select the circle at "C")*
Select object to trim: *(Select the circle at "D")*
Select object to trim: *(Strike Enter to exit this command)*

Step #8

Your drawing should be similar to the illustration at the right.

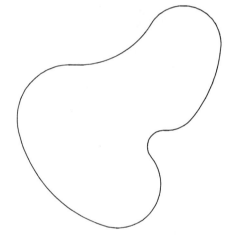

Step #9

Use the Circle command to place a circle of 2.00-unit diameter at the center of arc "A". Then, use the Copy command to duplicate the circle at the center of arcs "B" and "C".

Command: **Circle**
3P/2P/TTR/<Center point>: **Cen**
of *(Select the arc at "A")*
Diameter/<Radius>: **D**
Diameter: **2.00**

Command: **Copy**
Select objects: **L**
Select objects: *(Strike Enter to continue)*
<Base point or displacement>/Multiple: **M**
Base point: **Cen**
of *(Select the arc at "A")*
Second point of displacement: **Cen**
of *(Select the arc at "B")*
Second point of displacement: **Cen**
of *(Select the arc at "C")*
Second point of displacement: *(Strike Enter to exit this command)*

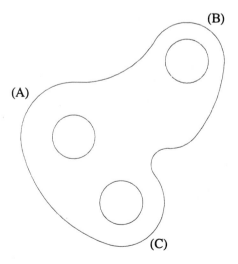

Step #10

Use the Layer command to set the new current layer to Center. Enter the dimensioning area of AutoCAD and change the variable Dimcen from a value of 0.09 to -0.12. This will place the center marker identifying the centers of circles and arcs when the Dim-Cen command is used.

Command: **Layer**
?/Make/Set/New/ON/OFF/Color/Ltype/Freeze/Thaw: **Set**
New current layer <0>: **Center**
?/Make/Set/New/ON/OFF/Color/Ltype/Freeze/Thaw:
    (Strike Enter to exit this command)

Command: **Dim**
Dim: **Dimcen**
Current value <0.09> New value: **-0.12**

Dim: **Center**
Select arc or circle: *(Select the arc at "A")*

Dim: **Center**
Select arc or circle: *(Select the arc at "B")*

Dim: **Center**
Select arc or circle: *(Select the arc at "C")*

Dim: **Center**
Select arc or circle: *(Select the arc at "D")*

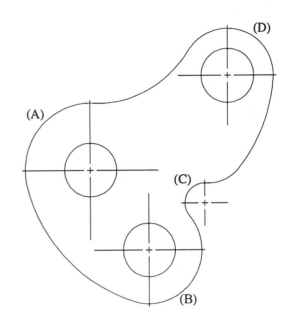

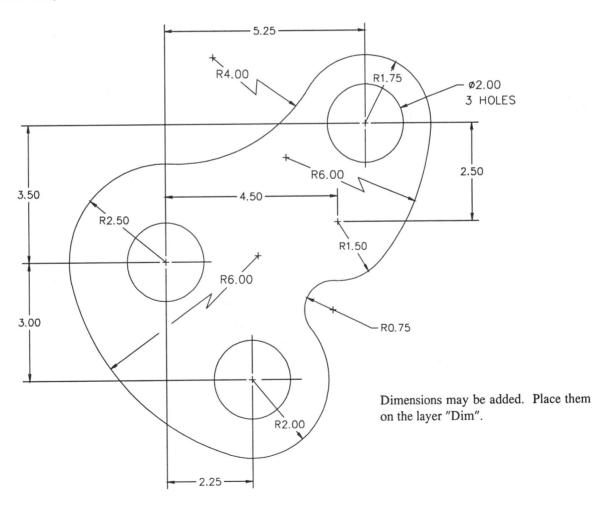

Dimensions may be added. Place them on the layer "Dim".

Tutorial Exercise #6
Gear-arm.Dwg

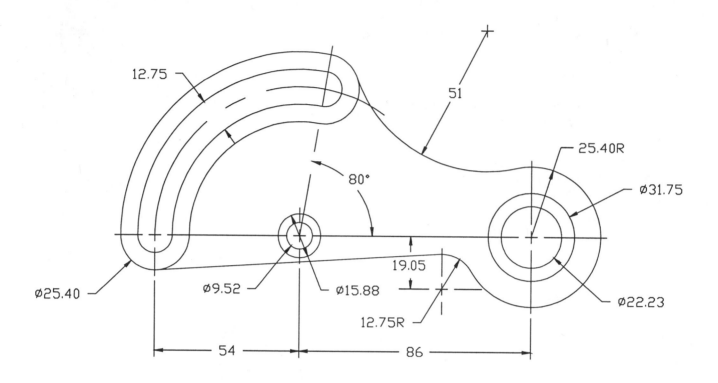

PURPOSE:

This tutorial is designed to use geometric commands to construct a one-view drawing of the Gear-arm. Follow the special system settings since this drawing is in metric.

SYSTEM SETTINGS:

Begin a new drawing called "Gear-arm." Use the Units command to change the number of decimal places past the zero from 4 to 2. Keep the remaining default unit values. Using the Limits command, keep 0,0 for the lower left corner and change the upper right corner from 12,9 to 265.00,200.00. Use the Grid command and change the grid spacing from 1.00 to 10.00 units. Do not turn the snap or ortho On. Since a layer called "Center" must be created to display center lines, use the Ltscale command and change the default value of 1.00 to 25.40. This will make the long and short dashes of the center lines appear on the display screen.

LAYERS:

Create the following layers with the format:
Name-Color-Linetype
Object - White - Continuous
Center - Yellow - Center
Dim - Yellow - Continuous

SUGGESTED COMMANDS:

The object consists of a combination of circles and arcs along with tangent lines and arcs. Use the Point command to identify and lay out the centers of all circles for construction purposes. Use the Arc command to construct a series of arcs for the left side of the Gear-arm. The Trim command will be used to trim circles, lines, and arcs to form the basic shape. Also, use the Circle-TTR command for tangent arcs to existing geometry. Since this object is metric, commands such as Ltscale and Dim-Dimscale need to be set to the metric-inch equivalent of 25.4 units. This value may be adjusted for better results.

DIMENSIONING:

This drawing may be dimensioned at a later date. Consult your instructor before continuing.

PLOTTING:

This tutorial exercise may be plotted on "A"-size paper (8.5" x 11"). Plot the Gear-arm as a metric drawing and a scale of full size, or 1=1.

VERSION OF AUTOCAD:

This tutorial exercise may be completed using either AutoCAD Release 10 or Release 11.

Step #1

Begin the gear-arm by drawing two circles of diameters 9.52 and 15.88 using the Circle command and coordinate 112.00,90.00 as the center of both circles.

Command: **Circle**
3P/2P/TTR/<Center point>: **112.00,90.00** *(Point "A")*
Diameter/<Radius>: **D**
Diameter: **9.52**

Command: **Circle**
3P/2P/TTR/<Center point>: **112.00,90.00** *(Point "A")*
Diameter/<Radius>: **D**
Diameter: **15.88**

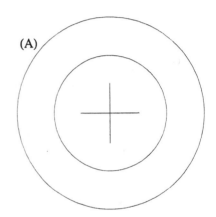

Step #2

Use the Setvar command and set the Pdmode system variable to a value of 2. This will form a "plus" when using the Point command. Again, use the Setvar command to set the Pdsize system variable to a value of 3. This variable controls the size of the point. Use the Point command and the Osnap-Center option to place a point at the center of the two circles.

Command: **Setvar**
Variable name or ?: **Pdmode**
New value for Pdmode <0>: **2**

Command: **Setvar**
Variable name or ? <Pdmode>: **Pdsize**
New value for Pdsize <0>: **3**

Command: **Point**
Point: **Cen**
of *(Select the large circle at "A")*

Step #3

Use the Copy command to duplicate the point using a polar coordinate distance of 86 units in the 0-degree direction.

Command: **Copy**
Select objects: **L** *(This will select the point)*
Select objects: *(Strike Enter to continue)*
<Base point or displacement>/Multiple: **Cen**
of *(Select the center of the large circle at "A")*
Second point of displacement: **@86<0**

Step #4

Use the Circle command to place three circles of different sizes from the same point at "A".

Command: **Circle**
3P/2P/TTR/<Center point>: **Node**
of *(Select the point at "A")*
Diameter/<Radius>: **25.40**

Command: **Circle**
3P/2P/TTR/<Center point>: **Node**
of *(Select the point at "A")*
Diameter/<Radius>: **D**
Diameter: **31.75**

Command: **Circle**
3P/2P/TTR/<Center point>: **Node**
of *(Select the point at "A")*
Diameter/<Radius>: **D**
Diameter: **22.23**

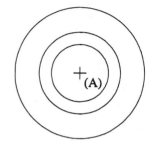

Step #5

Use the Copy command to duplicate point "A" using a polar coordinate distance of 54 units in the 180-degree direction.

Command: **Copy**
Select objects: *(Select the point at "A")*
Select objects: *(Strike Enter to continue)*
<Base point or displacement>/Multiple: **Cen**
of *(Select the center of the large circle at "A")*
Second point of displacement: **@54<180**

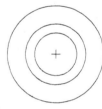

Step #6

Use the Circle command to place two circles of different sizes at point "A". These circles will be converted to arcs in later steps.

Command: **Circle**
3P/2P/TTR/<Center point>: **Node**
of *(Select the point at "A")*
Diameter/<Radius>: **D**
Diameter: **25.40**

Command: **Circle**
3P/2P/TTR/<Center point>: **Node**
of *(Select the point at "A")*
Diameter/<Radius>: **D**
Diameter: **12.75**

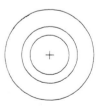

Step #7

Use the Copy command to duplicate point "A" using a polar coordinate distance of 54 units in the 80-degree direction.

Command: **Copy**
Select objects: *(Select the point at "A")*
Select objects: *(Strike Enter to continue)*
<Base point or displacement>/Multiple: **Cen**
of *(Select the center of the large circle at "A")*
Second point of displacement: **@54<80**

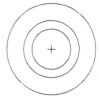

Step #8

Use the Line command to draw a line using a polar coordinate distance of 70 and a direction of 80 degrees. Start the line at point "A". Then use the Circle command to place two circles of different sizes at point "B". These circles will be converted to arcs in later steps.

Command: **Line**
From point: **Cen**
of *(Select the center of the large circle at "A")*
To point: **@70<80**
To point: *(Strike Enter to exit this command)*

Command: **Circle**
3P/2P/TTR/<Center point>: **Node**
of *(Select the point at "B")*
Diameter/<Radius>: **D**
Diameter: **25.40**

Command: **Circle**
3P/2P/TTR/<Center point>: **Node**
of *(Select the point at "B")*
Diameter/<Radius>: **D**
Diameter: **12.75**

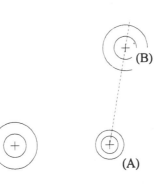
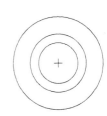

Step #9

Use the Arc command and draw an arc using point "A" as the center, point "B" as the start point, and point "C" as the endpoint.

Command: **Arc**
Center/<Start point>: **C**
Center: **Cen**
of *(Select the center of the circle at "A")*
Start point: **Int**
of *(Select the intersection of the line and circle at "B")*
Angle/Length of chord/<End point>: **Qua**
of *(Select the quadrant of the circle at "C")*

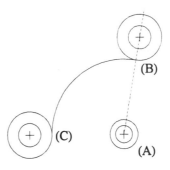
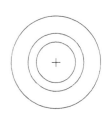

Step #10

Use the Arc command and draw an arc using point "A" as the center, point "B" as the start point, and point "C" as the endpoint.

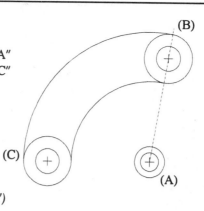
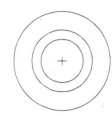

Command: **Arc**
Center/<Start point>: **C**
Center: **Cen**
of *(Select the center of the circle at "A")*
Start point: **Int**
of *(Select the intersection of the line and circle at "B")*
Angle/Length of chord/<End point>: **Qua**
of *(Select the quadrant of the circle at "C")*

Step #11

Use the Trim command, select the two dashed arcs at the right as cutting edges, and trim the two circles at points "C" and "D".

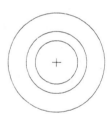

Command: **Trim**
Select cutting edge(s)...
Select objects: *(Select the two dashed arcs "A" and "B")*
Select objects: *(Strike Enter to continue)*
Select object to trim: *(Select the circle at "C")*
Select object to trim: *(Select the circle at "D")*
Select object to trim: *(Strike Enter to exit this command)*

The above prompts for the Trim command illustrate the Release 10 version of AutoCAD. Release 11 users have the following prompt sequence for Trim:

<Select objects to trim>Undo:

A built-in undo is provided in Release 11 if a mistake is made by trimming the wrong entity.

Step #12

Your drawing should be similar to the illustration at the right.

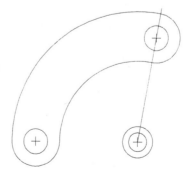
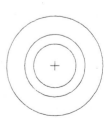

Step #13

Use the Arc command and draw an arc using point "A" as the center, point "B" as the start point, and point "C" as the endpoint.

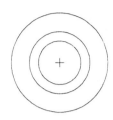
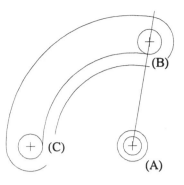

Command: **Arc**
Center/<Start point>: **C**
Center: **Cen**
of *(Select the center of the circle at "A")*
Start point: **Int**
of *(Select the intersection of the line and circle at "B")*
Angle/Length of chord/<End point>: **Qua**
of *(Select the quadrant of the circle at "C")*

Step #14

Use the Arc command and draw an arc using point "A" as the center, point "B" as the start point, and point "C" as the endpoint.

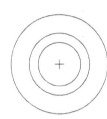
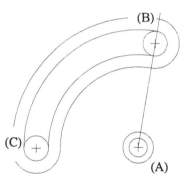

Command: **Arc**
Center/<Start point>: **C**
Center: **Cen**
of *(Select the center of the circle at "A")*
Start point: **Int**
of *(Select the intersection of the line and circle at "B")*
Angle/Length of chord/<End point>: **Qua**
of *(Select the quadrant of the circle at "C")*

Step #15

Use the Trim command, select the two dashed arcs at the right as cutting edges, and trim the two circles at points "C" and "D".

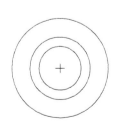
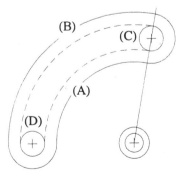

Command: **Trim**
Select cutting edge(s)...
Select objects: *(Select the two dashed arcs "A" and "B")*
Select objects: *(Strike Enter to continue)*
Select object to trim: *(Select the circle at "C")*
Select object to trim: *(Select the circle at "D")*
Select object to trim: *(Strike Enter to exit this command)*

Step #16

Your drawing should be similar to the illustration at the right. Always perform periodic screen redraws using the Redraw command to clean up the display screen.

Command: **Redraw**

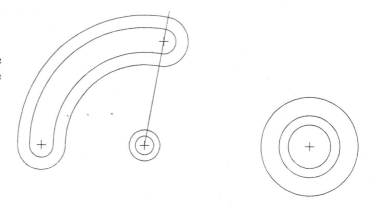

Step #17

Use the Line command and draw a line from the quadrant of the small circle to the quadrant of the large circle. This line is used only for construction purposes.

Command: **Line**
From point: **Qua**
of *(Select the circle at "A")*
To point: **Qua**
of *(Select the circle at "B")*
To point: *(Strike Enter to exit this command)*

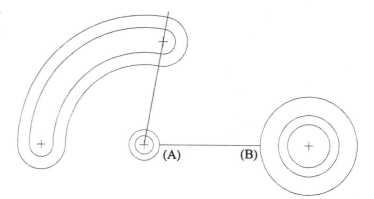

Step #18

Use the Move command to move the dashed line down the distance 19.05 units using the polar coordinate mode. Then, use the Offset command to offset the dashed circle the distance 12.75 units. The intersection of these two entities will be used to draw a 12.75 radius circle.

Command: **Move**
Select objects: *(Select the dashed line at the right)*
Select objects: *(Strike Enter to continue)*
Base point or displacement: **Endp**
of *(Select the endpoint of the line at "A")*
Second point of displacement: **@19.05<270**

Command: **Offset**
Offset distance or Through<Through>: **12.75**
Select object to offset: *(Select the circle at "B")*
Side to offset: *(Select a blank part of the screen at "C")*
Select object to offset: *(Strike Enter to exit this command)*

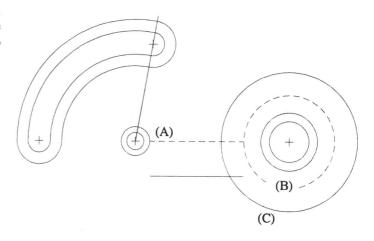

Step #19

Use the Circle command to draw a circle with a radius of 12.75. Use the center of the circle as the intersection of the large dashed circle and the dashed horizontal line illustrated at the right. Use the Erase command to erase the dashed circle and dashed line.

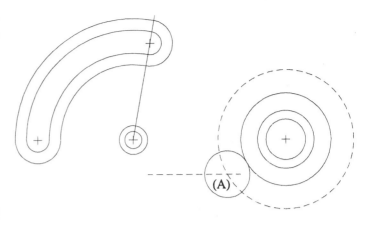

Command: **Circle**
3P/2P/TTR/<Center point>: **Int**
of *(Select the intersection of the line and circle at "A")*
Diameter/<Radius>: **12.75**

Command: **Erase**
Select objects: *(Select the dashed circle and dashed line)*
Select objects: *(Strike Enter to execute this command)*

Step #20

Use the Line command to draw a line from a point tangent to the arc at "A" to a point tangent to the circle at "B". Use the Osnap-Tan option to accomplish this.

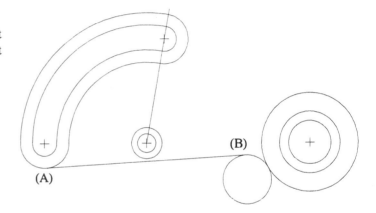

Command: **Line**
From point: **Tan**
to *(Select the arc at "A")*
To point: **Tan**
to *(Select the circle at "B")*
To point: *(Strike Enter to exit this command)*

Step #21

Use the Trim command, select the dashed line and dashed circle illustrated at the right as cutting edges, and trim the circle at "A".)

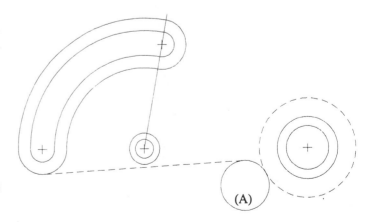

Command: **Trim**
Select cutting edge(s)...
Select objects: *(Select the dashed line at the right)*
Select objects: *(Select the dashed circle at the right)*
Select objects: *(Strike Enter to continue)*
Select object to trim: *(Select the circle at "A")*
Select object to trim: *(Strike Enter to exit this command)*

Step #22

Use the Circle-TTR command to draw a circle tangent
to the arc at "A" and tangent to the circle at "B" with
a radius of 51. When using the TTR option in the
Circle command, the Osnap-Tan option is automatically
invoked.

Command: **Circle**
3P/2P/TTR/<Center point>: **TTR**
Enter Tangent spec: *(Select the arc at "A")*
Enter second Tangent spec: *(Select the circle at "B")*
Radius: **51**

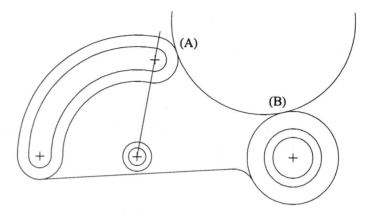

Step #23

Use the Trim command, select the dashed arc and dashed
circle illustrated at the right as cutting edges, and trim
the large 51 radius circle at "A".

Command: **Trim**
Select cutting edge(s)...
Select objects: *(Select the dashed arc at the right)*
Select objects: *(Select the dashed circle at the right)*
Select objects: *(Strike Enter to continue)*
Select object to trim: *(Select the circle at "A")*
Select object to trim: *(Strike Enter to exit this command)*

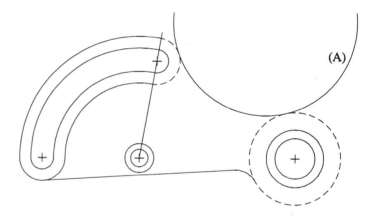

Step #24

Use the Trim command, select the two dashed arcs
illustrated at the right as cutting edges, and trim the
circle at "C". Use the Erase command to delete all four
points used to construct the circles.

Command: **Trim**
Select cutting edge(s)...
Select objects: *(Select the dashed arc "A")*
Select objects: *(Select the dashed arc "B")*
Select objects: *(Strike Enter to continue)*
Select object to trim: *(Select the circle at "C")*
Select object to trim: *(Strike Enter to exit this command)*

Command: **Erase**
Select objects: *(Select the four points illustrated at the
   right)*
Select objects: *(Strike Enter to execute this command)*

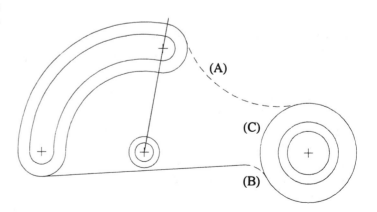

Step #25

Your display should appear similar to the illustration at the right. Notice the absence of the points that were erased in the previous step. Standard center lines will be placed to mark the center of all circles in the next series of steps. Use the Layer command to set the new current layer to Center. Change Ltscale to 25.40.

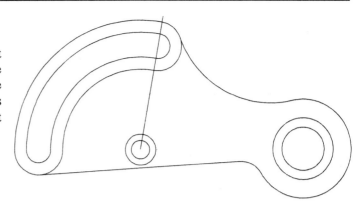

Command: **Ltscale**
New scale factor <1.0000>: **25.40**

Step #26

To place center lines for all circles and arcs, two dimension variables need to be changed. The dimension variable, Dimscale, is set to a value of 1. Since this is a metric drawing, the Dimscale variable needs to be changed to a value of 25.4. This will increase all variables by this value which is necessary because we are drawing in metric units. Also, the Dimcen variable needs to be changed from a value of 0.09 to -0.09. The negative value will extend the center lines past the edge of the circle when using the Dim-Cen command.

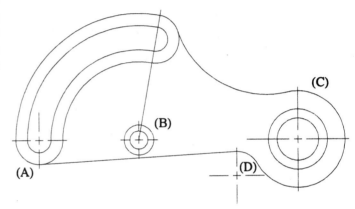

Command: **Dim**
Dim: **Dimscale**
Current value <1> New value: **25.4**

Dim: **Dimcen**
Current value <0.09> New value: **-0.09**

Dim: **Center**
Select arc or circle: *(Select the arc at "A")*
Dim: **Center**
Select arc or circle: *(Select the circle at "B")*
Dim: **Center**
Select arc or circle: *(Select the arc at "C")*
Dim: **Center**
Select arc or circle: *(Select the arc at "D")*
Dim: **Exit** *(To return to the "Command" prompt)*

Step #27

Use the Offset command to offset the inside arc at "A" the distance 6.375 units. Indicate a point in the vicinity of "B" for the side to perform the offset.

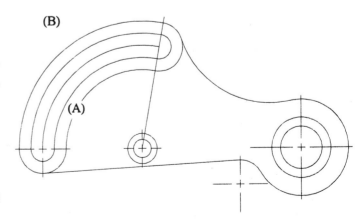

Command: **Offset**
Offset distance or Through <Through>: **6.375**
Select object to offset: *(Select the arc at "A")*
Side to offset? *(Select a point in the vicinity of "B")*
Select object to offset: *(Strike Enter to exit this command)*

Step #28

Use the Chprop command to change the arc at "A" and the line at "B" to the Center layer. This layer should have been created earlier for the arc and line to change to the proper color and linetype.

Command: **Chprop**
Select objects: *(Select the middle arc at "A")*
Select objects: *(Select the line at "B")*
Select objects: *(Strike Enter to continue)*
Change what property(Color/LAyer/LType/Thickness):
    LA
New layer <0>: **Center**
Change what property(Color/LAyer/LType/Thickness):
    (Strike Enter to exit this command)

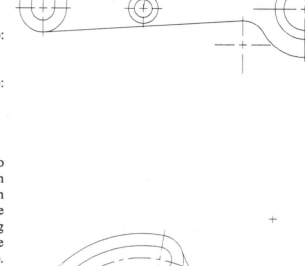

Step #29

Use the Extend command to extend the center line arc to intersect with the circular arc at "A"; see the illustration at the right. Next, reset the dimension variable Dimcen from -0.09 to a new value of 0.09. This will change the center point to a plus without the center lines extending beyond the arc when using the Dim-Cen command. The finished object may be dimensioned as an optional step.

Command: **Extend**
Select boundary edge(s)...
Select objects: *(Select the arc at "A")*
Select objects: *(Strike Enter to continue)*
Select object to extend: *(Select the center line arc)*
Select object to extend: *(Strike Enter to exit this command)*

Command: **Dim**
Dim: **Dimcen**
Current value <-0.09> New value: **0.09**

Dim: **Center**
Select arc or circle: *(Select the arc at "B")*

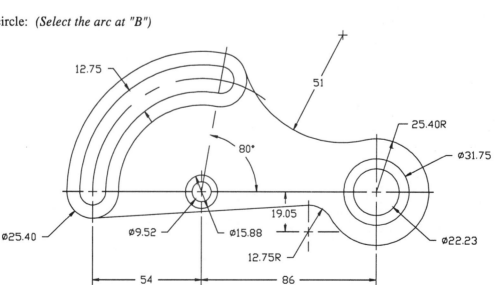

Problems for Unit 2

Directions for Problems 2-1 through 2-25:
Construct these geometric construction figures using existing AutoCAD commands.

Problem 2-1

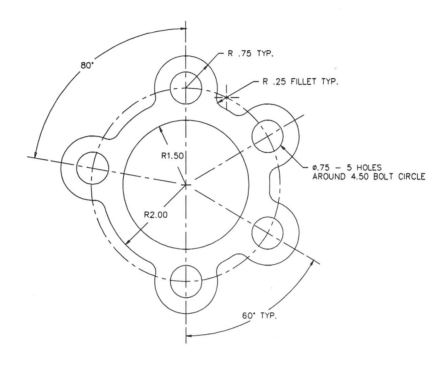

Problem 2-2

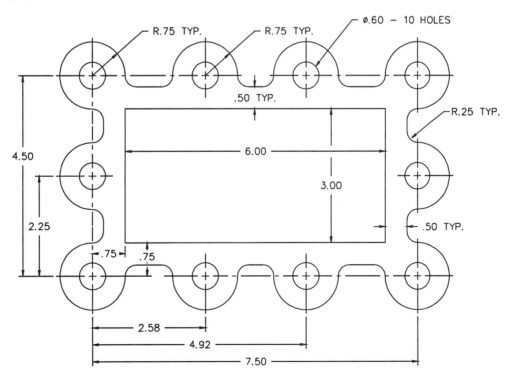

Problem 2-3

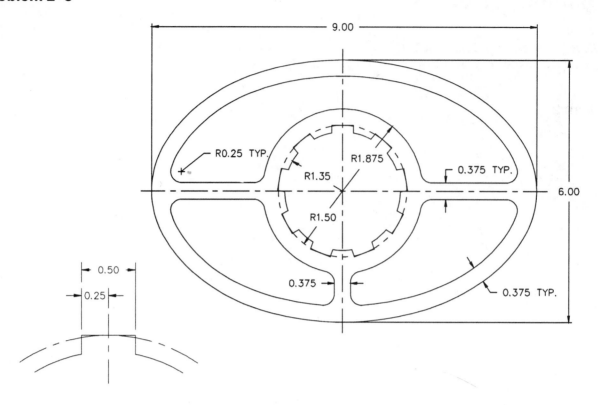

Problem 2-4

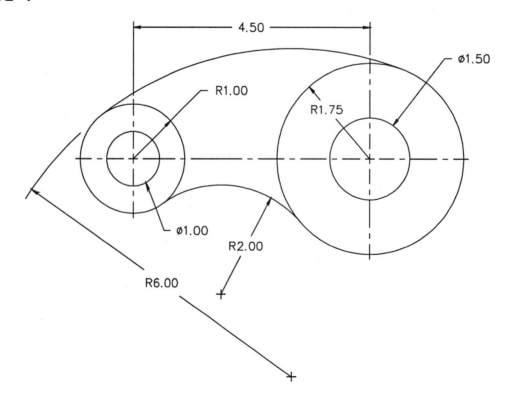

Problem 2-5

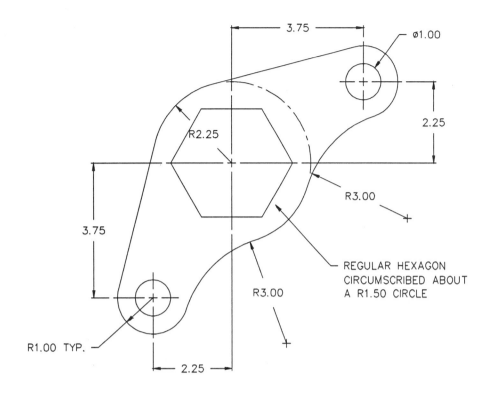

3.75

Ø1.00

R2.25

2.25

3.75

R3.00

REGULAR HEXAGON
CIRCUMSCRIBED ABOUT
A R1.50 CIRCLE

R3.00

R1.00 TYP.

2.25

Problem 2-6

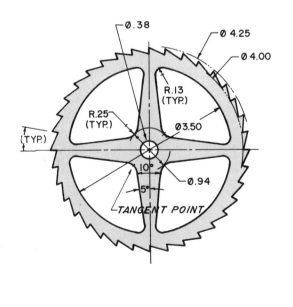

Ø.38

Ø4.25

Ø4.00

R.13
(TYP.)

R.25
(TYP.)

Ø3.50

(TYP.)

10°

Ø.94

5°

TANGENT POINT

Problem 2-7

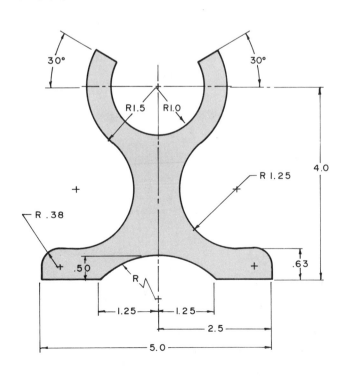

30°   30°

R1.5   R1.0

4.0

R1.25

R.38

.63

.50

R

1.25   1.25

2.5

5.0

Problem 2-8

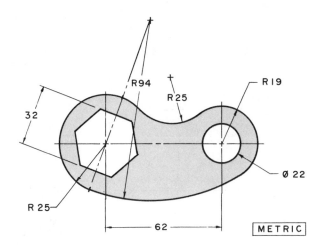

Problem 2-9

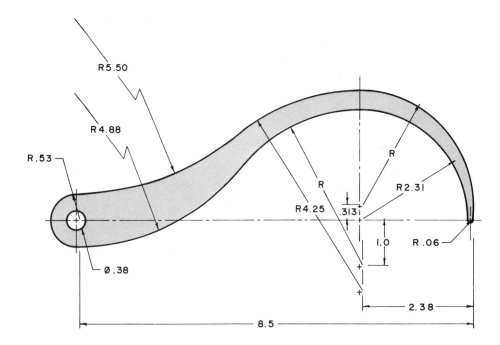

Problem 2-10

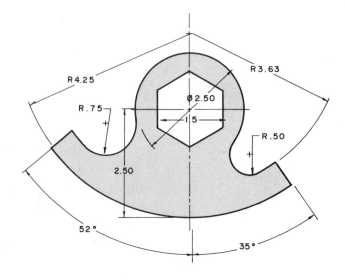

Problem 2-11

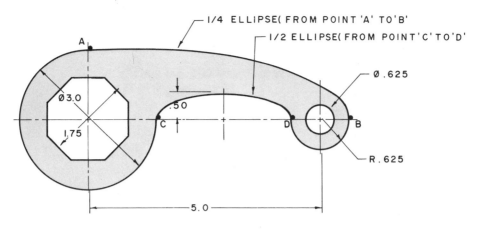

1/4 ELLIPSE(FROM POINT 'A' TO 'B'

1/2 ELLIPSE(FROM POINT 'C' TO 'D'

Ø.625

Ø3.0

1.75

.50

R.625

5.0

A

C

D

B

Problem 2-12

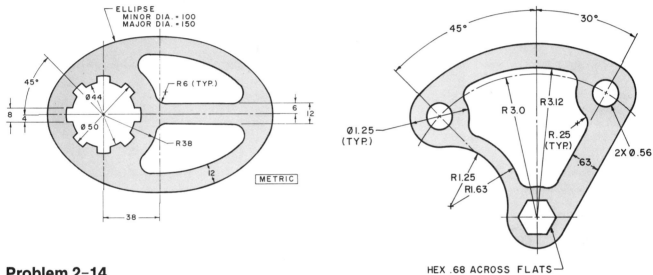

ELLIPSE
MINOR DIA. = 100
MAJOR DIA. = 150

R6 (TYP.)

45°

Ø44

Ø50

R38

8

4

6

12

12

38

METRIC

Problem 2-13

45°

30°

R 3.0

R 3.12

Ø1.25
(TYP.)

R.25
(TYP.)

R1.25

R1.63

.63

2X Ø.56

HEX .68 ACROSS FLATS

Problem 2-14

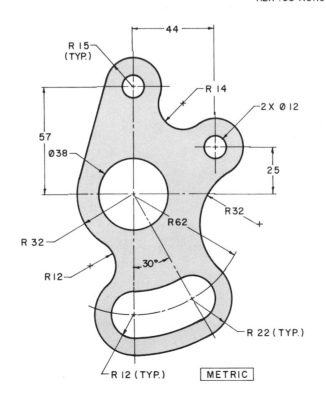

R 15
(TYP.)

44

R 14

2X Ø 12

57

Ø38

25

R 32

R62

R 32

R 12

30°

R 22 (TYP.)

R 12 (TYP.)

METRIC

Problem 2-15

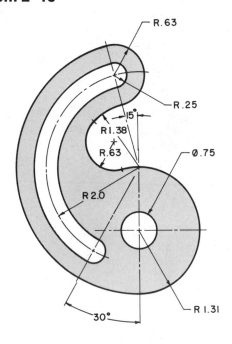

R.63
R.25
15°
R1.38
R.63
R2.0
Ø.75
30°
R1.31

Problem 2-17

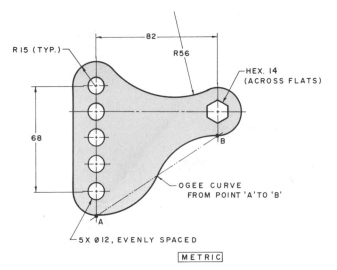

R15 (TYP.)
82
R56
HEX. 14
(ACROSS FLATS)
68
B
OGEE CURVE
FROM POINT 'A' TO 'B'
A
5X Ø12, EVENLY SPACED

METRIC

Problem 2-16

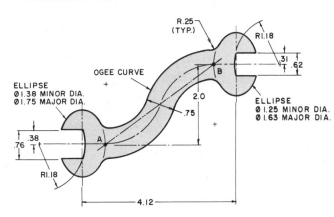

R.25
(TYP.)
R1.18
.31
.62
OGEE CURVE
B
2.0
ELLIPSE
Ø1.38 MINOR DIA.
Ø1.75 MAJOR DIA.
.75
ELLIPSE
Ø1.25 MINOR DIA.
Ø1.63 MAJOR DIA.
.38
.76
A
R1.18
4.12

Problem 2-18

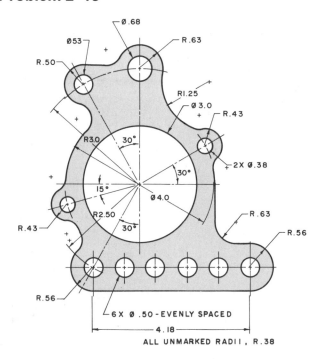

Ø.68
Ø53
R.63
R.50
R1.25
Ø3.0
R.43
R3.0
30°
2X Ø.38
30°
15°
Ø4.0
R.63
R2.50
30°
R.56
R.43
R.56
6X Ø.50 - EVENLY SPACED
4.18
ALL UNMARKED RADII, R.38

Problem 2-19

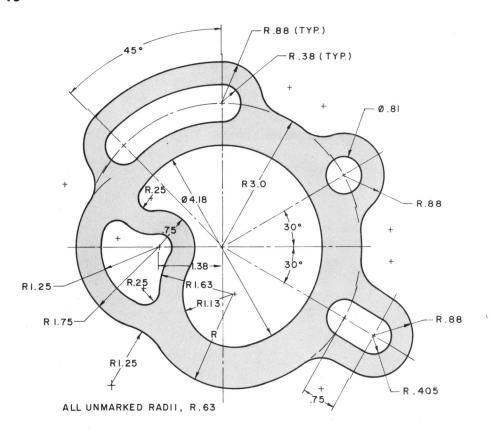

R.88 (TYP.)
R.38 (TYP.)
45°
Ø.81
R 3.0
R.88
R.25
Ø 4.18
30°
.75
30°
R1.25
1.38
R1.63
R.88
R.25
R1.13
R 1.75
R
R.405
R1.25
.75

ALL UNMARKED RADII, R.63

Problem 2-20

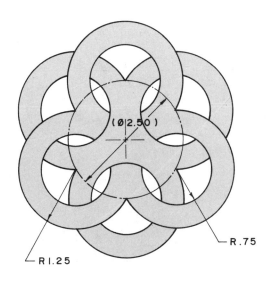

(Ø 2.50)

R.75

R 1.25

Problem 2-21

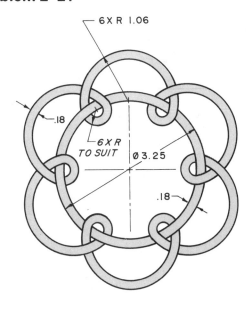

6X R 1.06

.18

6X R
TO SUIT

Ø 3.25

.18

Problem 2-22

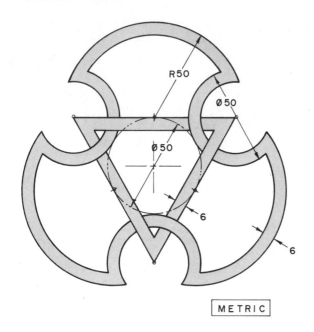

R50

Ø50

Ø50

6

6

METRIC

Problem 2-23

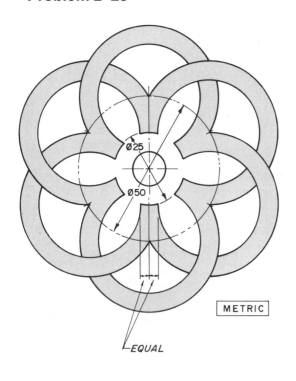

Ø25

Ø50

EQUAL

METRIC

Problem 2-24

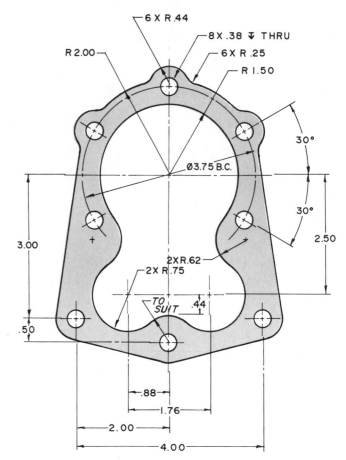

6 X R.44

8 X .38 ⊥ THRU

6 X R.25

R 1.50

R 2.00

30°

30°

Ø3.75 B.C.

3.00

2.50

2X R.62

2X R.75

TO SUIT

.44

.50

.88

1.76

2.00

4.00

Problem 2-25

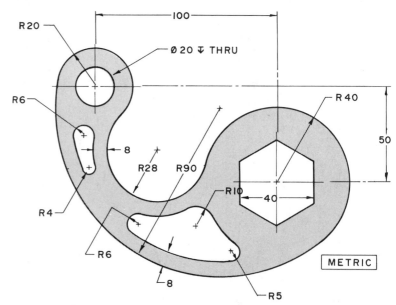

100

R20

Ø 20 ⊥ THRU

R6

R 40

50

8

R28

R90

R10

40

R4

R6

8

R5

METRIC

84

Unit 2

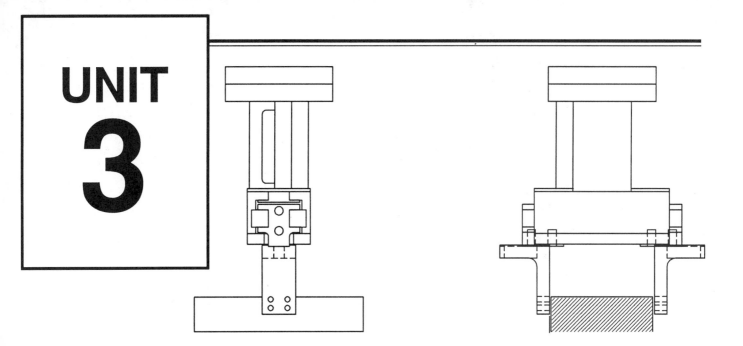

UNIT 3

Shape Description/ Multi-View Projection

Contents

Before any object is made in production, some type of drawing needs to be made. This is not just any drawing; rather, an engineering drawing consisting of overall sizes of the object and various views of the object is organized on the computer screen. This unit introduces the topic of shape description or how many views are really needed to describe an object, along with the art of multi-view projection with methods of constructing one-view, two-view, and three-view drawings using AutoCAD commands. Linetypes are explained as a method of communicating hidden features located in different views of a drawing.

Shape Description

Before performing engineering drawings, an analysis of the object being drawn must first be made. This takes the form of describing the object by views or how an observer looks at the object. In the illustration at the right of the simple wedge, it is no surprise that this object can be viewed at almost any angle to get a better idea of its basic shape. However, some standard method of determining how to and where to view the object must be exercised. This is to standardize how all objects are to be viewed in addition to limiting confusion that is usually associated with complex multi-view drawings.

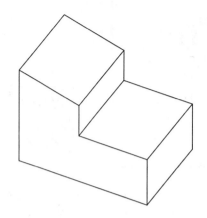

Even though the simple wedge is easy to understand because it is currently being displayed in picture or isometric form, it would be difficult to produce this object since it is unclear what the size of the front and top views are. In this way, the picture of the object is separated into six primary ways or directions to view an object as illustrated at the right. The front view begins the shape description followed by the top view and right side view. The left side view, back view, and bottom view complete the primary ways to view an object.

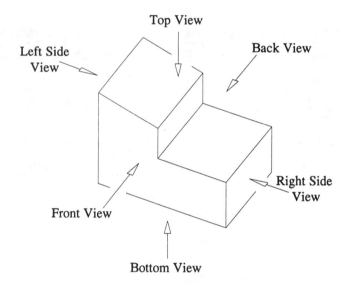

Now that the primary ways of viewing an object have been established, the views need to be organized in some manner to promote clarity and have the views reference themselves. Imagine the simple wedge positioned in a clear, transparent glass box similar to the illustration at the right. With the entire object at the center of the box, the sides of the box represent the ways to view the object. Images of the simple wedge are projected onto the glass surfaces of the box.

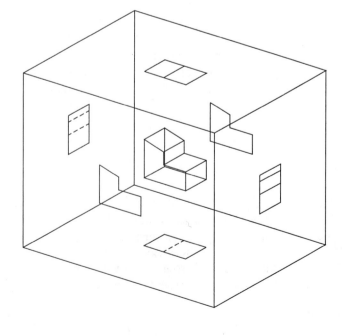

Shape Description Continued

With the views projected onto the sides of the glass box, we must now prepare the views to be placed on a two-dimensional (2D) drawing screen. To accomplish this, the glass box, which is hinged, is unfolded as in the illustration at the right. All folds occur from the front view which remains stationary.

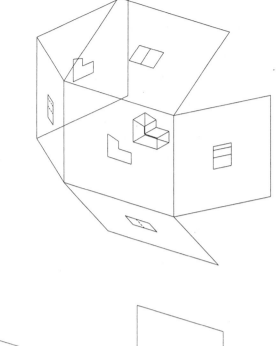

The illustration at the right shows the views in their proper alignment to one another; however, this illustration is still in a pictorial view. These views need to be placed flat before continuing.

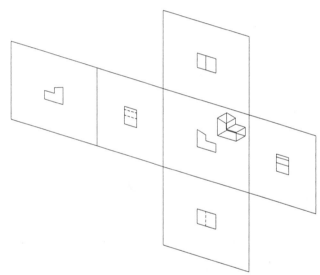

As the glass box is completely unfolded and laid flat, the result is illustrated at the right. The front view becomes the main view with other views being placed in relation to the front. Above the front view is the top view. To the right of the front is the right side view. To the left of the front is the left side view followed by the back view. Underneath the front view is the bottom view. This becomes the standard method of laying out the necessary views to describe an object. But are all views necessary? Upon closer inspection we find that, except for being a mirror image, the front and back views are identical. The top and bottom views appear similar as do the right and left side views. One very important rule to follow in multi-view objects is to only select those views that accurately describe the object and discard the remaining views.

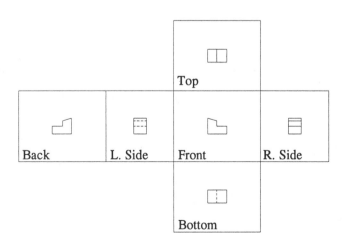

Shape Description Continued

The complete multi-view drawing of the simple wedge is illustrated at the right. Only the front, top, and right side views were needed to describe this object. Important information to remember when laying out views of an object is the front view is usually the most important view and holds the basic shape of the object being described. Directly above the front view is the top view and to the right of the front is the right side view. All three views are separated by spaces of various sizes. This space is commonly called a dimension space because it becomes a good area to place dimensions describing the size of the object. The space also acts as a separator between views; without it, the views would touch one another which would be difficult to read and interpret. The minimum distance of this space is usually 1.00 units.

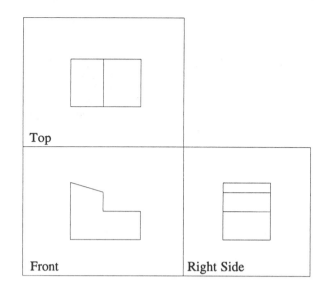

Relationships between Views

Some very interesting and important relationships are set up when placing the three views in the configuration illustrated at the right. Notice that since the top view is directly above the front view, and both views share the same width dimension. The front and right side views share the same height. The relationship of depth on the top and right side views can be explained by constructing a 45-degree projector line at "A" and projecting the lines over and down or vice versa to get the depth. Yet another principle illustrated by this example is that of projecting lines up, over, and across and down to create views. Editing commands such as Erase and Trim are then used to clean up unnecessary lines.

With the three views identified at the right, the final step is to annotate the drawing or add dimensions to the views. With dimensions, the object created in multi-view projection can now be produced. Even though this has been a simple example, the methods of multi-projection work even for the most difficult and complex objects.

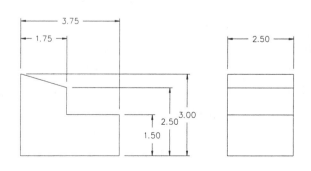

Linetypes and Conventions

The heart of any engineering drawing is the ability to assign different types of lines as a method of conveying some type of meaning to the drawing. All lines of a drawing are dark when output to a plotter; border and title block lines are the thickest lines of a drawing. Object lines outline visible features of a drawing and are made thick and dark (but not as thick as a border line). To identify features that are invisible in an adjacent view, a hidden line is used. This line is a series of dashes 0.12 units in length with a spacing of 0.06 units. Center lines are used to identify the centers of circular features such as holes. They are also used to show that a hidden feature in one view is circular in another. The center line consists of a series of long and short dashes. The short dash measures approximately 0.12 units while the long dash may vary from 0.75 to 1.50 units. A gap of 0.06 is placed in between dashes. Study the examples of these lines in the illustration at the right.

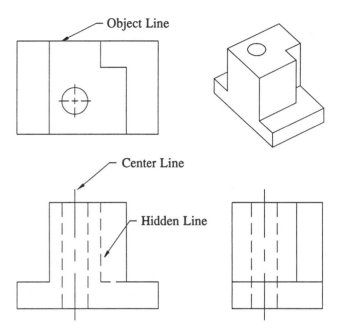

The object shown at the right is a good illustration of the use of phantom lines in a drawing. Phantom lines are especially useful where motion is applied. The arm on the right is shown using standard object lines consisting of the continuous linetype. To show that the arm rotates about a center pivot, the identical arm is duplicated using the Mirror command and all lines of this new element converted to phantom lines using the Change command. Notice the smaller segments are not shown as phantom lines due to their short sizes.

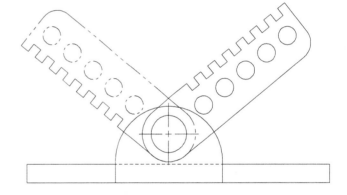

When using manual drafting techniques for drawing hidden and center lines before CAD systems were developed, the drafter first had to develop a skill of drawing a series of dashes as individual line segments along with spacing each dash equally. This was very tedious work but with practice and experience, the method proved acceptable in the field of engineering drawings. Since CAD has come along, the operator has the option of assigning a linetype to the drawing. This means that if the current linetype is called Hidden, the series of dashes and spaces will automatically be drawn once a line, circle, arc, or any drawing entity is drawn. Illustrated at the right were the standard linetypes supplied with AutoCAD ever since the software was introduced in 1982. The linetypes at the right were taken from the Release 10 version of AutoCAD.

Linetypes and Conventions Continued

The linetypes illustrated at the right are the current linetype definitions supplied with AutoCAD Release 11. These linetypes may be viewed by using the DOS "Type" command to type out a file called ACAD.LIN located in the AutoCAD subdirectory. The lines at the right are only a partial listing of the complete file. The Center linetype from AutoCAD Release 11 is identical to that of Release 10. There also exists two more types of center lines; Center2 is the same type of center linetype except that all dashes and spaces are half the size of the orginal Center linetype. CenterX2 has the orginal linetype doubled in size. All current linetypes of AutoCAD have three possible linetypes for the operator to choose from.

BORDER,____ .____ ____ .____ ____ .____
BORDER2,___.___.___.___.___.___.___.
BORDERX2,____ ____ . ____ ____ . ____ ____ . ____
CENTER,____ ____ ____ ____ ____ ____ ____
CENTER2,___ __ ___ __ ___ __ ___ __ ___ __
CENTERX2,____ ____ ____ ____ ____
DASHDOT,__ . __ . __ . __ . __ . __ . __
DASHDOT2,_._._._._._._._._._._._._._._._.
DASHDOTX2,____ . ____ . ____ . ____ . ____ .
DASHED,__ __ __ __ __ __ __ __ __ __ __ __
DASHED2,_ _ _ _ _ _ _ _ _ _ _ _ _ _ _ _ _ _
DASHEDX2,____ ____ ____ ____ ____ ____ ____

Use of linetypes in a drawing is crucial to the interpretation of the views and the final design before the object is actually made. Sometimes the linetype appears too long; in other cases the linetype does not appear at all even though using the List command on the entity will show the proper layer and linetype. The Ltscale command is used to manipulate the size of all linetypes loaded into a drawing. By default, all linetypes are assigned a scale factor of 1.00. This means that the actual dashes and/or spaces of the linetype are multiplied by this factor. The views illustrated show linetypes and use the default value of 1.00 from the Ltscale command.

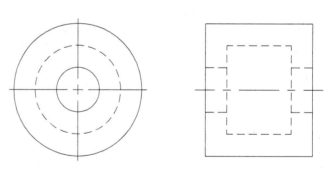

If a linetype appears too long, use the Ltscale command and set a new value to less than 1.00. If a linetype appears too short, use the Ltscale command and set a new value to greater than 1.00. The same views illustrated at the right show the affects of the Ltscale command set to a new value of 0.75. Notice the center in the right side view has one more series of dashes in its appearance than the same object illustrated above. The value 0.75 is the new multiplier that affects all dashes and spaces defined in the linetype.

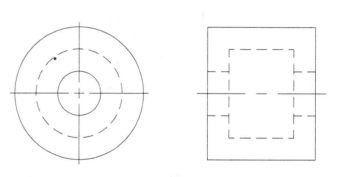

In this third example using the Ltscale command, a new value of 0.50 has been used to shorten the linetypes even more. Now, even the center mark identifying the circles in the front view have been changed into center lines. One other important note to remember when using the Ltscale command is that the new value, whether larger or smaller that 1.00, affects all linetypes visible on the display screen. In other words, it is not possible to affect a hidden line set to a certain scale without affecting a center line with the same scale. There are, however, the new sets of linetypes already defined as different scales supplied with Release 11 to make the scaling of individual linetypes much easier.

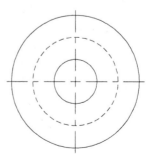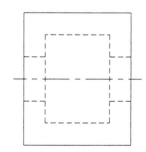

One-View Drawings

An important rule to remember concerning multi-view drawings is draw only enough views to accurately describe the object. In the drawing of the gasket at the right, a front and side view are shown; however, the side view is so narrow that it is difficult to interpret the hidden lines drawn inside. A better approach would be to leave out the side view and construct a one-view drawing consisting of just the front view.

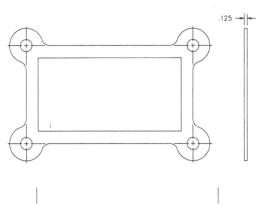

Begin the one-view drawing of the gasket by first laying out center lines marking the centers of all circles and arcs. A layer containing center lines could be used to show all lines as center lines.

.125 GASKET THICKNESS

Use the Circle command to layout all circles representing the bolt holes of the gasket. The Offset command could be used to form the large rectangle on the inside of the gasket. If lines of the rectangle extend past each other, use the Fillet command set to a value of 0. Selecting two lines of the rectangle will form a corner. Repeat this procedure for any other lines that do not form exact corners.

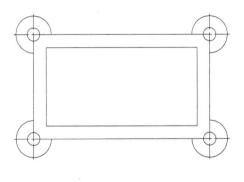

Use the Trim command to begin forming the outside arcs of the gasket.

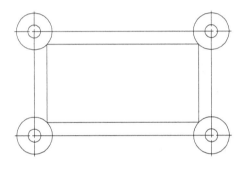

Use the Fillet command set to the desired radius to form a smooth transition from the arcs to the outer rectangle.

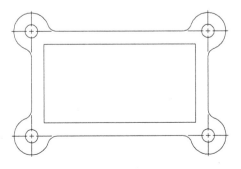

Two-View Drawings

Before attempting any drawing, some thought needs to be made in determining how many views need to be drawn. Only a minimum amount of views are needed to describe an object. Drawing extra views is not only time consuming, but may result in two identical views with mistakes in each view. The operator must interpret which is the correct set of views. The illustration at the right is a three-view, multi-view drawing of a coupler. The front view is identified by the circles and circular hidden circle. Except for their rotation angles, the top and right side views are identical. In this example or for other symetrical objects, only two views are needed to accurately describe the object being drawn. The top view has been deleted to leave the front and right side views. The side view could have easily been deleted in favor of leaving the front and top views. This decision is up to the designer depending on sheet size and which views are best suited for the particular application.

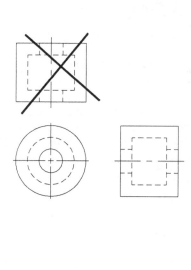

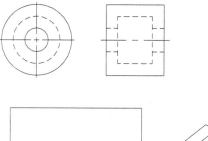

To illustrate how AutoCAD is used as the vehicle for creating a two-view engineering drawing, study the pictorial drawing illustrated at the right to get an idea of how the drawing will appear. Begin the two-view drawing by using the Line command to layout the front and side views. The width of the top view may be found by projecting lines up from the front since both views share the same width. Provide a space of 1.50 units in between views to act as a separator and allow for dimensions to be added at a later time.

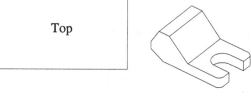

Begin adding visible details to the views such as circles, filleted corners, and angles. Use various editing commands such as Trim, Extend, and Offset to clean up unnecessary geometry.

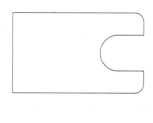

Two-View Drawings Continued

From the front view, project corners up into the top view. These corners will form visible edges in the top view. Use the same projection technique to project features from the top view into the front view.

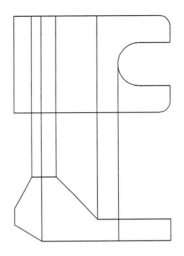

Use the Trim command to delete any geometry that appears in the 1.50 dimension space. The views now must conform to engineering standards by determining which lines are visible and which are invisible. The corner at "A" represents an area hidden in the top view. Use the Chprop command to convert the line in the top view from the continuous linetype to the hidden linetype. In the same manner, the slot visible in the top view is hidden in the front view. Again, use Chprop to convert the continuous line in the front view to the hidden linetype. Since the slot in the top view represents a circular feature, use the Dim-Cen command to place a center marker at the center of the semi-circle. To show in the front view that the hidden line represents a circular feature, add one center line consisting of one short dash and two short dashes. If the slot in the top view was square instead of circular, center lines would not be necessary.

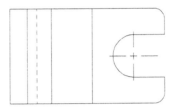

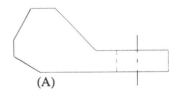

(A)

Use the spaces provided to properly add dimensions to the drawing. Once the dimension spaces are filled with numbers, use outside areas to call out distances. Placing dimensions will be discussed in a later chapter.

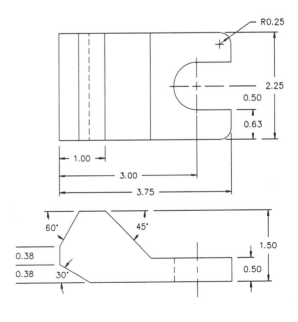

Three-View Drawings

If two views are not enough to describe an object, a three-view drawing of the object is drawn. This consists of front, top, and right side views. A three-view drawing of the guide block illustrated in pictorial format will be the focus of this segment. Notice the broken section exposing the spotfacing operation above a drill hole. Begin this drawing by laying out all views using overall dimensions of width, depth, and height. The Line command along with Offset are popular commands used to accomplish this. Provide a space in between views to accommodate dimensions at a later time.

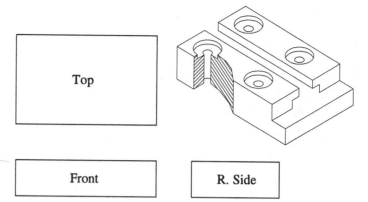

Begin drawing features in the views they appear visible in. Since the spotface holes appear above, draw these in the top view. The notch appears in the front view; draw it there. A slot is visible in the right side view and is drawn there.

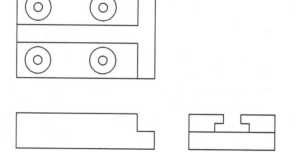

As in two-view drawings, all features are projected down from top to front view. To project depth measurements from top to right side views, construct a 45-degree line at "A" to accomplish this.

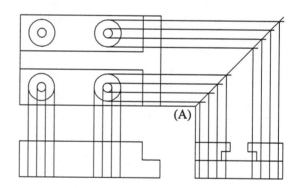

Use the 45-degree line to project the slot from the right side to the top view. Project the height of the slot from the right side to the front view. Use the Chprop command to convert continuous lines to hidden lines where features appear invisible such as the holes in the front and right side views.

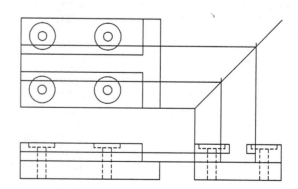

Three-View Drawings Continued

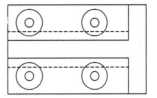

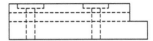

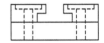

Use the Chprop command to change the remaining lines from continuous to hidden. Erase any construction lines including the 45-degree projection line.

Begin adding center lines to label circular features. The Dim-Cen command is used where the circles are visible. Where features are hidden but represent circular features, the single center line consisting of one short dash and two long dashes is used. In the illustration below, dimensions remain the final step in completing the engineering drawing before being checked and shipped off for production.

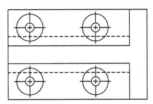

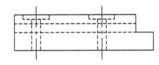

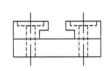

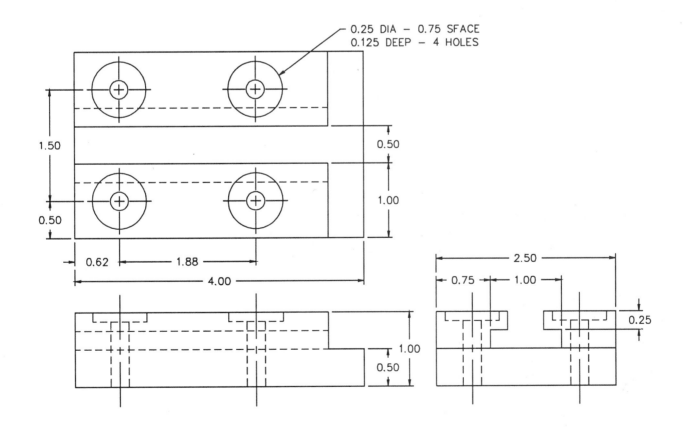

0.25 DIA — 0.75 SPACE
0.125 DEEP — 4 HOLES

1.50

0.50

0.50

1.00

0.62   1.88

4.00

2.50

0.75   1.00

0.25

1.00

0.50

1.00

0.50

Fillets and Rounds

Numerous objects require highly finished and polished surfaces consisting of extremely sharp corners. Fillets and Rounds represent the opposite case where corners are rounded off either for ornamental purposes or required by design. Generally, a fillet consists of a rounded edge formed in the corner of an object illustrated at the right at "A". A round is formed at an outside corner similar to "B". Fillets and rounds are primarily used where objects are cast or made from poured metal. The metal will form easier around a pattern that has rounded corners versus sharp corners which usually break away. Some drawings have so many fillets and rounds that a note is used to convey the size of all similar corners: "All Fillets and Rounds .125 Radius".

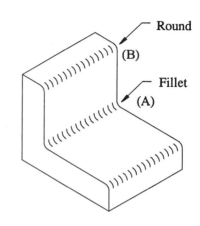

AutoCAD provides the Fillet command which allows the user to enter a radius followed by the selection of two lines. The result will be a fillet of the specified radius to the two lines selected. The two lines are also automatically trimmed leaving the radius drawn from the endpoint of one line to the endpoint of the other line. Illustrated to the right and the bottom are examples of using the Fillet command. Because sometimes the command is used repeatedly, the Multiple option automatically repeats the next command entered from the keyboard; in this case the Fillet command. Since Multiple continually repeats the command, use the CTRL-C keys to cancel the command and return back to the command prompt.

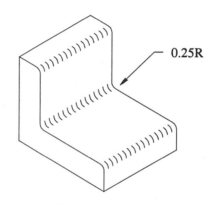

Command: **Multiple**
Fillet
Polyline/Radius/<Select two objects>: **R**
Enter fillet radius <0.0000>: **.25**
Polyline/Radius/<Select two objects>: *(Select at "A" and "B")*
Polyline/Radius/<Select two objects>: *(Select at "B" and "C")*
Polyline/Radius/<Select two objects>: *(Select at "C" and "D")*
Polyline/Radius/<Select two objects>: *(Type CTRL-C to cancel)*

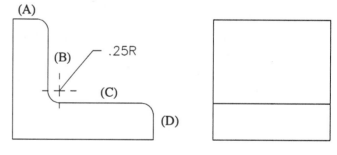

Yet another powerful feature of the Fillet command is to connect two lines at their intersections or corner the two lines. This is accomplished by setting the fillet radius to 0 and selecting the two lines. This is illustrated at the right.

Command: **Fillet**
Polyline/Radius/<Select two objects>: **R**
Enter fillet radius <0.2500>: **0**

Command: **Fillet**
Polyline/Radius/<Select two objects>: *(Select at "A" and "B")*

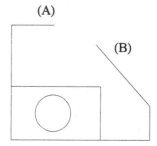
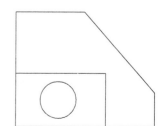

Chamfers

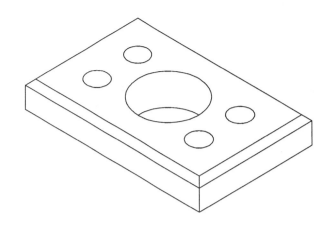

Chamfers represent yet another way to finish a sharp corner of an object. As fillets and rounds result from a pattern making operation and remain unfinished, a chamfer is a machining operation which may even result in a polishing operation. The illustration at the right is one example of an object that has been chamfered along its top edge.

As with the Fillet command, AutoCAD also provides a Chamfer command designed to draw an angle across a sharp corner given two chamfer distances. The most popular chamfer involves a 45-degree angle which is illustrated at the right. Even though this command does not allow the user to specify an angle, the operator may control the angle by the distances entered. In the example at the right, by specifying the same numeric value for both chamfer distances, a 45-degree chamfer will automatically be formed. As long as both distances are the same, a 45-degree chamfer will always be drawn. Study the illustration at the right and prompts below.

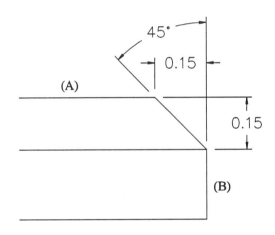

Command: **Chamfer**
Polyline/Distances/<Select first line>: **D**
Enter first chamfer distance <0.0000>: **.15**
Enter second chamfer distance <0.1500>: *(Strike Enter to accept)*

Command: **Chamfer**
Polyline/Distances/<Select first line>: *(Select the line at "A")*
Select second line: *(Select the line at "B")*

The illustration at the right is similar to the 45-degree chamfer above with the exception that the angle is different. This results from two different chamfer distances outlined in the prompts below. This type of edge is commonly called a bevel.

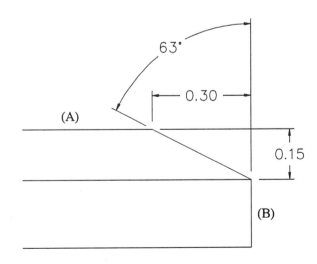

Command: **Chamfer**
Polyline/Distances/<Select first line>: **D**
Enter first chamfer distance <0.0000>: **.30**
Enter second chamfer distance <0.3000>: **.15**

Command: **Chamfer**
Polyline/Distances/<Select first line>: *(Select the line at "A")*
Select second line: *(Select the line at "B")*

Runouts

Where flat surfaces become tangent to cylinders, there must be some method of accurately representing this using fillets. In the object illustrated at the right, the front view shows two cylinders connected to each other by a tangent slab. The top view is complete; the front view has all geometry necessary to describe the object with the exception of the exact intersection of the slab with the cylinder.

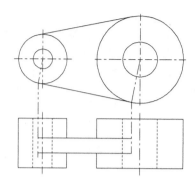

This illustration displays the correct method for finding intersections or runouts; areas where surfaces intersect others and blend in, disappear, or simply runout. A point of intersection is found at "A" in the top view with the intersecting slab and the cylinder. This actually forms a 90-degree angle with the line projected from the center of the cylinder and the angle made by the slab. A line is projected from "A" in the top view to intersect with the slab found in the front view.

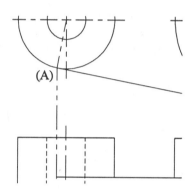

Fillets are drawn to represent the slab and cylinder intersections. This forms the runout.

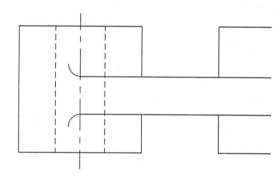

The resulting two-view drawing complete with runouts is illustrated at the right.

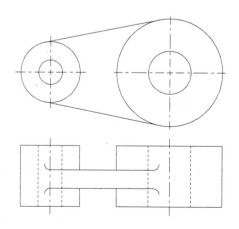

Tutorial Exercise #6
Shifter.Dwg

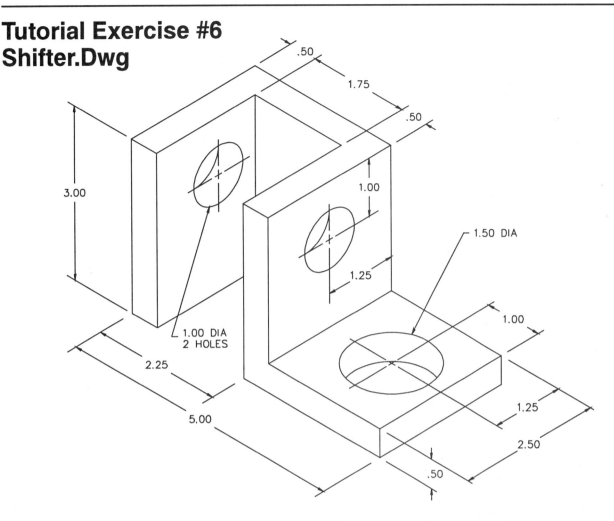

PURPOSE:
This tutorial is designed to allow the user to construct a three-view drawing of the Shifter.

SYSTEM SETTINGS:
Begin a new drawing called "Shifter." Use the Units command to change the number of decimal places past the zero from 4 to 2. Keep the remaining default unit values. Using the Limits command, keep 0,0 for the lower left corner and change the upper right corner from 12,9 to 10.50,8.00. The Grid command does not need to be set to any certain value. Do not turn snap or ortho On.

LAYERS:
Create the following layers with the format:

Name-Color-Linetype
Object - Green - Continuous
Hidden - Red - Hidden
Center - Yellow - Center
Dim-Yellow-Continuous

SUGGESTED COMMANDS:
The primary commands used during this tutorial are Offset and Trim. The Offset command is used for laying out all views before using the Trim command to clean up excess lines. Since different linetypes represent certain parts of a drawing, the Change command is used to convert to the desired linetype needed as set by the Layer command. Once all visible details are identified in the primary views, project the visible features into the other views using the Line command. A 45-degree inclined line is constructed to project lines from the top view to the right side view and vice versa.

DIMENSIONING:
Dimensions may be added to this problem at a later time. Consult your instructor.

PLOTTING:
This tutorial exercise may be plotted on "A"-size paper (8.5" x 11"). Use a plotting scale of 1=1 to produce a full size plot.

VERSION OF AUTOCAD:
This tutorial exercise may be completed using either AutoCAD Release 10 or Release 11.

Step #1

Begin the orthographic drawing of the shifter by constructing a right angle consisting of one horizontal and one vertical line. The corner formed by the two lines will be used to orientate the front view.

Command: **Line**
From point: **0.75,0.50**
To point: **@10.00<0**
To point: *(Strike Enter to exit this command)*

Command: **Line**
From point: **0.75,0.50**
To point: **@7.50<90**
To point: *(Strike Enter to exit this command)*

Step #2

Begin the layout of the primary views by using the Offset command to offset the vertical line at "A" the distance of 5.00 units which represents the length of the shifter.

Command: **Offset**
Offset distance or Through <Through>: **5.00**
Select object to offset: *(Select the vertical line at "A")*
Side to offset? *(Pick a point anywhere near "B")*
Select object to offset: *(Strike Enter to exit this command)*

Step #3

Use the Offset command to offset the horizontal line at "A" the distance of 3.00 units which represents the height of the shifter.

Command: **Offset**
Offset distance or Through <Through>: **3.00**
Select object to offset: *(Select the horizontal line at "A")*
Side to offset? *(Pick a point anywhere near "B")*
Select object to offset: *(Strike Enter to exit this command)*

Step #4

Begin laying out dimension spaces which will act as separators between views and allow for the placement of dimensions once the shifter is completed. A spacing of 1.50 units will be more than adequate for this purpose. Again, use the Offset command to accomplish this.

Command: **Offset**
Offset distance or Through <Through>: **1.50**
Select object to offset: *(Select the vertical line at "A")*
Side to offset? *(Pick a point anywhere near "B")*
Select object to offset: *(Select the horizontal line at "C")*
Side to offset? *(Pick a point anywhere near "D")*
Select object to offset: *(Strike Enter to exit this command)*

Step #5

Use the Offset command to layout the depth of the shifter at a distance of 2.50 units.

Command: **Offset**
Offset distance or Through <Through>: **2.50**
Select object to offset: *(Select the vertical line at "A")*
Side to offset? *(Pick a point anywhere near "B")*
Select object to offset: *(Select the horizontal line at "C")*
Side to offset? *(Pick a point anywhere near "D")*
Select object to offset: *(Strike Enter to exit this command)*

Step #6

Use the Trim command to trim away excess construction lines used when laying out the primary views of the shifter.

Command: **Trim**
Select cutting edge(s)...
Select objects: *(Select the lines at "A" and "B")*
Select objects: *(Strike Enter to continue)*
Select object to trim: *(Select the vertical line at "C")*
Select object to trim: *(Select the vertical line at "D")*
Select object to trim: *(Select the horizontal line at "E")*
Select object to trim: *(Select the horizontal line at "F")*
Select object to trim: *(Strike Enter to exit this command)*

Step #7

Use the Trim command again to complete trimming away excess construction lines used when laying out the primary views of the Shifter.

Command: **Trim**
Select cutting edge(s)...
Select objects: *(Select the lines at "A" and "B")*
Select objects: *(Strike Enter to continue)*
Select object to trim: *(Select the vertical line at "C")*
Select object to trim: *(Select the vertical line at "D")*
Select object to trim: *(Select the horizontal line at "E")*
Select object to trim: *(Select the horizontal line at "F")*
Select object to trim: *(Strike Enter to exit this command)*

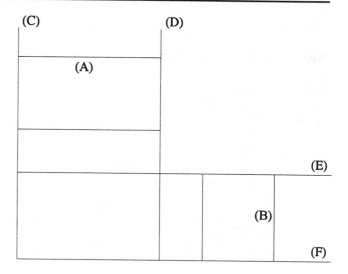

Step #8

Your display should appear similar to the illustration at the right with the layout of the front, top, and right side views. Begin adding details to all views through methods of projection. Use the Offset command to offset lines "A", "C", and "E" a distance of 0.50.

Command: **Offset**
Offset distance or Through<Through>: **0.50**
Select object to offset: *(Select the vertical line at "A")*
Side to offset? *(Pick a point anywhere near "B")*
Select object to offset: *(Select the horizontal line at "C")*
Side to offset? *(Pick a point anywhere near "D")*
Select object to offset: *(Select the horizontal line at "E")*
Side to offset? *(Pick a point anywhere near "F")*
Select object to offset: *(Strike Enter to exit this command)*

Step #9

Use the Offset command and offset the vertical line at "A" a distance of 1.75.

Command: **Offset**
Offset distance or Through<Through>: **1.75**
Select object to offset: *(Select the vertical line at "A")*
Side to offset? *(Pick a point anywhere near "B")*
Select object to offset: *(Strike Enter to exit this command)*

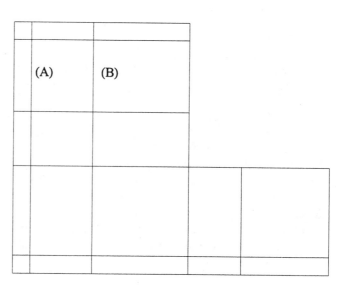

Step #10

Use the Offset command to offset the vertical line at "A" a distance of 0.50.

Command: **Offset**
Offset distance or Through<Through>: **0.50**
Select object to offset: *(Select the vertical line at "A")*
Side to offset? *(Pick a point anywhere near "B")*
Select object to offset: *(Strike Enter to exit this command)*

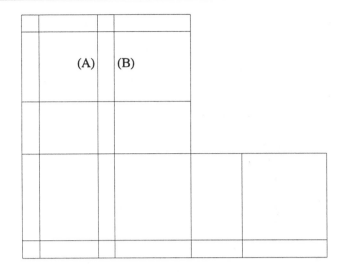

Step #11

Your display should appear similar to the illustration at the right. Next, use the Trim command to partially delete the line segments located in the spaces in between views.

Command: **Trim**
Select cutting edge(s)...
Select objects: *(Select the lines at "A", "B", "C", and "D")*
Select objects: *(Strike Enter to continue)*
Select object to trim: *(Select the horizontal line at "E")*
Select object to trim: *(Select the horizontal line at "F")*
Select object to trim: *(Select the horizontal line at "G")*
Select object to trim: *(Select the vertical line at "H")*
Select object to trim: *(Select the vertical line at "I")*
Select object to trim: *(Select the vertical line at "J")*
Select object to trim: *(Select the vertical line at "K")*
Select object to trim: *(Select the vertical line at "L")*
Select object to trim: *(Strike Enter to exit this command)*

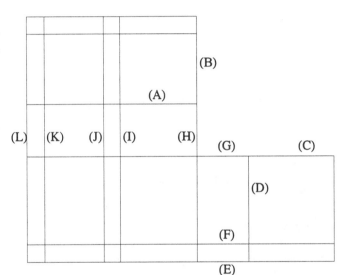

Step #12

Use the Zoom-Window option to magnify the top view similar to the illustration at the right. Then, use the Trim command to trim the excess entity labeled "B".

Command: **Trim**
Select cutting edge(s)...
Select objects: *(Select the vertical line at "A")*
Select objects: *(Strike Enter to continue)*
Select object to trim: *(Select the horizontal line at "B")*
Select object to trim: *(Strike Enter to exit this command)*

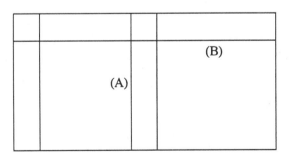

Step #13

Zoom back to the original display using the Zoom-Previous option. Use the Zoom-Window option to magnify the display to show the front view illustrated at the right. Use the Trim command to clean up the excess lines in the front view using the illustration at the right as a guide.

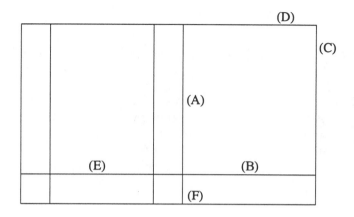

Command: **Trim**
Select cutting edge(s)...
Select objects: *(Select the lines at "A" and "B")*
Select objects: *(Strike Enter to continue)*
Select object to trim: *(Select the vertical line at "C")*
Select object to trim: *(Select the horizontal line at "D")*
Select object to trim: *(Select the horizontal line at "E")*
Select object to trim: *(Select the short vertical line at "F")*
Select object to trim: *(Strike Enter to exit this command)*

Step #14

Use the Zoom-Previous option to zoom back to the previous screen display containing the three views. Use the Fillet command to create a corner by selecting lines "A" and "B" illustrated at the right.

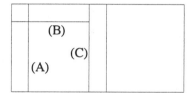

Command: **Fillet**
Polyline/Radius/<Select two objects>: *(Select the vertical line at "A" and horizontal line at "B")*

Command: **Fillet**
Polyline/Radius/<Select two objects>: *(Select the vertical line at "C" and horizontal line at "B")*

Step #15

Use the Trim command to partially delete the horizontal line at "C" and form the upside-down "U" shape illustrated at the right.

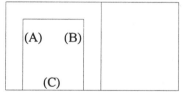

Command: **Trim**
Select cutting edge(s)...
Select objects: *(Select the vertical lines at "A" and "B")*
Select objects: *(Strike Enter to continue)*
Select object to trim: *(Select the horizontal line at "C")*
Select object to trim: *(Strike Enter to exit this command)*

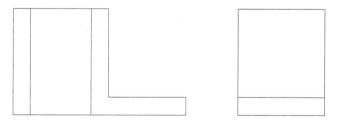

Step #16

Begin placing a circle in the top view representing the 1.50-diameter drill hole. Before accomplishing this, place a point of reference in the upper right corner of the top view. This point will then be referenced in the next step to accurately place the circle. In order to see the reference, use the Pdmode system variable to change the shape of the point to cross. Place the point using the Osnap-Endpoint option.

Command: **Setvar**
Variable name or ?: **Pdmode**
New value for pdmode <0>: **3**

Command: **Point**
Point: **Int**
of *(Select the intersection of the two lines at "A")*

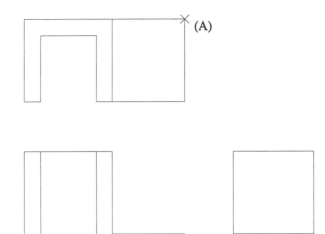

(A)

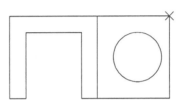

Step #17

Since the point was the last entity placed, immediately use the Circle command. Use coordinates to locate the center of the circle the distance @-1.00,-1.25 away from the reference point.

Command: **Circle**
3P/2P/TTR/<Center point>: **@-1.00,-1.25**
Diameter/<Radius>: **D**
Diameter: **1.50**

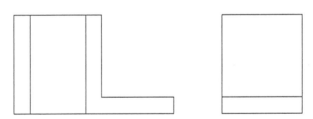

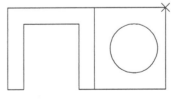

Step #18

Use the Point command again to place a reference point in the upper right corner of the right side view. This point will be used to help identify the center point of the 1.00-diameter drill holes. The current Pdmode of 3 will keep the shape of the point looking like a cross.

Command: **Point**
Point: **Int**
of *(Select the intersection of the right side view at "A")*

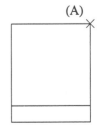

(A)

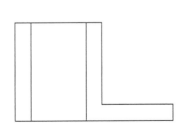

Step #19

Immediately use the Circle command and coordinates to locate the center of the circle the distance @-1.25,-1.00 away from the last reference point. Use the Erase command to delete the points.

Command: **Circle**
3P/2P/TTR/<Center point>: **@-1.25,-1.00**
Diameter/<Radius>: **D**
Diameter: **1.00**

Command: **Erase**
Select objects: *(Select both points)*
Select objects: *(Strike Enter to execute this command)*

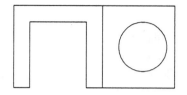

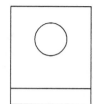

Step #20

Use the Line command and draw projection lines from both circles into the front view. Use the Osnap-Quadrant and Osnap-Perpend options to accomplish this. These segments will be converted into hidden lines in a later step.

Command: **Line**
From point: **Qua**
of *(Select the quadrant of the circle at "A")*
To point: **Per**
to *(Select the line at "B" to obtain the perpendicular point)*
To point: *(Strike Enter to exit this command)*

Repeat the above procedure for the quadrants "C", "D", and "E". Make "D" and "E" perpendicular to "F."

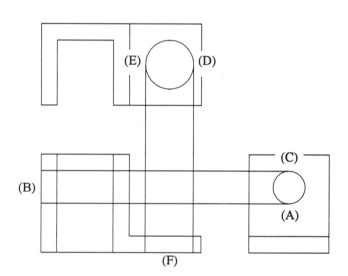

Step #21

Place center markers at the centers of both circles as illustrated at the right. Before proceeding with this operation, a system variable needs to be set to a certain value to achieve the desired results. Set the Dimcen system variable to a value of -0.12. This will not only place the center mark when using the Dim-Cen command but will also extend the center line a short distance outside of both circles.

Command: **Dim**
Dim: **Dimcen**
Current value <0.09> New value: **-0.12**

Dim: **Cen**
Select arc or circle: *(Select the circle at "A")*

Dim: **Cen**
Select the arc or circle: *(Select the circle at "B")*

Dim: **Exit**

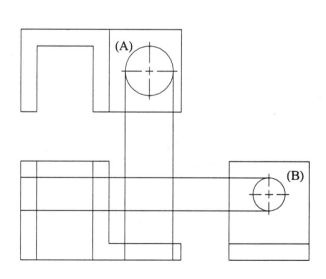

Step #22

Project two lines from the endpoints of both center marks using the Osnap-Endpoint option. Turn ortho mode on by striking the F8 function key. The lines will be converted to center lines at a later step. Draw the lines 0.50 units past the front view as illustrated at the right.

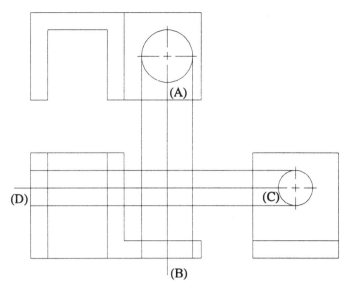

Command: **Line**
From point: **Endp**
of *(Select the endpoint of the center mark in the top view at "A")*
To point: *(Mark a point just below the front view at "B")*
To point: *(Strike Enter to exit this command)*

Command: **Line**
From point: **Endp**
of *(Select the endpoint of the center mark in the side view at "C")*
To point: *(Mark a point to the left of the front view at "D")*
To point: *(Strike Enter to exit this command)*

Step #23

Use the Trim command and the illustration at the right to trim away excess lines.

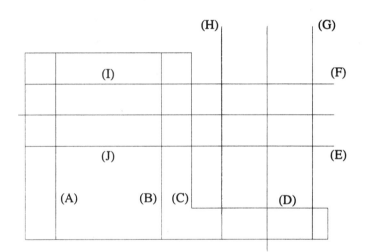

Command: **Trim**
Select cutting edge(s)...
Select objects: *(Select the lines at "A", "B", "C", and "D")*
Select objects: *(Strike Enter to continue)*
Select object to trim: *(Select the horizontal line at "E")*
Select object to trim: *(Select the horizontal line at "F")*
Select object to trim: *(Select the vertical line at "G")*
Select object to trim: *(Select the vertical line at "H")*
Select object to trim: *(Select the horizontal line at "I")*
Select object to trim: *(Select the horizontal line at "J")*
Select object to trim: *(Strike Enter to exit this command)*

Step #24

Use the Chprop command to change the six lines illustrated at the right from their current layer assignment of 0 to a new layer assignment of "Hidden" which will change the lines from object to hidden lines.

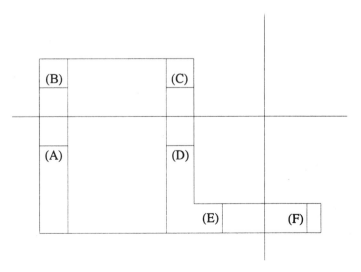

Command: **Chprop**
Select objects: *(Select all six lines labeled "A" to "F")*
Select objects: *(Strike Enter to continue)*
Change what property (Color/LAyer/LType/Thickness)?
    LA
New layer <0>: **Hidden**
Change what property(Color/LAyer/LType/Thickness)?
    (Strike Enter to exit this command)

Command: **Ltscale**
New scale factor <1.0000>: **0.50**

Step #25

Use the Break command to partially delete the lines illustrated at the right before converting them to center lines. Remember, center lines extend past the object lines when identifying hidden drill holes; it would be inappropriate to use the Trim command for this step.

Command: **Break**
Select object: *(Select the horizontal line at "A")*
Enter second point (or F for first point): *(Select the line at "B")*

Command: **Break**
Select object: *(Select the vertical line at "C")*
Enter second point (or F for first point): *(Select the line at "D")*

Command: **Break**
Select object: *(Select the horizontal line at "E")*
Enter second point (or F for first point): @

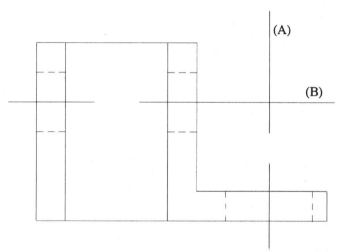

Step #26

The purpose of the @ option in the step above was to break a line into two segments without showing the break. The @ option means "the last known point" which completes the Break command by satisfying the "second point" prompt. To prove this, use the Erase command to delete the segments no longer needed.

Command: **Erase**
Select objects: *(Carefully select the lines at "A" and "B")*
Select objects: *(Strike Enter to execute this command)*

Step #27

Use the Chprop command to change the three lines illustrated at the right from their current layer assignment of 0 to a new layer assignment of "Center" which will change the lines from object to center lines.

Command: **Chprop**
Select objects: *(Select all three lines labeled "A" to "C")*
Select objects: *(Strike Enter to continue)*
Change what property(Color/LAyer/LType/Thickness)? **LA**
New layer <0>: **Center**
Change what property (Color/LAyer/LType/Thickness)?
    (Strike Enter to exit this command)

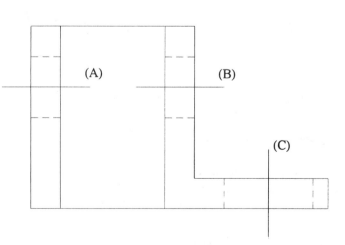

Step #28

A 45-degree angle needs to be constructed in order to begin projecting features from the top view to the right side view and then back again. This angle is formed by extending the bottom edge of the top view to intersect with the left edge of the side view. Use the Fillet command to accomplish this. Then, draw the 45-degree line; the length of this line is not important.

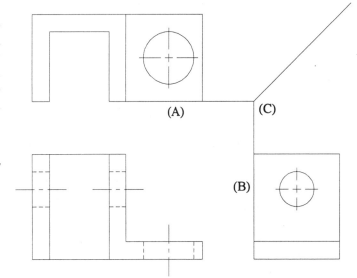

Command: **Fillet**
Polyline/Radius/<Select two objects>: *(Select lines "A" and "B")*

Command: **Line**
From point: **Int**
of *(Select the intersection at "C")*
To point: **@4<45**
To point: *(Strike Enter to exit this command)*

Step #29

Draw lines from points "A", "B", and "C" to intersect with the 45-degree angle projector. Be sure Ortho mode is on to draw horizontal lines. Use the Osnap-Intersec, -Endpoint, and -Quadrant options to assist in this operation.

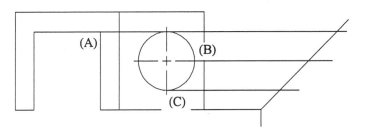

Command: **Line**
From point: **Int**
of *(Select the intersection of the corner at "A")*
To point: *(Draw a line just past the 45-degree angle)*
To point: *(Strike Enter to exit this command)*

Command: **Line**
From point: **Endp**
of *(Select the endpoint of the center line at "B")*
To point: *(Draw a line just past the 45-degree angle)*
To point: *(Strike Enter to exit this command)*

Command: **Line**
From point: **Qua**
of *(Select the quadrant of the circle at "C")*
To point: *(Draw a line just past the 45-degree angle)*
To point: *(Strike Enter to exit this command)*

Step #30

Draw lines from the intersection of the 45-degree angle projection line to points in the right side view. Follow the commands below to accomplish this.

Command: **Line**
From point: **Int**
of *(Select the intersection of the angle at "A")*
To point: **Per**
to *(Select the bottom horizontal line of the right side view at "B")*
To point: *(Strike Enter to exit this command)*

Command: **Line**
From point: **Int**
of *(Select the intersection of the angle at "C")*
To point: **Per**
to *(Select the bottom line of the right side view at "B")*
To point: *(Strike Enter to exit this command)*

Command: **Line**
From point: **Int**
of *(Select the intersection of the angle at "D")*
To point: *(Select a point below the bottom of the side view at "E")*
To point: *(Strike Enter to exit this command)*

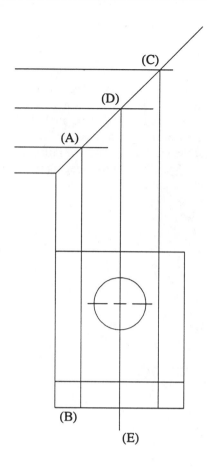

Step #31

Erase the three projection lines from the top view using the Erase command.

Command: **Erase**
Select objects: *(Select lines "A", "B", and "C")*
Select objects: *(Strike Enter to execute this command)*

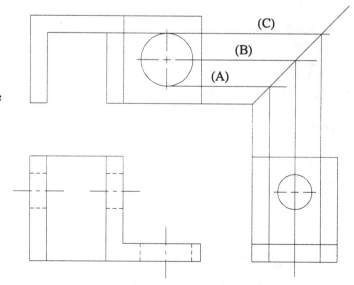

Step #32

Use the Trim command to trim away unnecessary geometry using the illustration at the right as a guide. The hidden hole and slot will be formed in the side view through this operation.

Command: **Trim**
Select cutting edges...
Select objects: *(Select the horizontal line at "A")*
Select objects: *(Strike Enter to continue)*
Object to trim: *(Select the vertical line at "B")*
Object to trim: *(Strike Enter to exit this command)*

Repeat the above procedure for the line illustrated at the right using the horizontal line "C" as the cutting edge and the vertical line "D" as the line to trim.

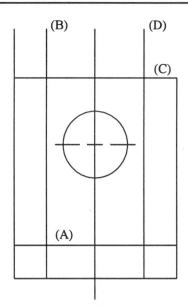

Step #33

Use the Break command to split the vertical line at "A" into two separate entities. This will be accomplished by typing @ in response to the prompt "Enter second point". This will split the line in two without noticing the break.

Command: **Break**
Select object: *(Select the vertical line at "A")*
Enter second point (or F for first point): @

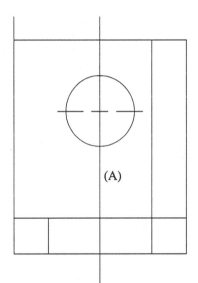

Step #34

Use the Erase command to delete the top half of the line broken previously in Step #33. This will leave a short line segment that will be changed or converted into a center line marking the center of a hidden hole. Redraw the screen to refresh the vertical center lines at the hole.

Command: **Erase**
Select objects: *(Select the vertical line at "A")*
Select objects: *(Strike Enter to execute this command)*

Command: **Redraw**

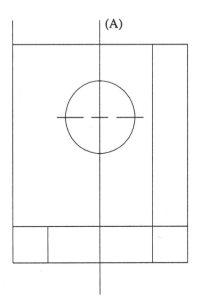

Step #35

Use the Chprop command to convert the two vertical lines labeled "A" and "B" at the right from the object layer to the layer named "Hidden". Do the same for the longer vertical line labeled "C" but change this line from the object layer to the layer named "Center."

Command: **Chprop**
Select objects: *(Select the two vertical lines labeled "A" and "B")*
Select objects: *(Strike Enter to continue)*
Change what property(Color/LAyer/LType/Thickness)? **LA**
New layer name <0>: **Hidden**
Change what property (Color/LAyer/LType/Thickness)? *(Strike Enter to exit this command)*

Command: **Chprop**
Select objects: *(Select the vertical line labeled "C")*
Select objects: *(Strike Enter to continue)*
Change what property(Color/LAyer/LType/Thickness)? **LA**

New layer <0>: **Center**
Change what property (Color/LAyer/LType/Thickness)? *(Strike Enter to exit this command)*

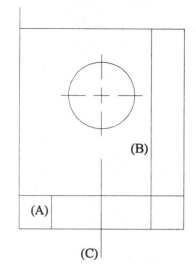

Step #36

Draw three lines from key features on the side view to intersect with the 45-degree angle projector. Use Osnap options whenever possible. Ortho mode must be On.

Command: **Line**
From point: **Qua**
of *(Select the quadrant of the circle at "A")*
To point: *(Identify a point past the 45-degree angle)*
To point: *(Strike Enter to exit this command)*

Repeat the above procedure for the other two projection lines. Use the Osnap-Endpoint option and begin the second projector line from the endpoint of the center marker at "B". Begin the third projector line from the quadrant of the circle at "C".

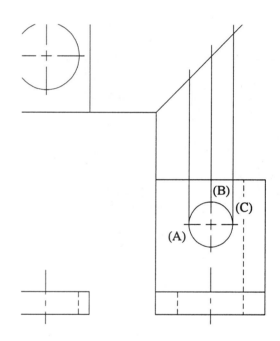

Step #37

Draw three lines from the intersection at the 45-degree projector across to the top view. Draw the middle line longer since it will be converted into a center line at a later step.

Command: **Line**
From point: **Int**
of *(Select the intersection at "A")*
To point: **Per**
to *(Select the vertical line at "B")*
To point: *(Strike Enter to exit this command)*

Repeat the above procedure for the other two lines. Do not use an Osnap option for the opposite end of the middle line but rather identify a point just to the left of the vertical line at "B".

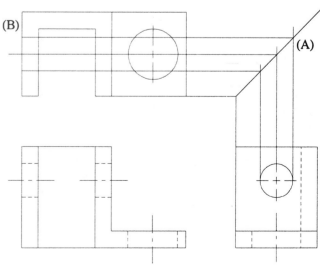

Step #38

Use the Erase command to erase the three vertical projector lines from the side view.

Command: **Erase**
Select objects: *(Select the vertical lines labeled "A", "B", and "C")*
Select objects: *(Strike Enter to execute this command)*

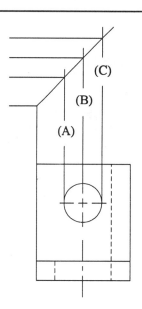

Step #39

Use the Trim command to trim away excess lines in the top view as illustrated at the right. Perform a redraw when completed.

Command: **Trim**
Select cutting edges...
Select objects: *(Select the vertical lines labeled "A", "B", and "C")*
Select objects: *(Strike Enter to continue)*
Select object to trim: *(Select the horizontal line at "D")*
Select object to trim: *(Select the horizontal line at "E")*
Select object to trim: *(Select the horizontal line at "F")*
Select object to trim: *(Select the horizontal line at "G")*
Select object to trim: *(Strike Enter to exit this command)*

Command: **Redraw**

Step #40

Use the Break command to partially delete the horizontal line segment from "A" to "B". Use the Break command with the @ option to break the line into two segments at "C". Use the Erase command to delete the trailing line segment at "D". Redraw the screen.

Command: **Break**
Select object: *(Select the horizontal line at "A")*
Enter second point (or F for first): *(Select the line at "B")*

Command: **Break**
Select object: *(Select the horizontal line at "C")*
Enter second point (or F for first): @

Command: **Erase**
Select objects: *(Select the horizontal line segment at "D")*
Select objects: *(Strike Enter to execute this command)*

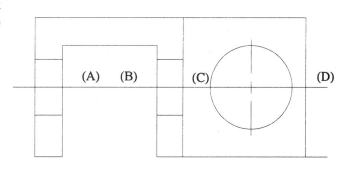

Step #41

Use the Chprop command to change the line segments labeled "A", "B", "C", and "D" to the new layer named "Hidden" which will display hidden lines. Change the line segments labeled "E" and "F" to the new layer named "Center" which will display center lines.

Command: **Chprop**
Select objects: *(Select the four lines labeled "A" to "D")*
Select objects: *(Strike Enter to continue)*
Change what property (Color/LAyer/LType/Thickness)?
   LA
New layer <0>: **Hidden**

Command: **Chprop**
Select objects: *(Select the two lines labeled "E" and "F")*
Select objects: *(Strike Enter to continue)*
Change what property(Color/LAyer/LType/Thickness)?
   LA
New layer <0>: **Center**
Change what property(Color/LAyer/LType/Thickness)?
   (Strike Enter to exit this command)

Step #42

Use the Erase command to delete the 45-degree angle line. Use the Fillet command to create corners in the top view and right side view.

Command: **Erase**
Select objects: *(Select the inclined line at "A")*
Select objects: *(Strike Enter to execute this command)*

Command: **Fillet**
Polyline/Radius/<Select two objects>: *(Select "B" and "C")*

Command: **Fillet**
Polyline/Radius/<Select two objects>: *(Select "D" and "E")*

Use the Chprop command to change the center marks at "A" and "B" to the Center layer. This completes this tutorial on performing a multi-view projection drawing. The steps have been numerous in order to detail every command sequence. In reality, the process is much faster, especially since a few basic commands were used most of the time such as Offset and Trim. Use this tutorial as a guide in completing the many multi-view drawing problems at the end of this chapter.

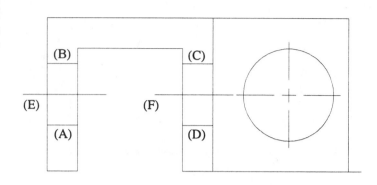

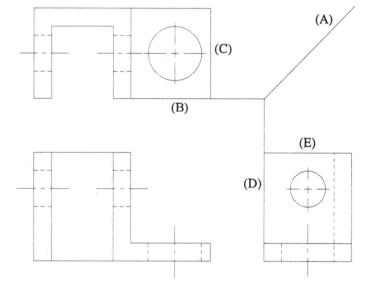

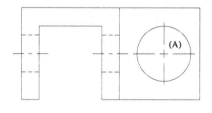

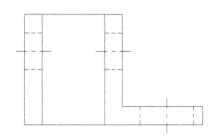

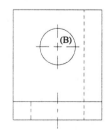

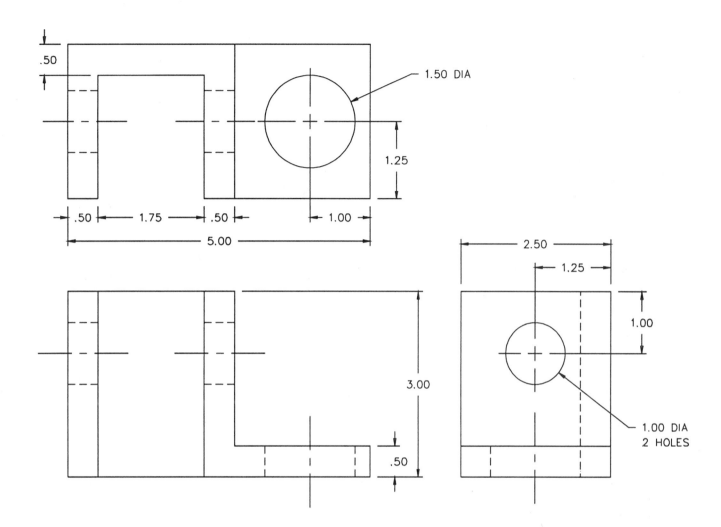

In keeping with the layers created, all object lines are to be changed from the layer they were created on to the object layer using the Chprop command.

Command: **Chprop**
Select objects: *(Select all twenty-four object lines)*
Select objects: *(Strike Enter to continue)*
Change what property(Color/LAyer/LType/Thickness)? **LA**
New layer <0>: **Object**
Change what property (Color/LAyer/LType/Thickness)?
  (Strike Enter to exit this command)

The final process in completing a multi-view drawing is to place dimensions to define size and locate features. This topic will be discussed in a later chapter.

Tutorial Exercise #8
XYZ.Dwg

It is possible to perform multi-view projections from one view to another without drawing construction lines and then trimming them to size. XYZ filters provide the means of accomplishing this along with a little practice. In the illustration at the right, the problem is to draw a circle using the center point of the circle exactly at the center of the rectangle. To complicate the issue, no other command or setting can be used to perform this operation. First, construct the rectangle.

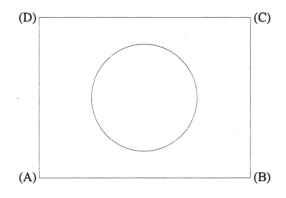

Command: **Line**
From point: **2,2** *(Point "A")*
To point: **@8<0** *(To Point "B")*
To point: **@6<90** *(To Point "C")*
To point: **@8<180** *(To Point "D")*
To point: **C** *(Back to Point "A" to close and exit the command)*

Use the Circle command and begin with the normal "Center point" prompt sequence. Begin finding the center of the rectangle by first filtering out the midpoint of the X value at "A" followed by the Y value at "B". Answer the prompt for Z by entering a value of 0. Complete the Circle command by supplying the radius value of 2.

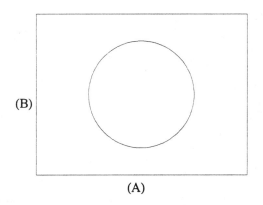

Command: **Circle**
3P/2P/TTR/<Center point>: **.X**
of **Mid**
of *(Select the horizontal line at "A")*
(Need YZ): **.Y**
of **Mid**
of *(Select the vertical line at "B")*
(Need Z): **0**
Diameter/<Radius>: **2**

To review, filtering out the midpoint of the horizontal line at "A" saved the point to satisfy the center point of the circle. However, a circle needs at least a second point; this is the reason for the prompt "Need YZ". A YZ point was needed to find the center point. Instead of selecting any point, the Y point was filtered out at the midpoint of the vertical line at "B". Now, the X and Y values were saved to satisfy the center of the circle. However, since AutoCAD now exists in a three-dimensional (3D) data base, a prompt "Need Z" appears. Since the entire drawing is located at an elevation of ∅, entering a value of 0 satisfies the centerpoint of the circle and marks a point. This may seem very tedious and too much trouble; with a little practice, however, filters become another drawing aid to work in your favor.

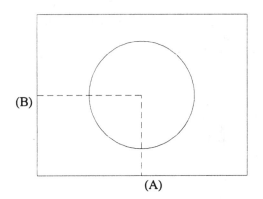

Tutorial Exercise #9
Gage.Dwg

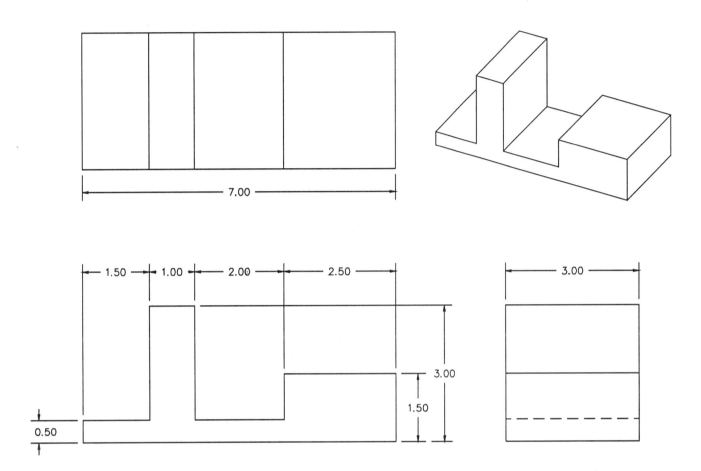

PURPOSE:
This tutorial is designed to allow the user to construct a three-view drawing of the Gage with the aid of XYZ filters.

SYSTEM SETTINGS:
Begin a new drawing called "Gage." Use the Units command to change the number of decimal places past the zero from 4 to 2. Keep the remaining default unit values. Using the Limits command, keep 0,0 for the lower left corner and change the upper right corner from 12,9 to 15.50,9.50. Use the Grid command and change the grid spacing from 1.00 to 0.25 units. Do not turn the snap or ortho On.

LAYERS:
Create the following layers with the format:

Name-Color-Linetype
Hidden-Red-Hidden

SUGGESTED COMMANDS:
Begin this tutorial by laying out the three primary views using the Line and Offset commands. Use the Trim command to clean up any excess line segments. As an alternate method used for projection, use XYZ filters in combination with Osnap options to add features in other views.

DIMENSIONING:
Dimensions may be added to this problem at a later time. Consult your instructor.

PLOTTING:
This tutorial exercise may be plotted on "B"-size paper (11" x 17"). Use a plotting scale of 1=1 to produce a full size plot.

VERSION OF AUTOCAD:
This tutorial exercise may be completed using either AutoCAD Release 10 or Release 11.

Step #1

Begin the multi-view drawing of the gage by constructing the front view using absolute and polar coordinates. Start the front view at coordinate 1.50,1.00.

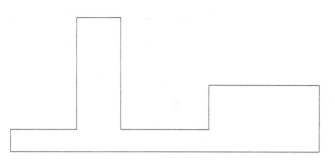

Command: **Line**
From point: **1.50,1.00**
To point: **@7.00<0**
To point: **@1.50<90**
To point: **@2.50<180**
To point: **@1.00<270**
To point: **@2.00<180**
To point: **@2.50<90**
To point: **@1.00<180**
To point: **@2.50<270**
To point: **@1.50<180**
To point: **C**

Step #2

Begin the construction of the top view by locating the lower left corner at coordinate 1.50,5.50.

Command: **Line**
From point: **1.50,5.50**
To point: **@7.00<0**
To point: **@3.00<90**
To point: **@7.00<180**
To point: **C**

Step #3

Begin the construction of the right side view by locating the lower left corner at coordinate 10.50,1.00.

Command: **Line**
From point: **10.50,1.00**
To point: **@3.00<0**
To point: **@3.00<90**
To point: **@3.00<180**
To point: **C**

Step #4

Your display should appear similar to the illustration at the right with the placement of the top view above the front view and the right side view directly to the right of the front. Filtering methods will now be used to complete the missing lines in the top and right side views.

Step #5

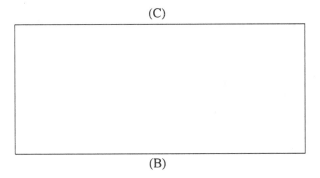

Zoom into the drawing so the front and top views appear similar to the illustration at the right. Follow the steps below for the proper use of XYZ filters when used in conjunction with the Line command.

Command: **Line**
From point: **.X**
of **Int**
of *(Select the intersection of the lines at "A")*
(Need YZ): **Nea**
to *(Select anywhere along the horizontal line "B")*
To point: **Per**
to *(Select anywhere along the horizontal line "C")*
To point: *(Strike Enter to exit this command)*

The X value identified by selecting the intersection on the front view was saved for later use by projecting the value to the horizontal line of the top view and completing the Line command with the Osnap-Perpendicular option.

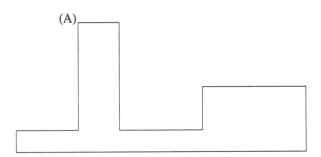

Step #6

Your display should appear similar to the illustration at the right. Rather than use filters for the next line, simply duplicate the last line drawn using the Copy command.

Command: **Copy**
Select objects: **L**
Select objects: *(Strike Enter to continue)*
<Base point or displacement>/Multiple: **Endp**
of *(Select the endpoint of the line at "A")*
Second point of displacement: **Endp**
of *(Select the endpoint of the line at "B")*

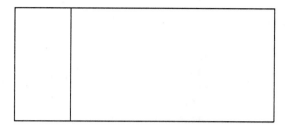

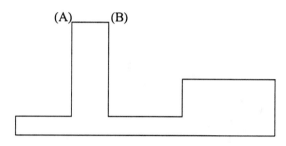

Step #7

After performing the Copy command, your display should appear similar to the illustration at the right. Use XYZ filters to project the last object line in the top view from the front view.

Command: **Line**
From point: **.X**
of **Int**
of *(Select the intersection of the lines at "A")*
(Need YZ): **Nea**
to *(Select anywhere along the horizontal line "B")*
To point: **Per**
to *(Select anywhere along the horizontal line "C")*
To point: *(Strike Enter to exit this command)*

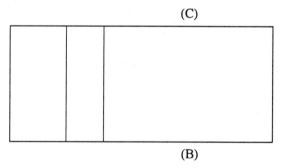

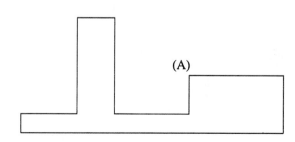

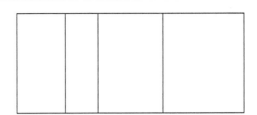

Step #8

The complete front and top views are illustrated at the right. Use the same procedure with XYZ filters for completing the right side view.

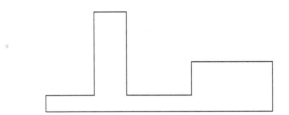

Step #9

Begin using XYZ filters to add missing lines to the right side view. First, use the Zoom-Window option to magnify the area illustrated at the right.

Command: **Zoom**
All/Center/Dynamic/Extents/Left/Previous/Window/
    <Scale(X)>: **W**
First corner: *(Select a point on the screen at "A")*
Second corner: *(Select a point on the screen at "B")*

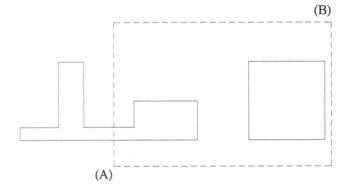

Step #10

Instead of filtering or saving the X coordinate, the same procedure can be used for saving the Y coordinate for later use.

Command: **Line**
From point: **.Y**
of **Int**
of *(Select the intersection of the lines at "A")*
(Need XZ): **Nea**
to *(Select anywhere along the vertical line "B")*
To point: **Per**
to *(Select anywhere along the vertical line "C")*
To point: *(Strike Enter to exit this command)*

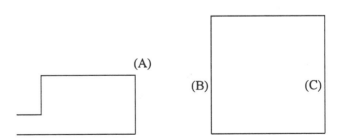

Step #11

Filter out the Y coordinate value of the intersecting area of the front view at the right to project the coordinate to the right side view.

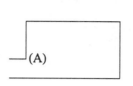 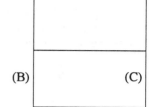

Command: **Line**
From point: **.Y**
of **Int**
of *(Select the intersection of the lines at "A")*
(Need XZ): **Nea**
to *(Select anywhere along the vertical line "B")*
To point: **Per**
to *(Select anywhere along the vertical line "C")*
To point: *(Strike Enter to exit this command)*

Step #12

Since the last line projected details a hidden surface, the object line at the right needs to be changed to a hidden line. The Chprop command will be used to accomplish this only if a layer identifying hidden lines, such as Hidden, has already been created.

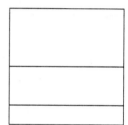

Command: **Chprop**
Select objects: **L**
Select objects: *(Strike Enter to continue)*
Change what property(Color/LAyer/LType/Thickness)? **LA**
New layer<0>: **Hidden**
Change what property(Color/LAyer/LType/Thickness)?
     (Strike Enter to exit this command)

Step #13

The completed right side view is illustrated at the right complete with visible and invisible surfaces. Use the Zoom-Previous or Zoom-All options to demagnify your display and show all three views of the gage.

 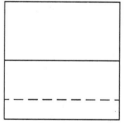

As a general rule of thumb, an X coordinate was filtered out and retrieved later for projecting lines from the front to the top view or vice versa. A Y coordinate was filtered out and retrieved later for projecting lines from the front to the right side view and vice versa. Filters may also be used for projecting such features as holes and slots between views. The key is to lock onto a significant part of an entity using one of the many options provided by the Osnap option.

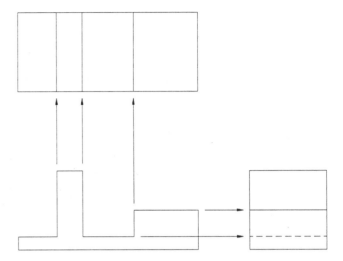

With the completed three-view drawing illustrated at the right, the next step would be to add dimensions to the views for manufacturing purposes. This topic will be discussed in a later chapter.

Problems for Unit 3

Directions for Problems 3-1 and 3-2:
Find the missing lines in these problems and sketch the correct solution.

Problem 3-1

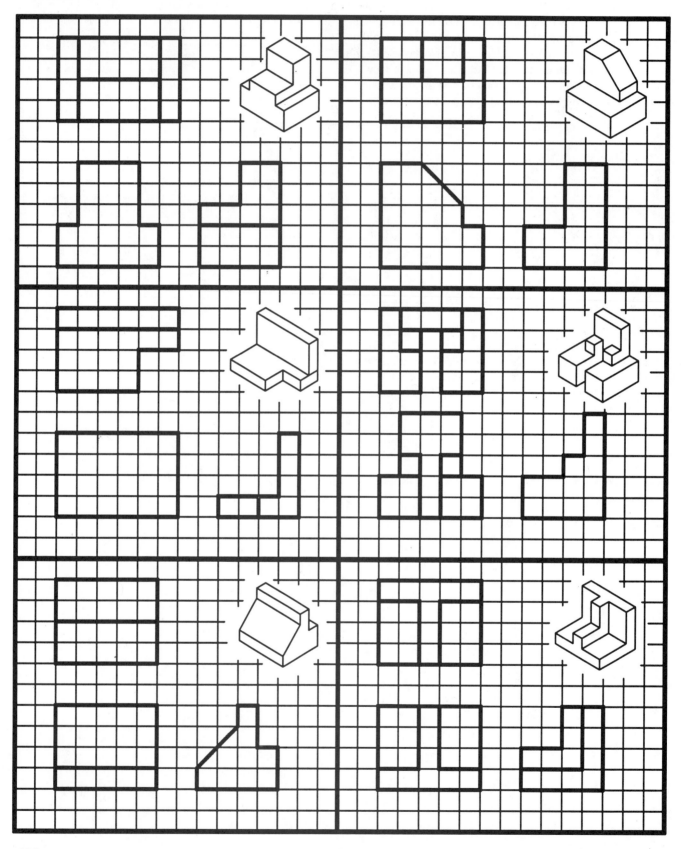

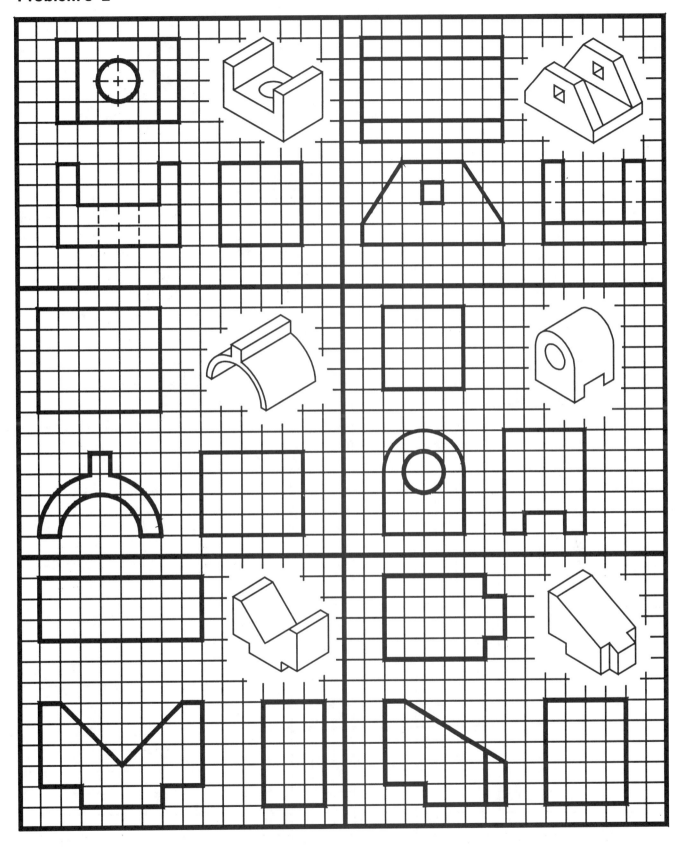

Directions for Problems 3–3 through 3–7:
Construct a multi-view drawing by sketching the front, top, and right side views in the grid provided below.

Problem 3-3

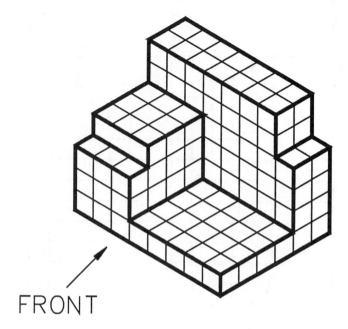

FRONT

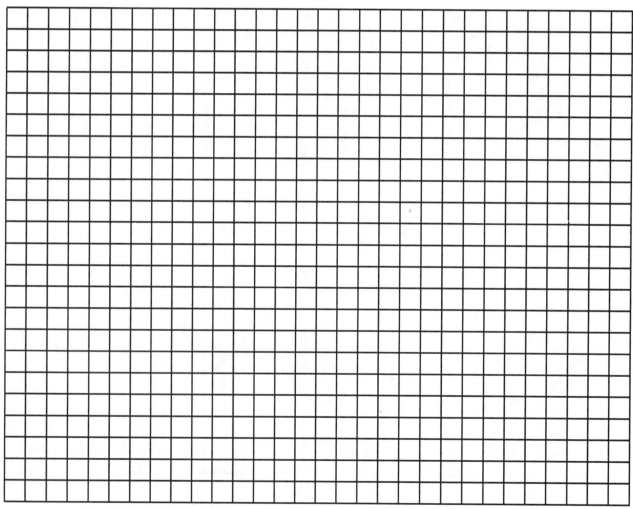

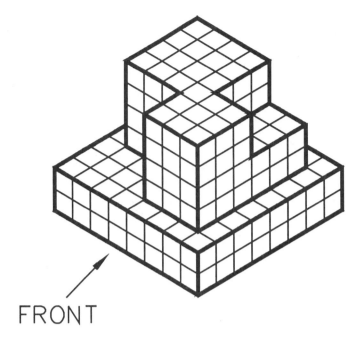

FRONT

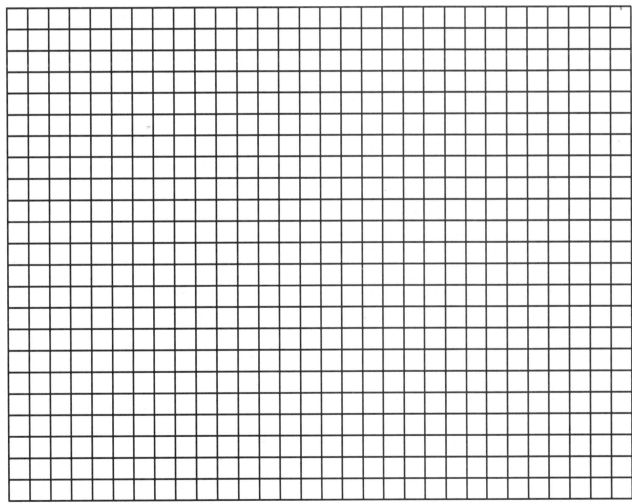

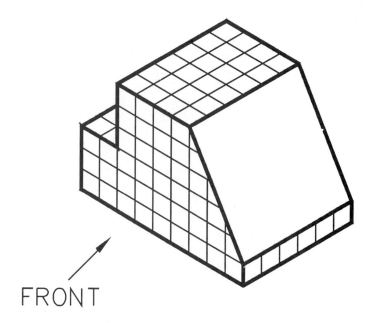

FRONT

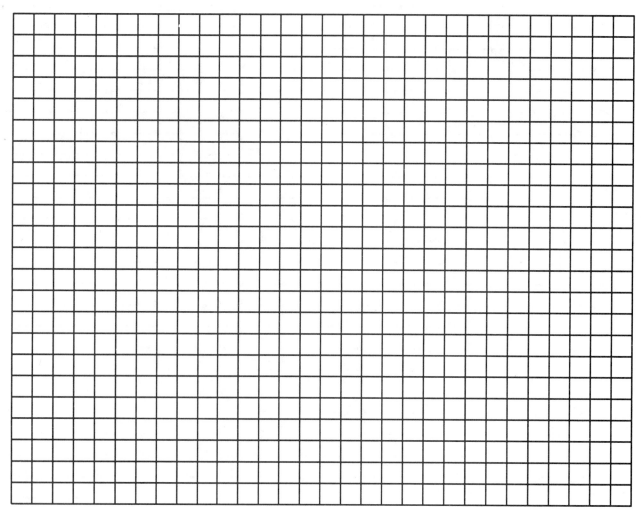

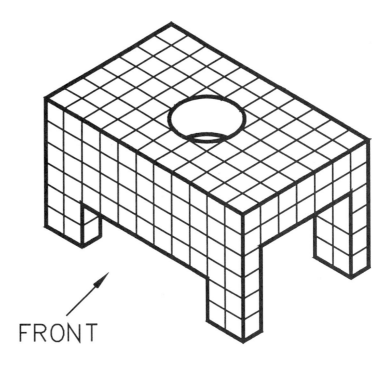

FRONT

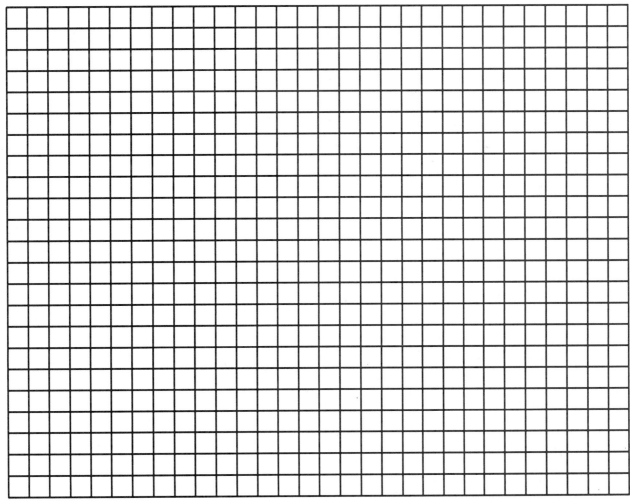

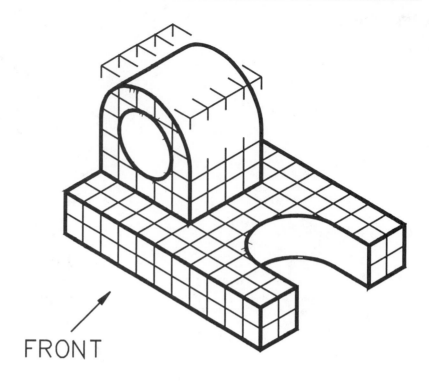

FRONT

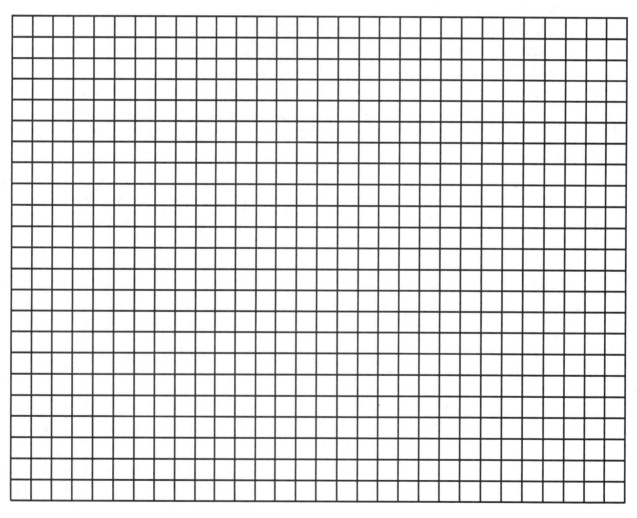

Problem 3–8

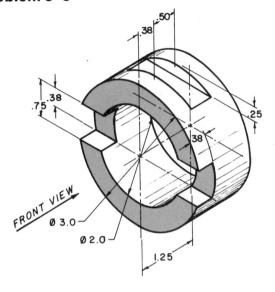

Problem 3–10

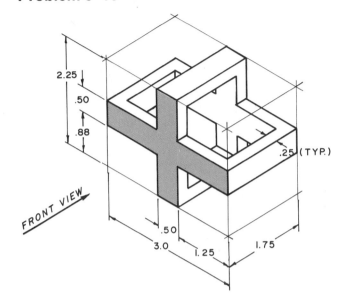

Problem 3–9

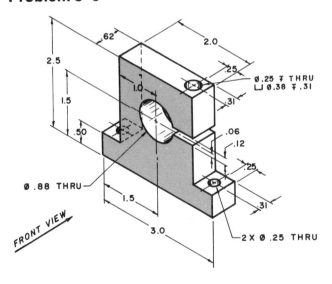

Problem 3–11

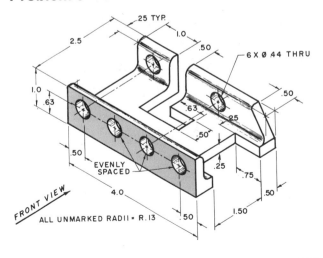

Shape Description/Multi-View Projection

Problem 3-12

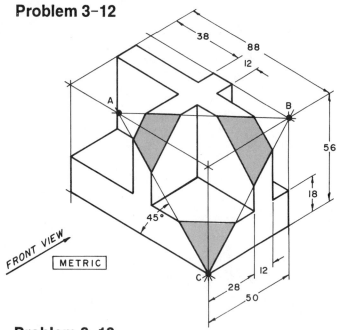

FRONT VIEW

METRIC

Problem 3-13

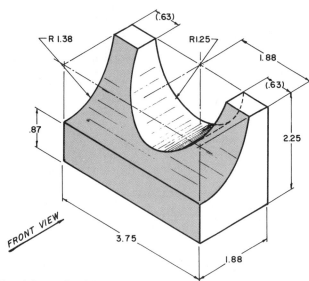

FRONT VIEW

Problem 3-14

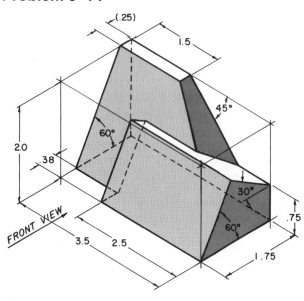

FRONT VIEW

Problem 3-15

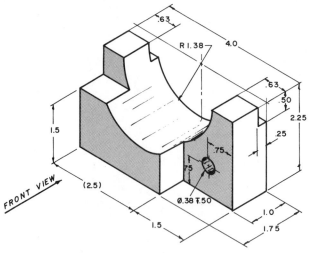

FRONT VIEW

Problem 3-16

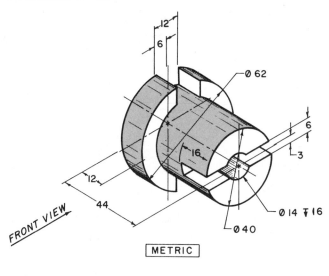

FRONT VIEW

METRIC

Directions for Problem 3-17
Construct a 4-view drawing of this object (front, top, right, and left-side views).

Problem 3-17

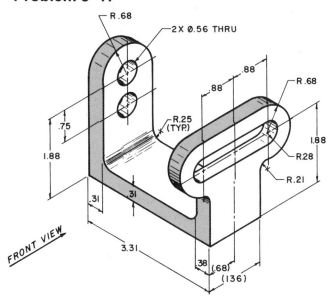

FRONT VIEW

Problem 3-18

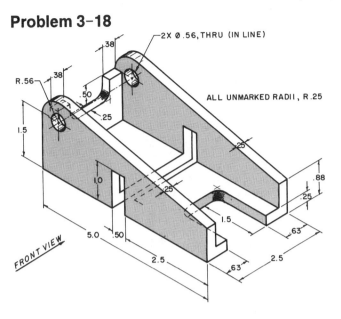

2X Ø.56, THRU (IN LINE)
38
R.56
38
.50
.25
ALL UNMARKED RADII, R.25
1.5
1.0
.25
.88
.25
1.5
FRONT VIEW
5.0   .50
2.5
.63
.63
2.5

Problem 3-19

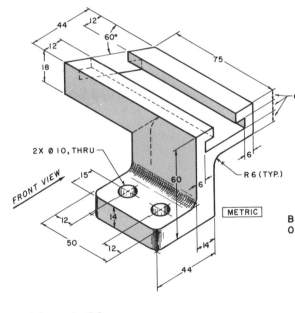

44   12
12
60°
18
75
6
2X Ø 10, THRU
FRONT VIEW   15
60
6
R 6 (TYP.)
12
14
6
METRIC
50
12
14
44

Problem 3-20

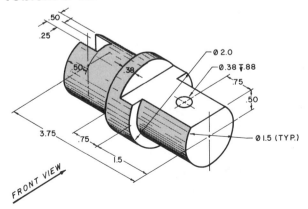

.50
.25
Ø 2.0
.50
.38
Ø.38 ⊤.88
.75
.50
3.75
.75
Ø 1.5 (TYP.)
1.5
FRONT VIEW

Problem 3-21

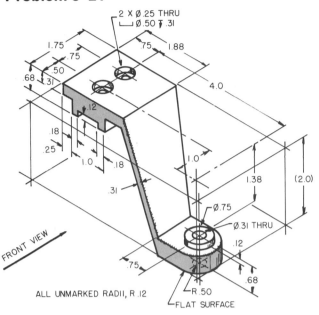

2 X Ø.25 THRU
⌴ Ø.50 ⊤ .31
1.75
.75   1.88
.75
.68
.50
.31
4.0
.12
.18
.25
.18
1.0
1.0
.31
1.38
(2.0)
Ø.75
Ø.31 THRU
.12
FRONT VIEW
.75
.68
R.50
ALL UNMARKED RADII, R.12
FLAT SURFACE

Problem 3-22

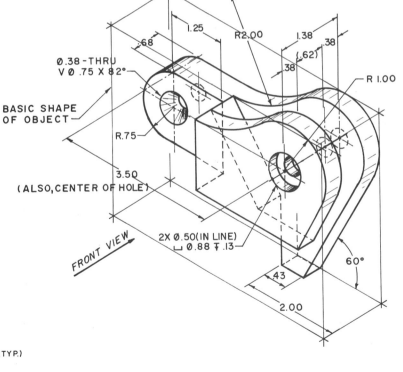

1.25
R2.00
1.38
.38
.68
.38
(.62)
Ø.38-THRU
V Ø.75 X 82°
38
R 1.00
BASIC SHAPE
OF OBJECT
R.75
3.50
(ALSO, CENTER OF HOLE)
2X Ø.50(IN LINE)
⌴ Ø.88 ⊤ .13
43
60°
2.00
FRONT VIEW

Problem 3-23

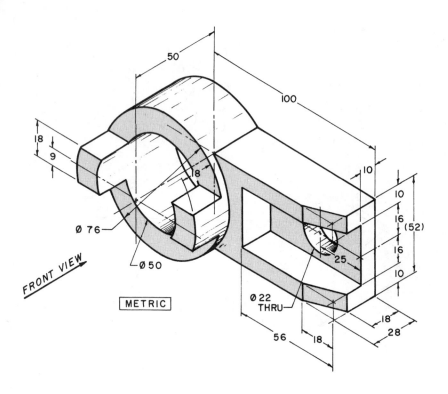

FRONT VIEW

METRIC

Problem 3-24

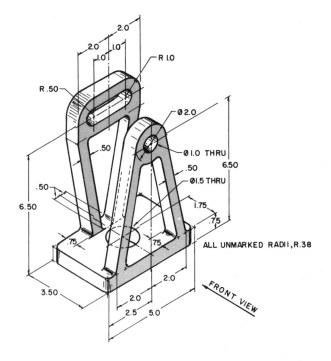

ALL UNMARKED RADII, R.38

FRONT VIEW

Problem 3-25

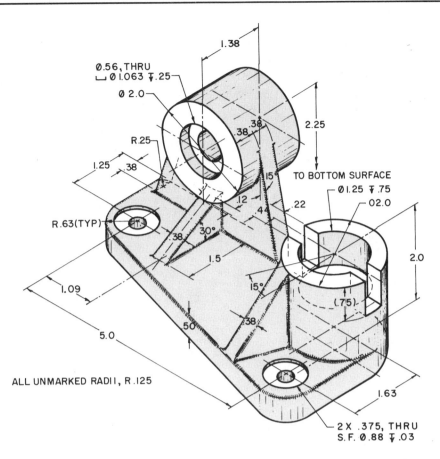

Ø.56, THRU
⌴ Ø1.063 ⍗.25

Ø 2.0

1.38

2.25

.38

.38

15°

TO BOTTOM SURFACE

R.25

1.25   .38

Ø1.25 ⍗ .75

02.0

.12

.44   .22

2.0

R.63(TYP)

30°

.38

15°

1.5

1.09

15°

5.0   .50   .38

(.75)

ALL UNMARKED RADII, R.125

1.63

2 X .375, THRU
S.F. Ø.88 ⍗.03

Problem 3-26

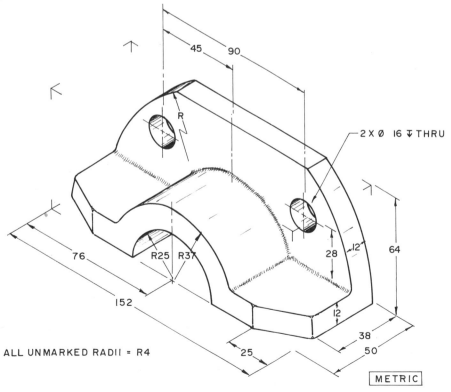

45   90

R

2 X Ø 16 ⍗ THRU

76   R25   R37

28   12

64

152

12

25

38

50

ALL UNMARKED RADII = R4

METRIC

Problem 3-27

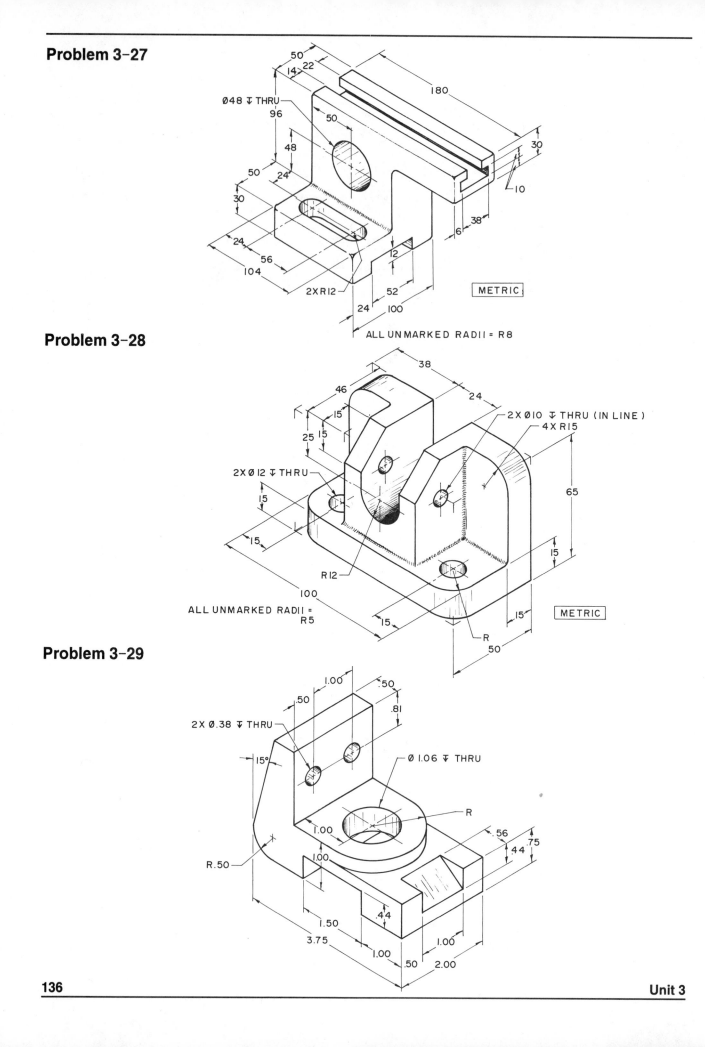

Ø48 ⍀ THRU
96
50
48
50
24
30
24
56
104
2X R12
24
100
52
12
6
38
50
14
22
180
30
10
METRIC

ALL UNMARKED RADII = R8

Problem 3-28

38
46
24
15
25   15
2X Ø12 ⍀ THRU
15
15
R12
100
ALL UNMARKED RADII =
R5
2X Ø10 ⍀ THRU (IN LINE)
4X R15
65
15
15
R
50
METRIC

Problem 3-29

1.00
.50
.50
.81
2X Ø.38 ⍀ THRU
15°
Ø 1.06 ⍀ THRU
R
1.00
R.50
1.00
.56
.75
.44
.44
1.50
3.75
1.00
.50
1.00
2.00

Problem 3-30

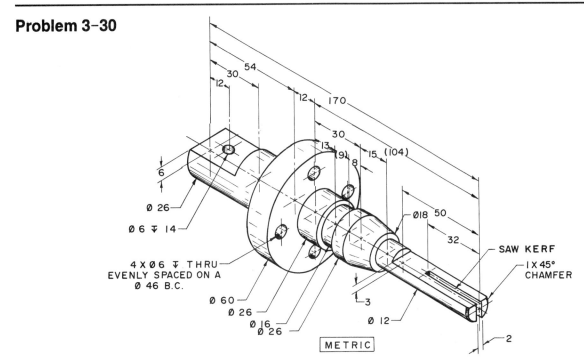

4 X Ø6 ⊥ THRU
EVENLY SPACED ON A
Ø 46 B.C.

Ø 26
Ø 6 ⊥ 14

Ø 60
Ø 26
Ø 16
Ø 26

Ø 12

Ø18 50

32

SAW KERF

I X 45°
CHAMFER

2

3

METRIC

54
30
12
170
30
12
13
(9) 8
15
(104)
6

Problem 3-31

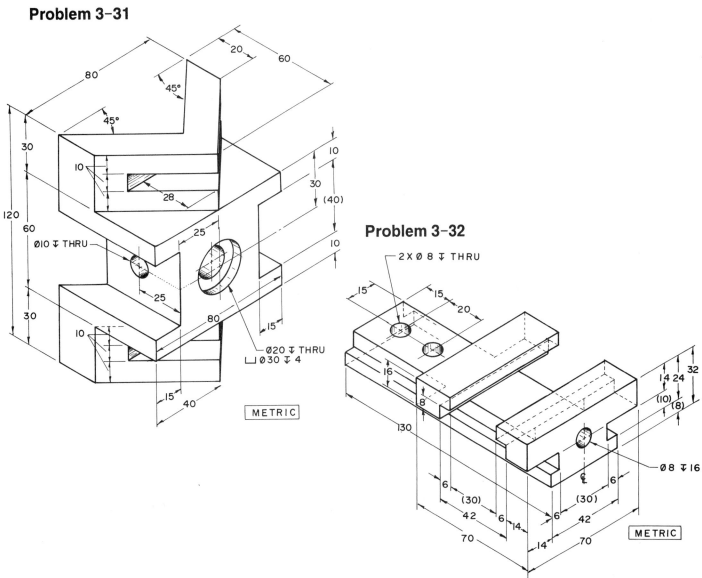

80
20
60
45°
45°
30
10
120
10
28
30
60
010 ⊥ THRU
25
25
30
10
80
15
020 ⊥ THRU
⊔ 030 ⊥ 4
10
15
40
10
30
(40)
10

METRIC

Problem 3-32

2 X Ø 8 ⊥ THRU

15
15
20
16
8
130

14 24 32
(10)
(8)

Ø 8 ⊥ 16

6
(30)
42
6
14
6
(30)
42
6
70
14
70

METRIC

Problem 3-33

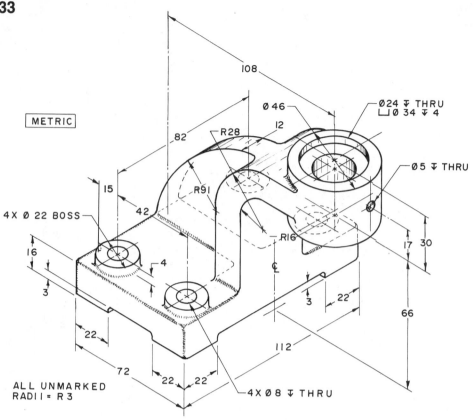

METRIC

108

82

Ø 46

R28

12

Ø24 ⟱ THRU
⊔ Ø 34 ⟱ 4

15

R91

R16

Ø5 ⟱ THRU

4X Ø 22 BOSS

42

C L

17   30

16

4

3

3   22

66

22

112

72

22   22

ALL UNMARKED
RADII = R 3

4X Ø8 ⟱ THRU

Problem 3-34

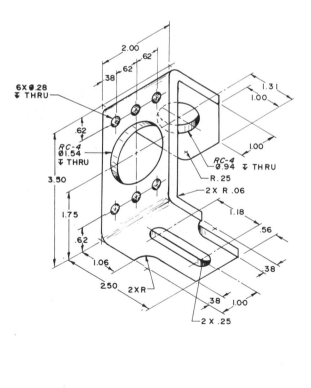

2 X R.52

2 X Ø.75 ⟱ THRU
(BOTH ENDS)

2 X R.75

.09, BOTH ENDS

Ø2.50

18
TYP

2X Ø1.25
TOP/BOTTOM

6.38

Ø.66

RC-4
Ø2.00
⟱ THRU

1.12

1.06

Ø.22 ⟱ THRU

3.19

2.00   .25

4.50

18
TYP

(1.50)

RC-6
2 X Ø.75
⟱ THRU

ALL FILLETS/ROUNDS =
R .09

Problem 3-35

2.00

62

38   62

6X Ø.28
⟱ THRU

1.31

1.00

.62

1.00

RC-4
Ø1.54
⟱ THRU

RC-4
Ø.94 ⟱ THRU

3.50

R. 25

2 X R .06

1.75

1.18

.62

.56

1.06

.38

2.50   2XR

.38   1.00

2 X .25

Problem 3-36

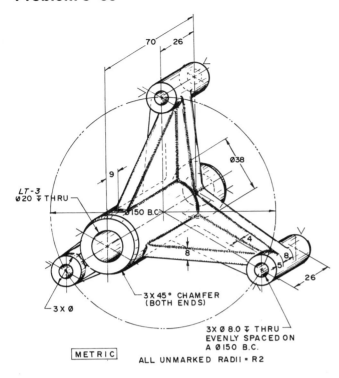

70   26°

Ø38

9

LT-3
Ø20 ↧ THRU

Ø150 B.C.

4

8

8

26

5°

3X Ø

3X 45° CHAMFER
(BOTH ENDS)

3X Ø 8.0 ↧ THRU
EVENLY SPACED ON
A Ø150 B.C.

METRIC

ALL UNMARKED RADII = R2

Problem 3-38

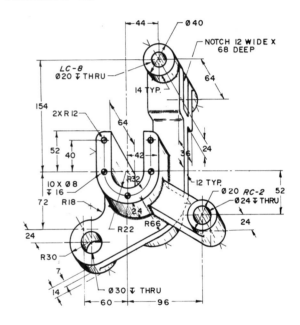

44

Ø40

NOTCH 12 WIDE X
68 DEEP

LC-8
Ø20 ↧ THRU

64

14 TYP.

154

2X R12

64

52

40

42

36

24

10 X Ø8
↧ 16

R32

12 TYP.

Ø20 RC-2
Ø24 ↧ THRU

52

72   R18

24

R66

R22

24

R30

7

14

Ø30 ↧ THRU

60

96

Problem 3-37

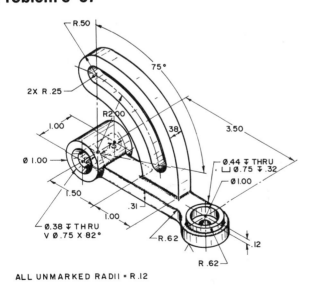

R.50

75°

2X R.25

R2.00

1.00

38

3.50

75°

Ø 1.00

Ø.44 ↧ THRU
⊔ Ø.75 ↧ .32

Ø1.00

1.50

.31

1.00

.12

Ø.38 ↧ THRU
V Ø.75 X 82°

R.62

R .62

ALL UNMARKED RADII = R.12

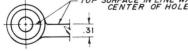

TOP SURFACE IN LINE W/
CENTER OF HOLE

.31

Problem 3-39

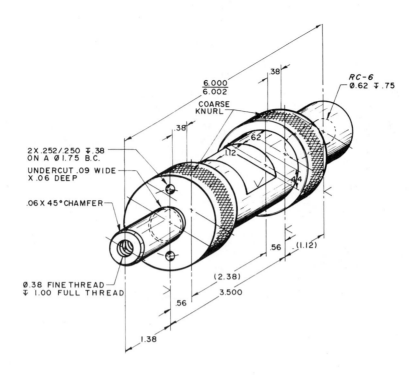

2X.252/250 ⊽.38
ON A Ø1.75 B.C.

UNDERCUT .09 WIDE
X .06 DEEP

.06 X 45° CHAMFER

Ø.38 FINE THREAD
⊽ 1.00 FULL THREAD

6.000
6.002

COARSE
KNURL

RC-6
Ø.62 ⊽ .75

.38

.62

1.12

.44

.56

(1.12)

.56

(2.38)

3.500

.56

1.38

.38

Directions for Problems 3-40 through 3-48
Using the background grid as a guide, reproduce each problem on a CAD system.

Problem 3-40

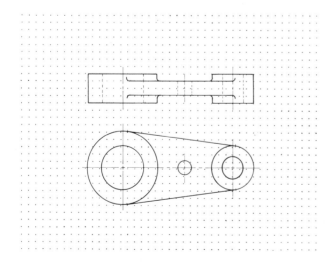

Problem 3-41

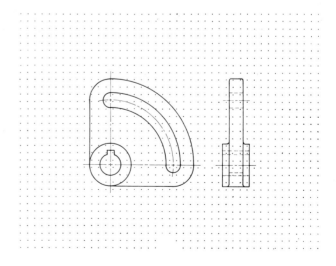

Problem 3-42

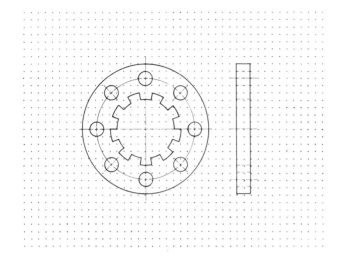

Problem 3-43

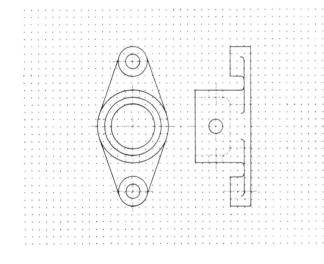

Problem 3-44

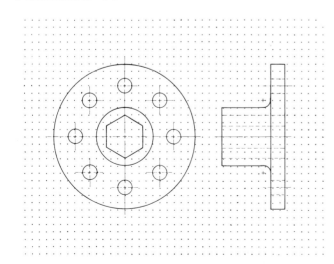

Problem 3-45

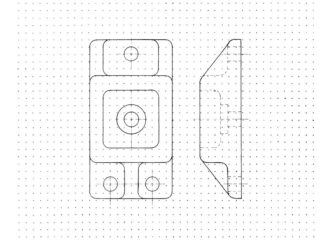

Problem 3-46

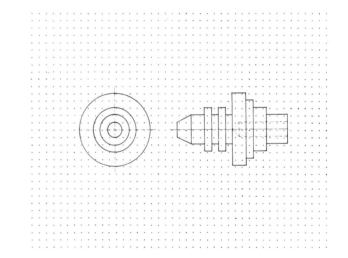

Problem 3-47

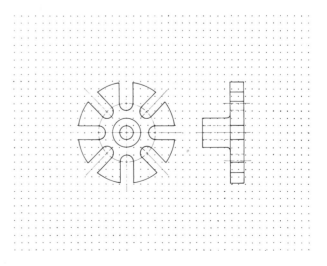

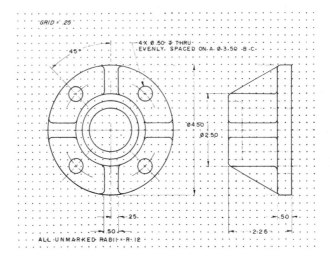

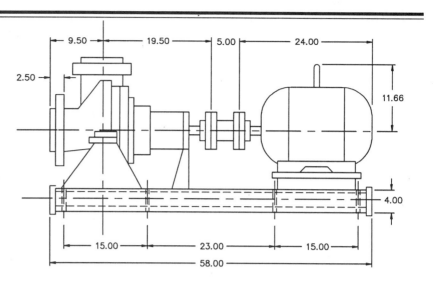

Dimensioning Techniques

Contents

Once views have been laid out, a design is not ready for the production line until numbers describing how wide, tall, or deep the object is are added to the drawing. However, these numbers must be added in a certain organized fashion; otherwise, the drawing becomes difficult to read. This may lead to confusion and the possible production of a part that is incorrect by the original design. This unit will focus on the basics of dimensioning and includes a few rules to proper dimensioning practices with numerous examples. Dimensioning techniques using AutoCAD will also be discussed in great detail including linear dimensioning, radius and diameter dimensioning, dimensioning angles, leader line usage, and a complete listing of all dimension variables with an explanation of their purpose. Special topics include dimensioning isometric drawings and ordinate or datum dimensioning. Three tutorials follow the main body of text; all three tutorials are complete regarding geometry. To complete them, dimensions need to be added or edited in some cases.

Dimension Basics

Before discussing the components of a dimension, remember that object lines (at "A") continue to be the thickest lines of a drawing with the exception of border or title blocks. To promote contrasting lines, dimensions become visible, yet thin lines. The heart of a dimension is the dimension line, (at "B") which is easily identified by the location of arrow terminators at both ends (at "D"). In mechanical cases, the dimension line is broken in the middle which provides an excellent location for the dimension text (at "E"). For architectural applications, dimension text is usually placed above an unbroken dimension line. The extremities of the dimension lines are limited by placing lines that act as stops for the arrow terminators. These lines, called extension lines (at "C"), begin close to the object without touching the object. Extension lines will be highlighted in greater detail in the pages that follow. For placing diameter and radius dimensions, a leader line consisting of an inclined line with a short horizontal shoulder is used (at "F"). Other applications of leader lines are for adding notes to drawings.

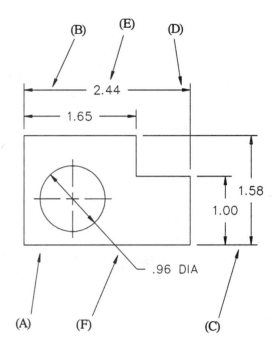

When placing dimensions in a drawing, it is recommended to provide a spacing of at least 0.38 units between the first dimension line and object being dimensioned (at "A"). If placing stacked or baseline dimensions, provide a minimum spacing of at least 0.25 units between the first and second dimension line at "B" or any other dimension lines placed thereafter. This will prevent dimensions from being placed too close to each other.

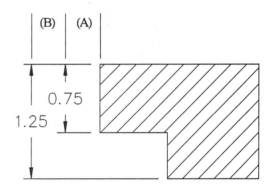

It is recommended that extensions never touch the object being dimensioned and begin approximately between 0.03 and 0.06 units away from the object at "A". As dimension lines are added, extension lines should extend no further than 0.12 beyond the arrow or any other terminator (at "B"). The height of dimension text is usually 0.125 units (at "C"). This value also applies to notes placed on objects with leader lines. Certain standards may require a taller lettering height. Become familiar with office practices that may deviate from these recommended values.

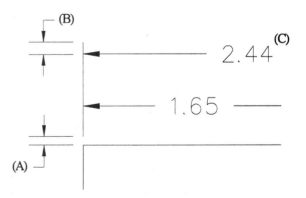

Placement of Dimensions

When placing multiple dimensions on one side of an object, place the shorter dimension closest to the object followed by the next larger dimension. When placing multiple horizontal and vertical dimensions involving extension lines that cross other extension lines, do not place gaps in the extension lines at their intersection points.

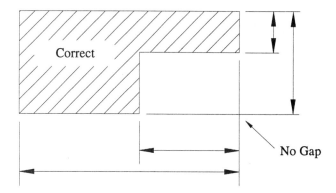

As it is acceptable for extension lines to cross each other, it is considered unacceptable practice for extension lines to cross dimension lines as in the example at the right. The shorter dimension is placed closest to the object followed by the next larger dimension.

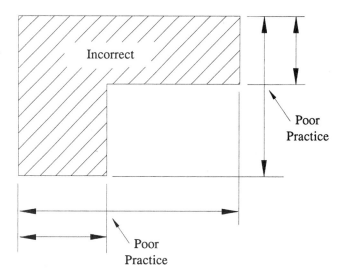

It is considered poor practice to dimension on the inside of an object when there is sufficient room to place dimensions on the outside. There may be exceptions to this rule, however. It is also considered poor practice to cross dimension lines since this may render the drawing confusing and possibly result in the inaccurate interpretation of the drawing.

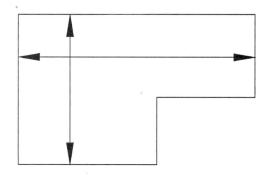

Placement of Extension Lines

As two extension lines may intersect without providing a gap, so also may extension lines and object lines intersect with each other without the need for a gap in between them. This is the same rule practiced when using center lines that extend beyond the object without gapping.

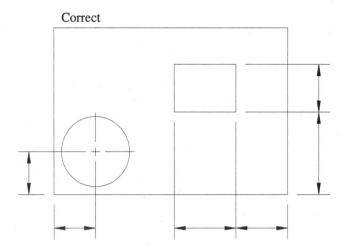

Correct

In the example at the right, the gaps in the extension lines may appear acceptable; however, in a very complex drawing, gaps in extension lines would render a drawing confusing. Draw extension lines as continuous lines without providing breaks in the lines.

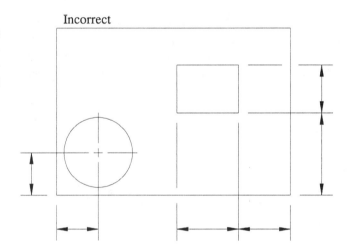

Incorrect

In the same manner, when center lines are used as extension lines for dimension purposes, no gap is provided at the intersection of the center line and the object. As with extension lines, the center line should extend no further than 0.125 units past the arrow terminator.

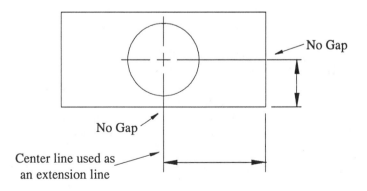

No Gap

No Gap

Center line used as an extension line

Grouping Dimensions

To promote ease of reading and interpretation, it is considered good practice to group dimensions whenever possible as in the example at the right. This promotes good organizational skills and techniques in addition to making the drawing and dimensions easier to read. As in previous examples, always place the shorter dimensions closest to the drawing followed by any larger or overall dimensions.

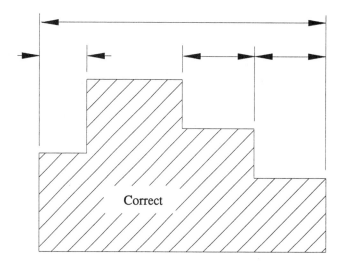

Avoid placing dimensions to with an object line substituting for an extension line as in the illustration at the right. The drawing is more difficult to follow with the dimensions being placed at different levels instead of being grouped. It must be pointed out at this time however, that there may be cases where even this practice of dimensioning is unavoidable.

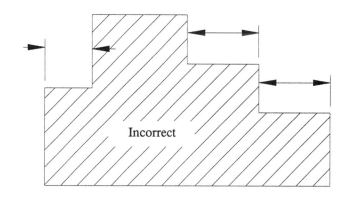

For tight spaces, arrange dimensions as in the illustration at the right. Extra care needs to be exercised to follow proper dimension rules without sacrificing clarity. AutoCAD dimension variables Dimtix and Dimsoxd may aid in the placing of dimensions in small spaces.

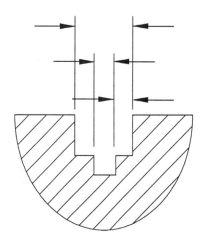

Dimensioning to Visible Features

The object is dimensioned correctly, however, the problem is that hidden lines are used to dimension to. As there are always exceptions, try to avoid dimensioning to any hidden surfaces or features.

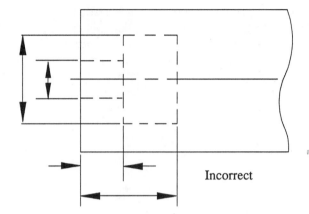

The object illustrated at the right is almost identical to the previous figure with the exception that it has been converted into a full section. Surfaces that were previously hidden are now exposed. This example illustrates a better way to dimension details that were previously invisible.

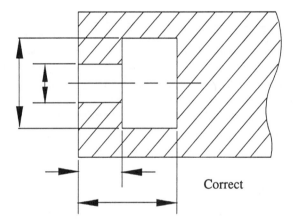

Dimensioning to Center Lines

Center lines are used to identify the center of circular features as in "A". The AutoCAD dimension variable Dimcen may be used to control the size of the center marker and whether the center marker extends beyond the largest circle. Center lines can also be used to indicate an axis of symmetry as in "B". Here, the center line consisting of a short dash flanked by two long dashes signifies the feature is circular in shape and form. Center lines may take the place of extension lines when placing dimensions in drawings.

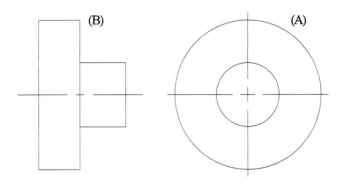

The illustration at the right represents the top view of a "U"-shaped object with two holes placed at the base of the "U". It also represents the correct way of utilizing center lines as extension lines when dimensioning to holes. What makes this example correct is the rule of always dimensioning to visible features. The example at the right uses center lines to dimension to holes that appear as circles. This is in direct contrast to the next example.

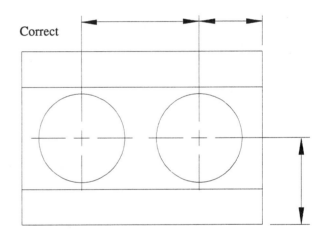

This illustration represents the front view of the "U"-shaped object. The hidden lines display the circular holes passing through the object along with center lines. Center lines are being used as extension lines for dimensioning purposes; however, it is considered poor practice to dimension to hidden features or surfaces. Always attempt to dimension to a view where the features are visible before dimensioning to hidden areas.

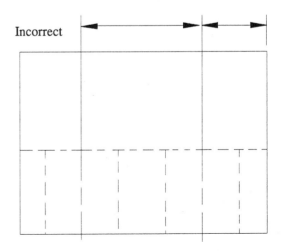

Arrowheads

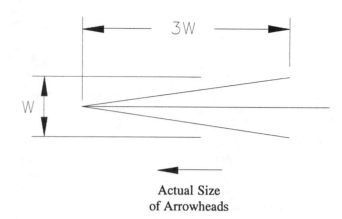

Arrowheads are generally made three times as long as they are wide, or very long and narrow. The actual size of an arrowhead would measure approximately 0.125 units in length. This size is controlled by the AutoCAD dimension variable Dimasz.

Actual Size
of Arrowheads

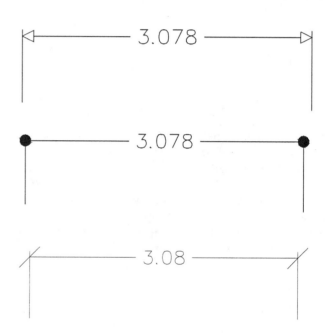

Dimension line terminators may take the form of shapes other than filled-in arrowheads. Open arrowheads and filled-in dots are controlled by first defining the shape as a symbol using the Block command and then identifying the name in the AutoCAD dimension variable Dimblk. The 45-degree slash or "tick" is controlled by the dimension variable Dimtsz. This is a favorite dimension line terminator used by architects although they are sometimes seen in mechanical applications.

Linear Dimensions

Horizontal Dimensioning

This linear dimensioning mode generates a dimension line that is horizontal in appearance. The following prompts illustrate generation of a horizontal dimension using the Dim-Horizontal option:

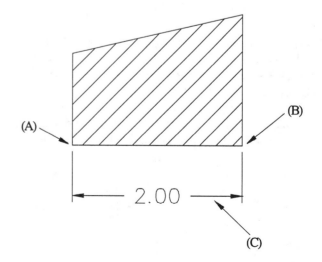

Dim: **Hor**
First extension line origin or RETURN to select: **Endp**
of *(Select the endpoint of the horizontal line at "A")*
Second extension line origin: **Endp**
of *(Select the other endpoint of the horizontal line at "B")*
Dimension line location: *(Select a point at "C")*
Dimension text <2.00>: *(Strike Enter to accept the default value)*

Linear Dimensions Continued

Vertical Dimensioning

This linear dimensioning mode generates a dimension line that is vertical in appearance. The following prompts illustrate generation of a horizontal dimension using the Dim-Vertical option:

Dim: **Ver**
First extension line origin or RETURN to select: **Endp**
of *(Select the endpoint of the vertical line at "A")*
Second extension line origin: **Endp**
of *(Select the other endpoint of the vertical line at "B")*
Dimension line location: *(Select a point at "C")*
Dimension text <2.10>: *(Strike Enter to accept the default value)*

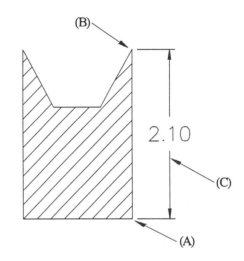

Aligned Dimensioning

This linear dimensioning mode generates a dimension line that is parallel to the distance specified by the two extension line origins. The following prompts illustrate generation of an aligned dimension using the Dim-Aligned option:

Dim: **Ali**
First extension line origin or RETURN to select: **Endp**
of *(Select the endpoint of the line at "A")*
Second extension line origin: **Endp**
of *(Select the other endpoint of the line at "B")*
Dimension line location: *(Select a point at "C")*
Dimension text <5.6569>: *(Strike Enter to accept the default value)*

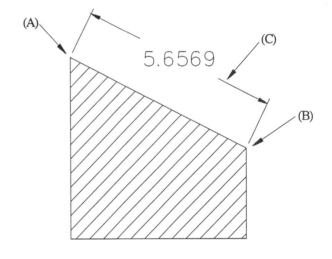

Rotated Dimensioning

This linear dimensioning mode generates a dimension line that is rotated at a specified angle. The following prompts illustrate generation of a rotated dimension using the Dim-Rotated option:

Dim: **Rot**
Dimension line angle <0>: **45**
First extension line origin or RETURN to select: **Endp**
of *(Select the endpoint of the line at "A")*
Second extension line origin: **Endp**
of *(Select the other endpoint of the line at "B")*
Dimension line location: *(Select a point at "C")*
Dimension text <2.8284>: *(Strike Enter to accept the default value)*

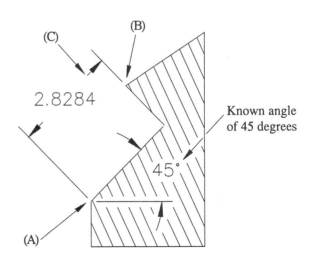

Known angle of 45 degrees

Radius and Diameter Dimensioning

Arcs and circles are to be dimensioned in the view where their true shape is visible. The mark in the center of the circle or arc indicate its center point. The dimension text may be placed either inside or outside of the circle or arc depending on the current values of two dimension variables, namely Dimtix and Dimtofl. Both of these variables will be discussed in detail at a later time. The prompts for Diameter and Radius dimensions are as follow:

Dim: **Dia**
Select arc or circle: *(Select the edge of an arc or circle)*
Dimension text <value>: *(Strike Enter to accept the default)*

Dim: **Rad**
Select arc or circle: *(Select the edge of an arc or circle)*
Dimension text <value>: *(Strike Enter to accept the default)*

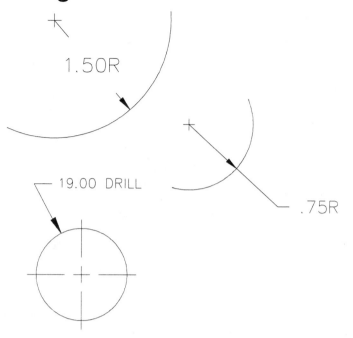

Leader Lines

A leader line is a thin, solid line leading from a note or dimension ending with an arrowhead illustrated at "A". The arrowhead should always terminate on an object line such as the edge of a hole or arc. A leader to a circle or arc should be radial; this means it is drawn so that if extended it would pass through the center of the circle illustrated at "B". Leaders should cross as few object lines as possible and should never cross each other. The short horizontal shoulder of a leader should meet the dimension illustrated at "A". It is considered poor practice to underline the dimension with the horizontal shoulder illustrated at "C". Example "C" also illustrates a leader not lined up with the center or radial. This may affect the appearance of the leader. Again, check for the standard office practices to ensure this example is acceptable. Yet another function of a leader is to attach notes to a drawing illustrated at "D". Notice the two notes attached to the view have different terminators, arrows, and dots. It is considered good practice to adopt only one terminator for the duration of the drawing. The Dimblk dimension variable may be used for defining different terminators such as dots. The prompt sequence for the AutoCAD Leader command is as follows:

Dim: **Lea**
From point: *(Select a starting point for the leader)*
To point: *(Select an ending point for the leader)*
To point: *(Strike Enter to place the short horizontal shoulder)*
Dimension text < >: *(Enter the desired text for the leader)*

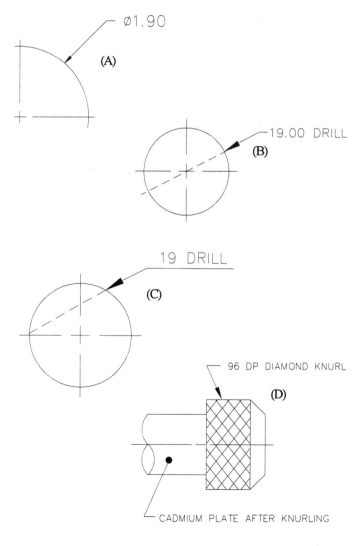

Dimensioning Angles

Dimensioning angles requires two lines forming the angle in addition to the location of the vertex of the angle. Other important information needed before dimensioning an angle includes the dimension arc location, the dimension text or what the angle actually measures, and the location of the dimension text. Before going any further, understand where the curved arc for the angular dimension is derived from. In illustration "A" at the right, the dimension arc is struck from an imaginary center or vertex of the arc. The following prompts are taken from the AutoCAD Angular dimension command:

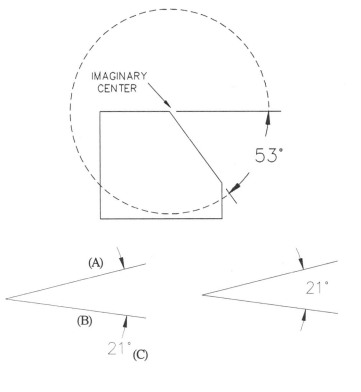

Dim: **Ang**
Select arc, circle, line or RETURN: *(Select the line at "A")*
Second line: *(Select the line at "B")*
Enter dimension line arc location: *(Select a point at "C")*
Dimension text <21>: *(Strike Enter to accept the default value)*
Enter text location: *(Strike Enter to place text in the center of the dimension arc)*

Dimensioning Slots

For slots, first select the view where the slot is visible. Two methods of dimensioning the slot are illustrated at the right. A slot may be called out by locating the center-to-center distance of the two semi-circles followed by a radius dimension to one of the semi-circles; which radius dimension selected depends on the available room to dimension. A second method involves the same center-to-center distance followed by an overall distance designating the width of the slot. This dimension happens to be the same as the diameter of the semi-circles. It is also considered good practice to place this dimension inside of the slot. A more complex example involves slots formed by curves and angles. Here, the radius of the circular center arc is called out. Angles reference each other for accuracy. As in the previous example, the overall width of the slot is dimensioned which happens to be the diameter of the semi-circles at opposite ends of the slot.

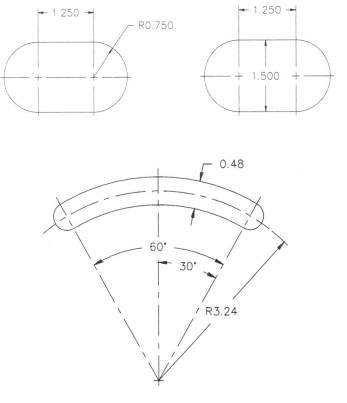

Dimensioning Systems

The Unidirectional System

When placing dimensions using AutoCAD, a typical example is illustrated at the right. Here, all text items are right-reading or horizontal. This goes for all vertical, aligned, angular, and diameter dimensions. When dimension text can be read right-reading, this is the Unidirectional Dimensioning System. By default, all AutoCAD dimension variables are set to dimension in the Unidirectional System.

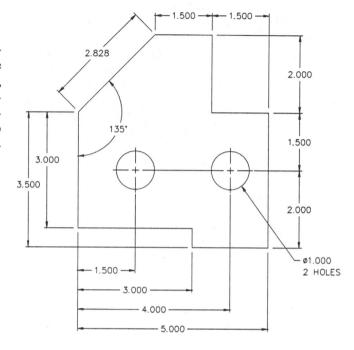

The Aligned System

The identical object from the previous example is illustrated at the right. Notice that all horizontal dimensions have the text positioned in the horizontal direction as in the previous example. However, vertical and aligned dimension text is rotated or aligned with the direction being dimensioned. This is the most notable feature of the Aligned Dimensioning System. Text along vertical dimensions is rotated in such a way that the drawing must be read from the right. Angular dimensions remain unaffected in the Aligned System; however, aligned dimension text is rotated parallel with the feature being dimensioned. Dimtih and Dimtoh are the two dimension variables that control whether the dimension text is horizontal. Both are currently in an Off mode. Once switched On, text for vertical dimensions will appear similar to the example at the right.

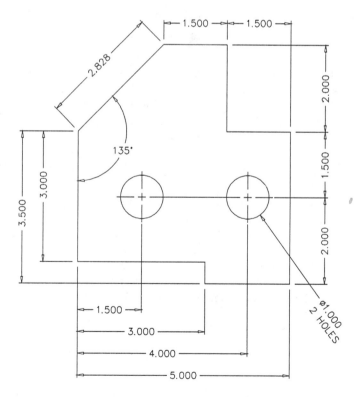

Continuous and Baseline Dimensions

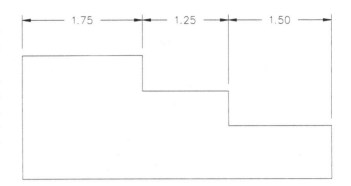

The power of grouping dimensions for ease of reading has already been explained. The illustration at the right shows yet another feature while dimensioning in AutoCAD; namely, the practice of using Continuous dimensions. With one dimension already placed, the Continuous subcommand of DIM: is selected which prompts the user for the second extension line location. Picking the second extension line location strings the dimensions next to each other or continues the dimension.

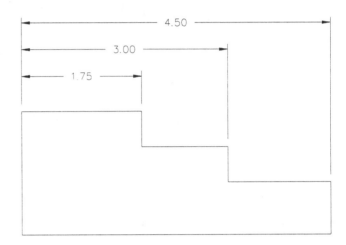

Yet another aid in grouping dimensions is using the Baseline mode which is also located in the DIM: area of AutoCAD. Continuous dimensions place dimensions next to each other; Baseline dimensions establish a base or starting point for the first dimension. Any dimensions that follow in Baseline mode are calculated from the common base point already established. This is a very popular mode to use when one end of an object acts as a reference edge. As dimensions are placed in Baseline mode, the AutoCAD dimension variable Dimdli controls the spacing of the dimensions away from each other. This variable has a default spacing of 0.38 units.

Tolerances

Interchangeability of parts requires replacement parts to fit in an assembly of an object no matter where the replacement part comes from. In the example at the right, the "U"-shaped channel of 2.000 units in width is to accept a mating part of 1.995 units. Under normal situations, there is no problem with this drawing or callout. However, what if the person cannot make the channel piece exactly at 2.000? What if he is close and the final product measures 1.997? Again, the mating part will have no problems fitting in the 1.997 slot. What if the mating part is not made exactly 1.995 units but is instead 1.997? You see the problem. As easy as it is to attach a dimension to a drawing, some thought needs to go into the possibility that based on the numbers, maybe the part cannot be easily made. Instead of locking dimensions in using one number, a range of numbers would allow the production individual the flexability to vary in any direction and still have the parts fit together. This is the purpose of converting some basic dimensions to tolerances.

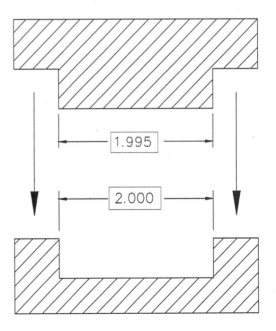

Tolerances Continued

The example at the right shows the same two mating parts; this time two sets of numbers for each part are assigned. For the lower part, the machinist must make the part anywhere in between 2.002 and 1.998, which creates a range of .004 units of variance. The upper mating part must be made within 1.997 and 1.993 units in order for the two to fit correctly. The range for the upper part is also .004 units. As we will soon see, no matter if upper or lower numbers are used, the parts will always fit together. If the bottom part is made to 1.002 and the top part is made to 1.993, the parts will fit. If the bottom part is made to 1.998 and the upper part is made to 1.997, the parts will fit. In any case or combination, if the dimensions are followed exactly as stated by the tolerances, the pieces will always fit. If the bottom part is made to 1.998 and the upper part is made to 1.999, the upper piece is rejected. The method of assigning upper and lower values to dimensions is called limit dimensioning. Here, the larger value in all cases is placed above the smaller value. This is also called a clearance fit since any combinations of numbers may be used and the parts will still fit together. The AutoCAD dimension variable Dimlim controls the display of limit dimensions. Variables Dimtp and Dimtm assign positive and negative ranges for the tolerance. These will be discussed later in this chapter.

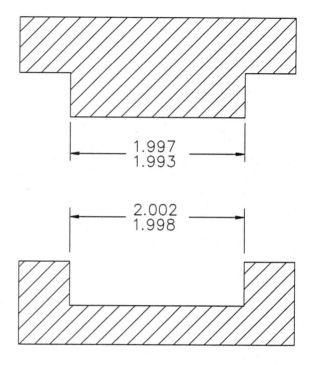

The object at the right has a different tolerance value assigned to it. The basic dimension is 2.000; in order for this part to be accepted, the width of this object may go as high as 2.002 units or as low as 2.001 units giving a range of .003 units the part may vary. This type of tolerance is called a plus/minus dimension with an upper limit of .002 different from the lower limit of -.001. The AutoCAD dimension Dimtol controls the display of plus/minus dimensions. Variables Dimtp and Dimtm assign positive and negative ranges for the tolerance. These will be discussed later in this chapter.

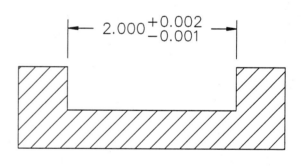

Illustrated at the right is yet another way to display tolerances. It is very similar to the plus/minus method except that both upper and lower limits are the same; for this reason, this method is called plus and minus dimensions. The basic dimension is still 2.000 with upper and lower tolerance limits of .002 units. The dimension variable Dimtol again controls the display of plus and minus dimensions. When both the Dimtp and Dimtm variables have the same values, the result is both tolerances listed together with the plus and minus symbol (±) placed in front of the tolerance. All variables will be discussed later in this chapter.

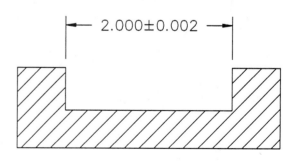

Repetitive Dimensions

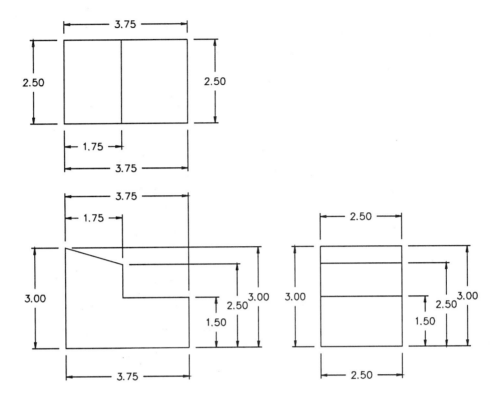

Throughout this chapter, numerous methods of dimensioning have been discussed supported by many examples. Just as it was important in multi-view projection to draw only those views that accurately described the object, so also is important to dimension these views. In this manner, the actual production of the part may start. However, care needs to be taken when placing dimensions; dimensioning takes planning to better place a dimension. The problem with the illustration above is that even though the views are correct and the dimensions call out the overall sizes of the object, there are too many cases where dimensions are duplicated. Once a feature has been dimensioned, such as the overall width of 3.75 units, this number does not need to be placed in the top view. This is the purpose of understanding the relationship between views; or what dimensions the views have in common with each other. Adding unnecessary dimensions also makes the drawing very busy and cluttered in addition to being very confusing to read. Compare the illustration above with the illustration at the right which shows just those dimensions needed to describe the size of the object. Do not be concerned that the top view has no dimension; the designer should interpret the width of the top as 3.75 units from the front view and the depth as 2.50 from the right side view.

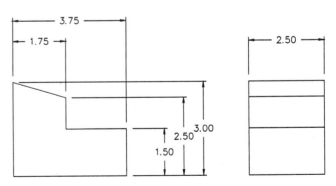

Dimension Variables

Dimension variables control the manner in which dimensions are placed in an AutoCAD drawing. These variables range from specifying a new size for arrowheads or text to turning On or Off variables that control things like if dimension text is placed horizontal or vertical. Illustrated at the right is a partial list of variables with default values. Each variable will be explained in detail in the next series of pages that follow.

DIMTIH	On
DIMTIX	Off
DIMTM	0.00
DIMTOFL	Off
DIMTOH	Off
DIMTOL	Off
DIMTP	0.00
DIMTSZ	0.00
DIMTVP	0.00
DIMTXT	0.18
DIMZIN	0

DIMALT
Alternate Dimension Units
DIMALTF
Alternate Dimension Unit Scale Factor
DIMALTD
Alternate Dimension Unit Decimal Places

All three variables operate together in controlling an alternate unit display. If Dimalt is On, as at "B" in the illustration at the right, a second dimension text string is placed next to the existing dimension text. The alternate text string is placed in brackets to separate it from the orginal string. The value of the alternate dimension is calculated from the current value held in the Dimaltf variable. This value is multiplied by the current dimension text string to arrive at the alternate string. Illustrated at "C" is the Dimaltd which controls the number of decimal places held in the alternate dimension string. The prompt for this dimension variable is:

Dim: DIMALT
Current value <Off> New value: *(Enter On or accept default)*

Dim: DIMALTF
Current value <25.4000> New value: *(Enter a numeric value)*

Dim: DIMALTD
Current value <2> New value: *(Enter a numeric value)*

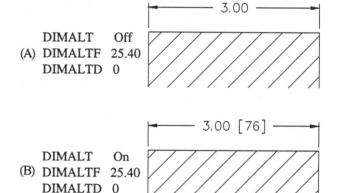

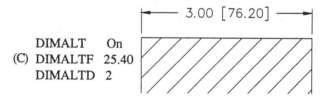

DIMAPOST
The Default Suffix for Alternate Text

When Dimalt is On and Dimapost is set to the text string, "mm", the alternate dimension text value, will have the Dimapost string added to its end. To change this value back to a null response, type a "." for the new value. The prompt for this dimension variable is:

Dim: DIMAPOST
Current value < > New value: *(Enter a numeric value)*

DIMALT On
DIMAPOST Set to "mm"

DIMASO

The Associative Dimensioning Control Variable

Use this variable to turn associative dimensioning On or Off. If On, all entities that make up the dimension will be considered one entity. If Off, all dimension components such as arrowheads, extension lines, dimension lines, and dimension text will be considered single entities. This is the same effect as using the Explode command on an associative dimension. With Dimaso set to Off, commands that normally affect dimensions that are associative will have no affect. The prompt for this dimension variable is:

Dim: DIMASO
Current value <On> New value: *(Enter Off or keep the default)*

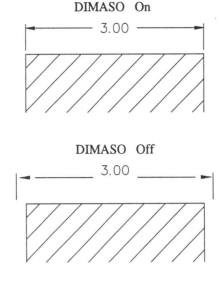

DIMASZ

Dimension Arrow Size

Use this variable to control the size of the arrow terminator. This arrow is solid controlled by the Fill command. The prompt for this dimension variable is:

Dim: DIMASZ
Current value <0.18> New value: *(Enter a numeric value)*

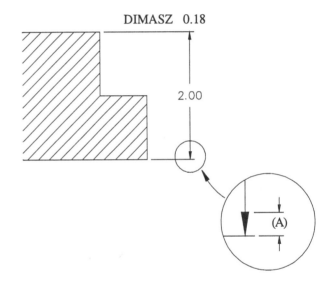

DIMBLK

New Dimension Line Terminator Name

This variable controls the shape of the terminator at the two ends of the dimension line. Dimblk waits for the name of a Block that has already been defined in the drawing data base such as the "Dot" at the right. Once the dot has been called out in the Dimblk variable, all dimension lines will be terminated at the extension lines by dots at both ends. There is some skill involved in creating the block of the arrow. Generally the block is constructed inside a 1-unit by 1-unit grid area. A short horizontal shoulder is drawn to complete the block; this shoulder will be added to the dimension line preventing a gap between the end of the dimension line and the new terminator. To disable this terminator and return to standard arrows, reply to the new value of the variable with a period, "." The prompt for this dimension variable is:

Dim: DIMBLK
Current value < > New value: *(Enter the name of a block)*

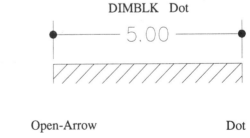

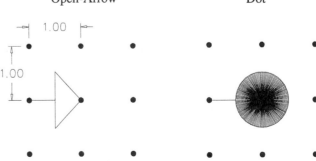

DIMBLK1
First Dimension Terminator Block Name
DIMBLK2
Second Dimension Terminator Block Name
DIMSAH
Separate Dimension Terminators

Some applications require two different dimension line terminators. All of the above variables work in combination with each other when specifying two different dimension line terminators. When a Block called "Dot" is assigned to Dimblk1 and open "Arrow" is assigned to Dimblk2, the user may place a dot at the first extension line origin and an open arrow at the second extension line origin. This is only possible if two different blocks have been defined and called out in Dimblk1 and Dimblk2 in addition to having Dimsah set to On. The prompts for Dimblk1 and Dimblk2 are identical to Dimblk. The prompt for Dimsah is:

Dim: DIMSAH
Current value <Off> New value: *(Enter On or keep the default)*

DIMBLK1 Dot
DIMBLK2 Arrow
DIMSAH   Off

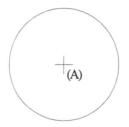

DIMBLK1 Dot
DIMBLK2 Arrow
DIMSAH   On

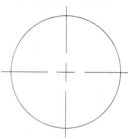

DIMCEN
Center Mark Size for Circles and Arcs

This dimension variable controls the size of the center mark that is placed when using the Dim-Cen command. Selecting a circle or arc places a mark similar to "A" when the value of Dimcen is 0.09. With a new value of -0.09 for Dimcen places the center mark at the center in addition to drawing extender lines beyond the perimeter of the circle. The prompt for this dimension variable is:

Dim: DIMCEN
Current value <0.09> New value: *(Enter a numeric value)*

DIMCEN 0.12

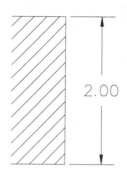

DIMCEN -0.12

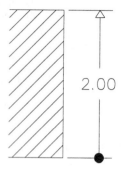

DIMCLRE
Color for Extension Lines
DIMCLRD
Color for Dimension Lines
DIMCLRT
Color for Dimension Text

These three variables control the color of the following dimension components, extension lines, dimension lines and arrowheads, and dimension text. It may seem that these variables merely give the operator control of colorfully displaying dimensions. However, a more practical use of these variables is to assign colors to the different dimension components to control line quality during plotting by assigning different pen weights. The prompt for all three variables is:

Dim: DIMCLRE
Current value < > New value: *(Enter the name of a color)*

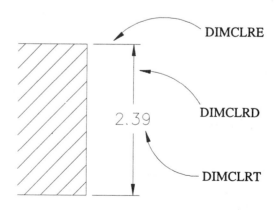

DIMDLE
Dimension Line Extension

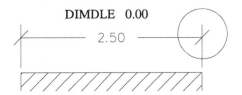

Some professions prefer the dimension line be drawn past the current arrow terminator. Setting the variable Dimdle to a value of 0.20 extends the dimension line past the terminator at that distance. The prompt for this dimension variable is:

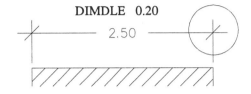

Dim: DIMDLE
Current value <0.00> New value: *(Enter a numeric value)*

DIMDLI
Dimension Line Increment for Continuation

This variable controls the spacing when multiple dimensions are placed away from each other similar to the illustration at the right. The default value of Dimdli (0.38) satisfies the minimum value on the spacing of dimensions especially while in Baseline mode. The prompt for this dimension variable is:

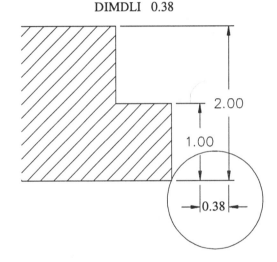

Dim: DIMDLI
Current value <0.38> New value: *(Enter a numeric value)*

DIMEXE
Extension Above Dimension Line

Use this variable to control how far the extension line extends past the dimension line. In keeping with dimension basics, this variable may be changed from the default value of 0.18 to a new value of 0.12. The prompt for this dimension variable is:

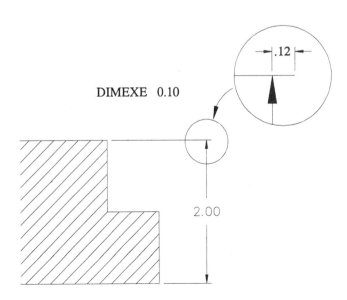

Dim: DIMEXE
Current value <0.18> New value: *(Enter a numeric value)*

Dimensioning Techniques

DIMEXO
Extension Line Origin Offset

This variable controls how far away from the object the extension will start (at "A"). The prompt for this dimension variable is:

Dim: DIMEXO
Current value <0.0625> New value: *(Enter a numeric value)*

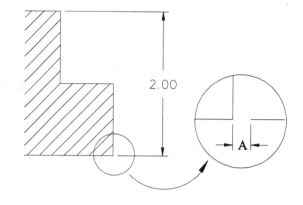

DIMLFAC
Linear Unit Scale Factor

This variable acts as a multiplier for all linear dimensions including radius and diameter dimensions. As a dimension is calculated by AutoCAD, the current Dimlfac value is multiplied by the dimension to arrive at a new dimension value. With a value of 1.00, all dimensions are taken as default. With a value of 2.00, all dimension values are first multiplied by 2.00 before being placed. In the same manner, a Dimlfac value of 0.50 would cut all dimensions in half. The prompt for this dimension variable is:

Dim: DIMLFAC
Current value <1.00> New value: *(Enter a numeric value)*

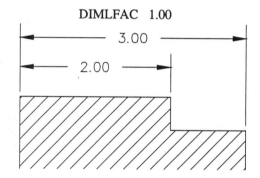

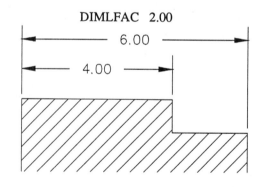

DIMGAP
Dimension Line Gap Increment

This variable maintains the area around dimension text by providing the gap between the text and the dimension line ends. The prompt for this dimension variable is:

Dim: DIMGAP
Current value <0.09> New value: *(Enter a numeric value)*

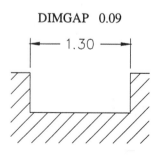

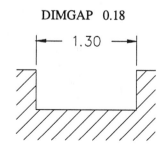

DIMPOST
Default Suffix for Dimension Text

This variable adds a character string immediately after all dimension values except for angular dimensions. The text string may be disabled by entering a single period "." at the "New value:" prompt. The prompt for this dimension variable is:

Dim: DIMPOST
Current value <0> New value: *(Enter a numeric value)*

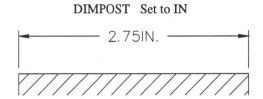

DIMRND
Dimension Round-Off Value

This variable will round-off all dimension distances based on the current rounding value. If the Dimrnd value is set to 0.25, all distances will be rounded to the nearest 0.25 unit. This variable does not affect angular dimensions. The prompt for this dimension variable is:

Dim: DIMRND
Current value <0> New value: *(Enter a numeric value)*

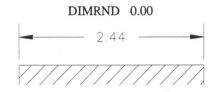

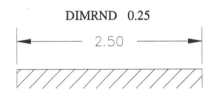

DIMSCALE
Overall Dimension Scale Factor

This variable acts as a multiplier and globally affects all current dimension variables that are specified by sizes or distances. This means that if the current Dimscale value is 1.00, variables such as Dimtxt set to 0.18 will remain unchanged. If the Dimscale value is changed to 2.00, the dimension text visible on the display screen will be changed from 0.18 to 0.36. The prompt for this dimension variable is:

Dim: DIMSCALE
Current value <1.0000> New value: *(Enter a numeric value)*

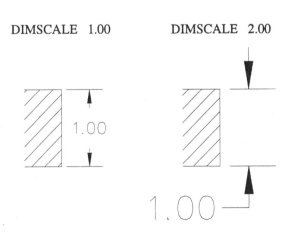

DIMSE1
Suppress the First Extension Line
DIMSE2
Suppress the Second Extension Line

These two variables control the display of extension lines. They are useful when dimensioning to an object line and to avoid placing the extension line on top of the object line. Dimse1, if turned On, suppresses the first extension. This is another way of saying that if Dimse1 is On, all first extension lines from that point on will be turned off. The same is true for Dimse2 which will turn off or suppress the second extension line if set to On. Study the many examples illustrated at the right. The prompt for this dimension variable is:

Dim: DIMSE1
Current value <Off> New value: *(Enter On or keep the default)*

Dim: DIMSE2
Current value <Off> New value: *(Enter On of keep the default)*

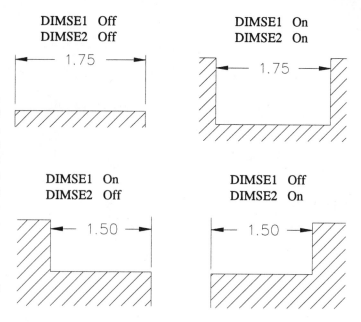

DIMTAD
Place Text above the Dimension Line

This variable controls whether dimension text will be placed inside of the dimension line or placed above the dimension. The default value is Off which means the dimension line will break to allow text to be placed in between. If turned On, text will be placed above the dimension line as in the illustration at the right. This variable is very important to architectural office practices although some mechanical applications use this variable On. The prompt for this dimension variable is:

Dim: DIMTAD
Current value <Off> New value: *(Enter On or keep the default)*

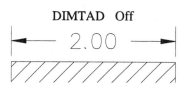

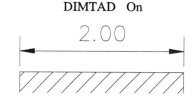

DIMTIX
Place Text Inside Extension Lines

This variable controls whether text is placed outside of extension lines or forced in between extension lines. By default, the variable is turned Off; depending on the current value of Dimasz and Dimtxt, AutoCAD will calculate whether the dimension will fit or be placed outside of extension lines. Turning this variable On forces the text to be placed inside the extension lines even if the text is so large that it goes beyond the extensions The prompt for this dimension variable is:

Dim: DIMTIX
Current value <Off> New value: *(Enter On or keep the default)*

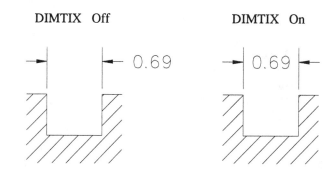

DIMTIH
Text Inside Extension Lines is Horizontal
DIMTOH
Text Outside Extension Lines is Horizontal

Both of these variables control whether dimension text is placed horizontally or is placed parallel to the distance dimensioned. Both variables are either On or Off; the default setting for both is On. This means that if text can fit inside of extension lines or is placed outside of extension lines, the text is placed horizontally as in "A" at the right. If both variables are turned Off, the results are illustrated at "B". Here, dimension text placed outside of extension lines, such as the 0.50 distance, is no longer horizontal but vertical. The same is true for the 1.00 vertical dimension. Notice that the 2.06 aligned dimension is placed parallel to the dimension line. One variable may be turned On while the other remains Off as in "C" at the right. Here, the variable Dimtih is turned Off while Dimtoh is turned On. Dimension text that falls outside of extension lines will be placed horizontally as in the 0.50 dimension while dimension text that falls inside extension lines will not be horizontal as with the 1.00 and 2.06 dimensions. The prompts for these dimension variables are:

Dim: DIMTIH
Current value <On> New value:

Dim: DIMTOH
Current value <On> New value:

DIMTOFL
Force the Dimension Line Inside the Extension Lines

This variable forces the dimension line to be drawn inside of extension lines even if the text is placed outside of the extension lines. The prompt for this dimension variable is:

Dim: DIMTOFL
Current value <Off> New value: *(Enter On or keep the default)*

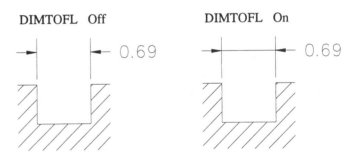

DIMTSZ
Tick Terminator Size

Use this variable to convert arrow terminators into 45-degree slashes or "ticks". With the Dimtsz value 0.00, arrowheads will be drawn. Setting the variable to a value of 0.12 overrides the current arrow setting and places ticks at the ends of the dimension line. Set Dimtsz back to 0.00 to return to arrowheads. The prompt for this dimension variable is:

Dim: DIMTSZ
Current value <0.00> New value: *(Enter a numeric value)*

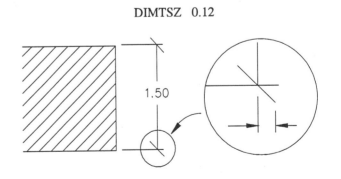

DIMSOXD

Suppress Dimension Line Outside of Extension Lines

This variable controls the dimension line for dimension text placed outside of extension lines. The default value is Off which means the dimension line with arrowheads will be drawn whether the text is placed inside or outside. With this variable turned On, the dimension line is suppressed or turned off leaving the text and extension lines.
The prompt for this dimension variable is:

Dim: DIMSOXD
Current value <Off> New value: *(Enter Off or keep the default)*

DIMTIX On
DIMSOXD Off

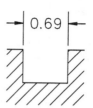

DIMTIX On
DIMSOXD On

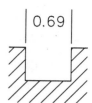

DIMTOL

Generate Tolerance Dimensions
DIMLIM
Generate Limit Dimensions
DIMTP
Positive Tolerance Value
DIMTM
Minus Tolerance Value

These variables work together in converting normal dimensions to limit or tolerance dimensions. Dimtp and Dimtm require numeric values; Dimtol and Dimlim are On/Off switches. An example of a tolerance dimension is illustrated at "A" complete with "plus" and "minus" values. An example of a limit dimension is illustrated at "B". Here the upper limit is placed over the lower limit. As both Dimtol and Dimlim can be switched On or Off, only one variable may remain On. This means if Dimlim is On and Dimtol is changed from Off to On, the Dimlim variable automatically switches itself Off. With Dimtol On in example "A" and different values entered for Dimtp and Dimtm, the base dimension of 4.000 is placed along with the upper limit, (Dimtp) and lower limit, (Dimtm). In example "B", Dimlim is On with Dimtp and Dimtm the same values. The limit dimension takes the basic dimension size (4.000), adds the Dimtp value, subtracts the Dimtm value, and places the upper limit and lower limit as the new dimension. If Dimtp and Dimtm have values assigned but both Dimtol and Dimlim are Off, the result is the basic size at "C". If Dimtp and Dimtm are the same size and Dimtol is On, a plus/minus dimension is placed as in "D". The prompts for these variables are:

Dim: DIMTOL
Current value <Off> New value:

Dim: DIMLIM
Current value <Off> New value:

Dim: DIMTP
Current value <0.00> New value:

Dim: DIMTM
Current value <0.00> New value:

If...
DIMTP = 0.005
DIMTM = 0.003

And...
DIMTOL is On
DIMLIM is Off

Then...

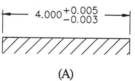

(A)

If...
DIMTP = 0.003
DIMTM = 0.003

And...
DIMTOL is Off
DIMLIM is On

Then...

4.003
3.997

(B)

If...
DIMTP = 0.005
DIMTM = 0.005

And...
DIMTOL is Off
DIMLIM is Off

Then...

4.000

(C)

If...
DIMTP = 0.005
DIMTM = 0.005

And...
DIMTOL is On
DIMLIM is Off

Then...

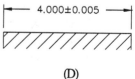

(D)

DIMTFAC
Dimension Tolerance Text Size Factor

This variable acts as a multiplier and affects the height of tolerance text. Dimtfac will multiply its current value by the value set in Dimtxt. This variable is designed to work when Dimtm and Dimtp are not equal and Dimtol and Dimlim are On, but not at the same time. The prompt for this dimension variable is:

Dim: DIMTFAC
Current value <1.00> New value: *(Enter a numeric value)*

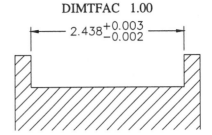

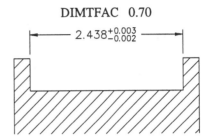

DIMTXT

This variable controls the height of text added to dimensions. To control the text font, use the Style command. The prompt for this dimension variable is:

Dim: DIMTXT
Current value <0.18> New value: *(Enter a numeric value)*

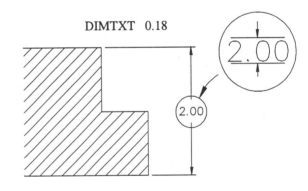

DIMTVP
Vertical Dimension Text Position

This variable allows you to control the position of the dimension text either above or below the dimension line. With Dimtad Off and if Dimtvp is set to 1.00 in the illustration at the right, AutoCAD will multiply this value by the current Dimtxt value and place the text above the dimension line. If Dimtvp is set to -1.00, the negative multiplier places the text below the dimension line. The prompt for this dimension variable is:

Dim: DIMTVP
Current value <0> New value: *(Enter a numeric value)*

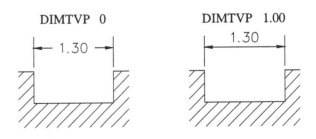

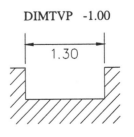

DIMZIN
Zero Inches/Feet Control

This variable controls the display of feet and inches in a drawing depending on what the current Dimzin value is set to. This variable may also suppress the leading zero or trailing zeros of decimal units depending on the setting. See the table at the right for the affects of Dimzin on different types of units. The prompt for this dimension variable is:

Dim: DIMZIN
Current value <0> New value: *(Enter a supported numeric value)*

DIMZIN Value				
0	3/8"	5"	2'	0.7500
1	0'-0 3/8"	0'-5"	2'-0"	
2	0'-0 3/8"	0'-5"	2'	
3	3/8"	5"	2'-0"	
4				.7500
8				0.75
12				.75

Diameter and radius dimensioning can be affected by the Dimtix and Dimtofl system variables depending on the results desired. With both variables Off in example "A", the dimension text is placed on the outside of the circle resembling a leader line. This dimension, however, remains associative as long as Dimaso is On. If Dimtix is turned On as in example "B", the dimension text will be forced inside of the circle. Turning Dimtix Off and Dimtofl On places the dimension text outside of the circle and forces the dimension line to be drawn through the entire circle.

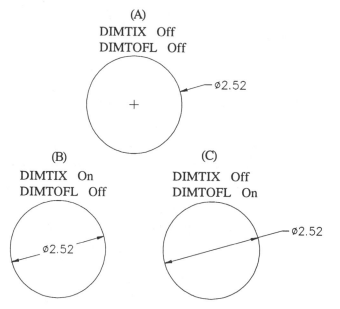

The affects of Dimtix and Dimtofl on diameter dimensions is identical to radius dimensions.

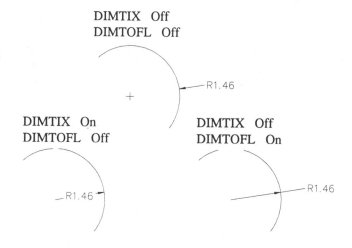

Dimensioning Isometric Drawings

As of AutoCAD Release 11, a form of isometric dimensioning is now supported. In the past, it was possible to use the Aligned dimensioning mode to make the dimension line parallel with the surface being dimensioned. Arrowheads were also drawn parallel; however, extension lines were drawn perpendicular to the dimension lines and not at an isometric angle. The results of this type of dimensioning technique are illustrated at the right. One of the only ways to simulate isometric dimensions was to turn off the extension lines, manually draw new extension lines at isometric angles, and move the dimension to a new location because of the position of the extension lines.

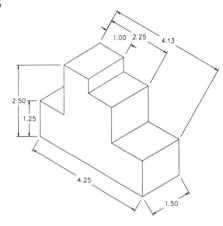

The Oblique command allows the user to enter an obliquing angle which will rotate the extension lines and reposition the dimension line.

Dim: **Obl**
Select objects: *(Select the 2.50 and 1.25 dimensions at the right)*
Select objects: *(Strike Enter to continue with this command)*
Enter obliquing angle (Return for none): **150**

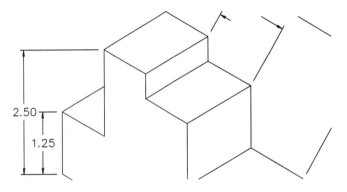

Both dimensions at the right were repositioned using the Oblique command. Notice that the extension lines and dimension line were affected; however, the text remains in the horizontal position as defined by the Dimtih and Dimtoh dimension variables. The Tedit and Trotate commands will allow text to be rotated at an angle; however, the text will not be in true isometric form.

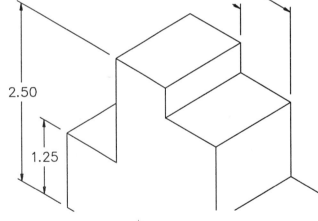

The Oblique command was used to rotate the dimension at "A" at an obliquing angle of 210 degrees. An obliquing angle of -30 degrees was used to rotate the dimension at "B" and the dimensions at "C" required an obliquing angle of 90. This represents proper isometric dimensions except for the orientation of the text.

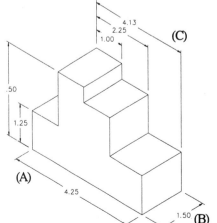

Ordinate Dimensioning

The plate at the right consists of numerous drill holes with a few slots in addition to numerous 90-degree angle cuts along the perimeter. This object is not considered difficult to draw or make since it mainly consists of drill holes. However, conventional dimensioning techniques make the plate appear complex since a dimension is required for the location of every hole and slot in both the X and Y directions. Add to that standard dimension components such as extension lines, dimension lines, and arrowheads and it is easy to get lost in the complexity of the dimensions even on this simple object.

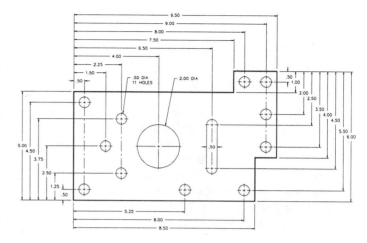

A better dimensioning method to use is illustrated at the right, called Ordinate or Datum dimensioning. Here, no dimension lines or arrowheads are drawn; instead, one extension line is constructed from the selected feature to a location specified by the user. A dimension is added to identify this feature in either the X or Y directions. It is important to understand that all dimension calculations occur in relation to the current User Coordinate System (UCS), or the current 0,0 origin. In the example at the right, with the 0,0 origin located in the lower left corner of the plate, all dimensions in the horizontal and vertical directions are calculated in relation to this 0,0 location. Holes and slots are called out using the Dim-Diameter option. The following illustrates a typical ordinate dimensioning prompt sequence:

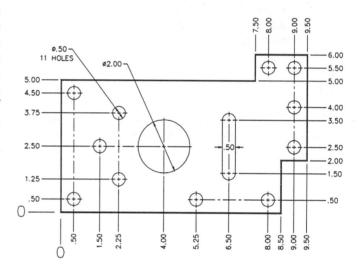

Command: **Dim**
Dim: **Ord**
Select feature: *(Select a feature using an Osnap option)*
Leader endpoint (Xdatum/Ydatum): *(Locate an outside point)*
Dimension text < >: *(Strike Enter to accept the default value)*

To illustrate how to place ordinate dimensions, see the example at the right and the prompt sequence below. Before placing any dimensions, a new UCS must be moved to a convenient location on the object using the UCS-Origin option. All ordinate dimensions will reference this new origin since it is located at coordinate 0,0. Enter the dimensioning area by picking from a menu or typing "Dim". Once in dimensioning mode, select or type Ordinate to enter ordinate dimensioning. Select the Quadrant of the arc at "A" as the feature. For the leader endpoint, pick a point at "B". Be sure Ortho mode is On. It is also helpful to snap to a convenient grid point for this and other dimensions along this direction. Follow the prompt sequence below:

Command: **Dim**
Dim: **Ord**
Select feature: *(Select the Quadrant of the arc at "A")*
Leader endpoint (Xdatum/Ydatum): *(Select a point at "B")*
Dimension text <1.50>: *(Strike Enter to accept this value)*

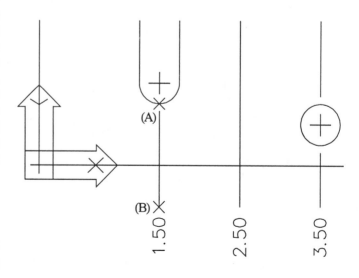

Ordinate Dimensioning Continued

With the previous example highlighting horizontal ordinate dimensions, placing vertical ordinate dimensions is identical. With the location of the UCS still located in the lower left corner of the object, select the feature at "A" using either the Endpoint or Quadrant modes. Pick a point at "B" in a convenient location on the drawing. Accept the default value and the dimension is placed. Again, it is helpful if Ortho is On and a grid dot is snapped to.

Command: **Dim**
Dim: **Ord**
Select feature: *(Select the Endpoint of the line or arc at "A")*
Leader endpoint (Xdatum/Ydatum): *(Select a point at "B")*
Dimension text <3.00>: *(Strike Enter to accept this value)*

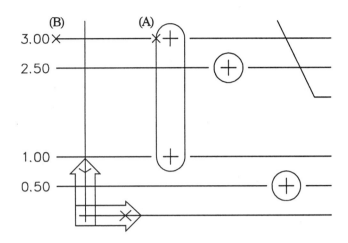

When spaces are tight to dimension to, two points not parallel to the X or Y axis will result in a "jog" being drawn. The jog always occurs in the middle of the extension line and always at a 90-degree angle to the extension line. It is still helpful to snap to a grid dot when performing this operation; however, be sure Ortho is Off.

Command: **Dim**
Dim: **Ord**
Select feature: *(Select the Endpoint of the line at "A")*
Leader endpoint (Xdatum/Ydatum): *(Select the point at "B")*
Dimension text <2.00>: *(Strike Enter to accept this default value)*

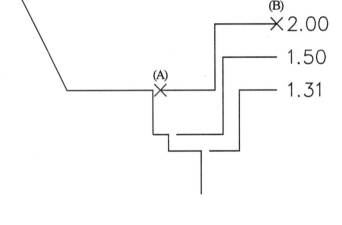

Ordinate dimensioning provides a neat and easy way of organizing dimensions for drawings where geometry leads to applications involving numerical control. Only two points are required to place the dimension which references the current location of the UCS. Certain dimension variables have no affect on ordinate dimensions. Dimension text along the X direction will never be drawn horizontally regardless of what Dimtih and Dimtoh are set to. However, dimension text may be placed above the extension line controlled by Dimtad or at specified distances above or below the extension line controlled by Dimtvp. The gap between the dimension text and the end of the extension line can be manipulated with Dimgap. The spacing between the selected feature and the beginning of the extension line is controlled by Dimexo.

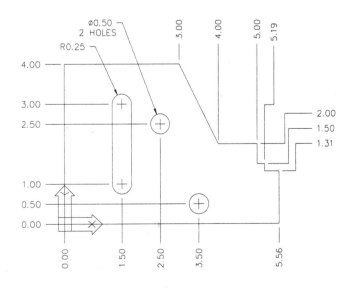

Tutorial Exercise #10
Dimex.Dwg

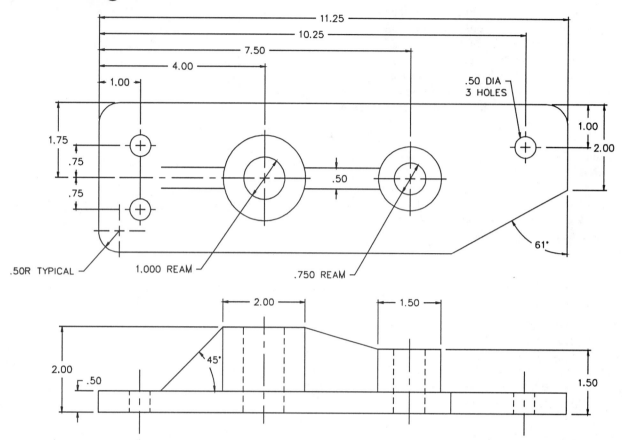

PURPOSE:
The purpose of this tutorial is to draw the two-view object illustrated above and place dimensions on the drawing. This drawing is available on diskette under the file name "Dimex.Dwg".

SYSTEM SETTINGS:
Either copy "Dimex.Dwg" from the diskette provided or begin a new drawing called "Dimex." Use the Units command to change the number of decimal places past the zero from 4 to 2. Keep the remaining default unit values. Using the Limits command, keep 0,0 for the lower left corner and change the upper right corner from 12,9 to 21.00,16.00. Use the Grid command and change the grid spacing from 1.00 to 0.25 units to aid in the placement of dimension lines. Do not turn the snap or ortho On.

LAYERS:
If the drawing file "Dimex.Dwg" was copied from the diskette provided, the following layers are already created for this tutorial. If starting a new drawing called "Dimex," create the following layers with the format:

Name-Color-Linetype
Object - Magenta - Continuous
Hidden - Red - Hidden
Center - Yellow - Center
Dim - Yellow - Continuous

SUGGESTED COMMANDS:
All commands for this tutorial deal with dimensioning; all dimensioning commands and options begin at the prompt "Dim:". The following options of the Dim: command will be used: Horizontal, Vertical, Continuous, Baseline, Center, Leader, Radius, Diameter, and Angular. All dimension options may be picked from the digitizing pad, screen menu, or entered from the keyboard. When entering dimension options from the keyboard, the first three letters are all that are required (e.g., Horizontal=Hor). Use the Zoom command to get a closer look at details and features that are being dimensioned.

DIMENSIONING:
Follow the tutorial for manipulating and setting dimension variables.

PLOTTING:
This tutorial exercise may be plotted on "C"-size paper (18" x 24"). Use a plotting scale of 1=1 to produce a scaled size plot.

VERSION OF AUTOCAD:
This tutorial exercise may be completed using either AutoCAD Release 10 or Release 11.

Step #1

Construct the front and top views. Starting the front view at absolute coordinate 4.50,4.25 and spacing the views a distance 1.75 units away from each other will ensure the dimensions will fit on the defined limits. Next, use the DIM: command, enter the dimensioning mode, and use the Center option to place center marks at the center of all circles. Be sure the dimension variable Dimcen is set to a value of -.09. This value will place the markers at the circle center in addition to the lines that extend to the outside of the circle.

Command: **Dim**
Dim: **Dimcen**
Current value <0.09>   New value: **-.09**
Dim: **Center**
Select arc or circle: *(Select the circle at "A")*
Dim: **Center**
Select arc or circle: *(Repeat for circles "B", "C", "D", and "E")*

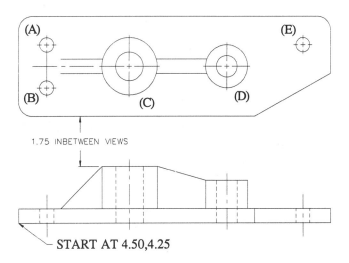

1.75 INBETWEEN VIEWS

START AT 4.50,4.25

Step #2

Change the following dimension variables:

Dim: **Dimtxt**
Current value <0.18>   New value: **0.12**
Dim: **Dimasz**
Current value <0.18>   New value: **0.12**
Dim: **Dimexo**
Current value <0.06>   New value: **0.12**
Dim: **Dimexe**
Current value <0.18>   New value: **0.07**
Dim: **Dimzin**
Current value <0>   New value: **4**

Use the Dim-Horizontal option and place the 1.00 dimension as illustrated below. All dimensioning options can be entered by the first three letters. This will be used throughout this tutorial.

Dim: **Hor**
First extension line origin or RETURN to select: **Endp**
of *(Select the endpoint of the line at "A")*
Second extension line origin: **Endp**
of *(Select the endpoint of the line at "B")*
Dimension line location: *(Select a point at "C")*
Dimension text <1.00>: *(Strike Enter to accept this default value)*

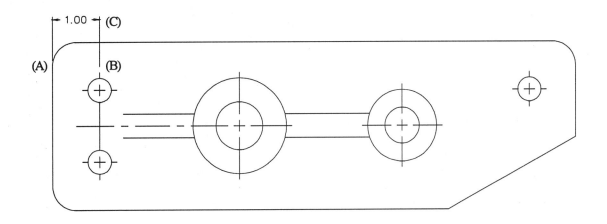

Step #3

The first extension line location of the last dimension, 1.00, will be used to establish a point of reference for the next four dimensions. This is accomplished with the Dim-Baseline option. Again, only the first three letters will be used to begin the command.

Dim: **Bas**
Second extension line origin: **Endp**
of *(Select the endpoint of the line at "A")*
Dimension text <4.00>: *(Strike Enter to accept this default value)*

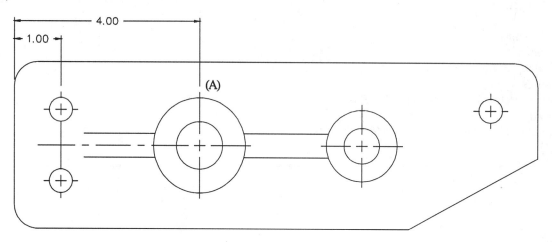

Step #4

Repeat the Dim-Baseline option for the 7.50 dimension.

Dim: **Bas**
Second extension line origin: **Endp**
of *(Select the endpoint of the line at "A")*
Dimension text <7.50>: *(Strike Enter to accept this default value)*

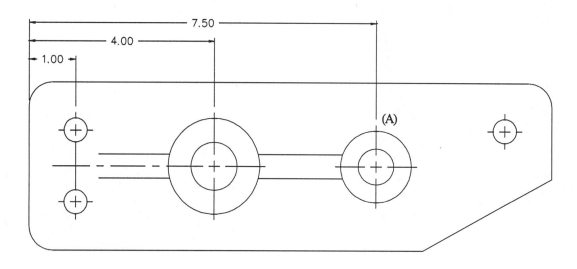

Step #5

Repeat the Dim-Baseline option for the 10.25 dimension.

Dim: **Bas**
Second extension line origin: **Endp**
of *(Select the endpoint of the line at "A")*
Dimension text <10.25>: *(Strike Enter to accept this de-*
fault value)

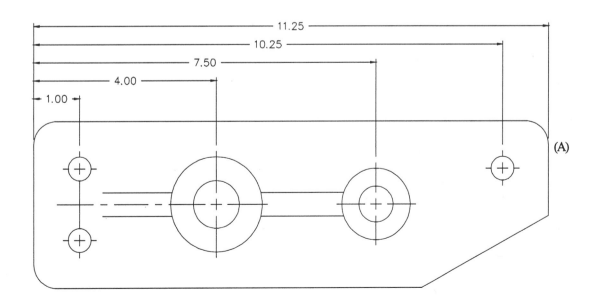

Step #6

Repeat the Dim-Baseline option for the 11.25 dimension.

Dim: **Bas**
Second extension line origin: **Endp**
of *(Select the endpoint of the line at "A")*
Dimension text <11.25>: *(Strike Enter to accept this de-*
fault value)

Step #7

Use the Zoom command to get a closer look at the left side of the top view as illustrated to the right. Then use the Dim-Vertical option to place the .75 dimension. Note that the value .75 does not display as 0.75; this is the purpose of the dimension variable Dimzin. Setting this value from 0 to 4 eliminates the leading zero before the decimal point.

Dim: **Ver**
First extension line of RETURN to select: **Endp**
of *(Select the endpoint of the line at "A")*
Second extension line origin: **Endp**
of *(Select the endpoint of the line at "B")*
Dimension line location: *(Select a point at "C")*
Dimension text <.75>: *(Strike Enter to accept this default value)*

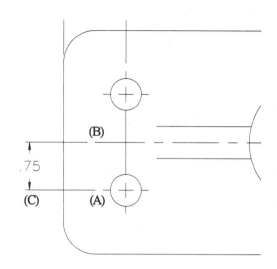

Step #8

The Dim-Continuous option is used to place the next dimension in line or along side the previous dimension. When using the Continuous option, the placement of the dimension line is remembered from the previous dimension.

Dim: **Con**
Second extension line origin: **Endp**
of *(Select the endpoint of the line at "A")*
Dimension text <.75>: *(Strike Enter to accept this default value)*

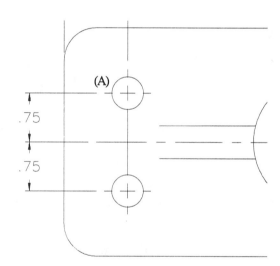

Step #9

Use the Dim-Vertical option to place the 1.75 dimension illustrated at the right.

Dim: **Ver**
First extension line of RETURN to select: **Endp**
of *(Select the endpoint of the line at "A")*
Second extension line origin: **Endp**
of *(Select the endpoint of the line at "B")*
Dimension line location: *(Select a point at "C")*
Dimension text <1.75>: *(Strike Enter to accept this default value)*

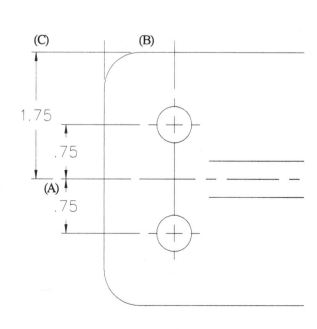

Step #10

Use the Zoom command to get a closer look at the right side of the top view as illustrated to the right. Then use the Dim-Vertical option to place the 1.00 dimension.

Dim: **Ver**
First extension line of RETURN to select: **Endp**
of *(Select the endpoint of the line at "A")*
Second extension line origin: **Endp**
of *(Select the endpoint of the line at "B")*
Dimension line location: *(Select a point at "C")*
Dimension text <1.00>: *(Strike Enter to accept this default value)*

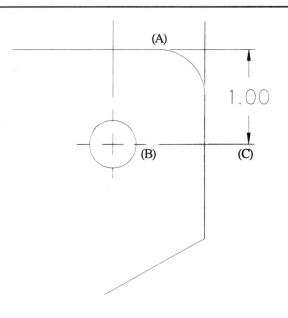

Step #11

Repeat the Dim-Baseline option for the 2.00 dimension.

Dim: **Bas**
Second extension line origin: **Endp**
of *(Select the endpoint of the line at "A")*
Dimension text <2.00>: *(Strike Enter to accept this default value)*

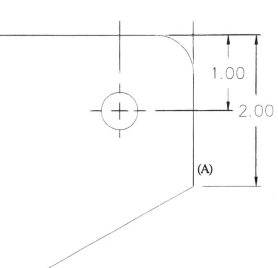

Step #12

Use the Dim-Angular option to place the 61-degree dimension illustrated to the right. Follow the system prompts carefully when performing angular dimensioning. If the results are not satisfactory, type "U" at the "Dim:" prompt to erase the last dimension and try again.

Dim: **Ang**
Select first line: *(Select the line at "A")*
Second line: *(Select the line at "B")*
Enter the dimension line arc location: *(Select near "C")*
Dimension text <61>: *(Strike Enter to accept this default value)*
Enter text location: *(Strike the Enter key)*

Striking the Enter key in response to the "Enter text location:" prompt automatically places the text in the center of the arc.

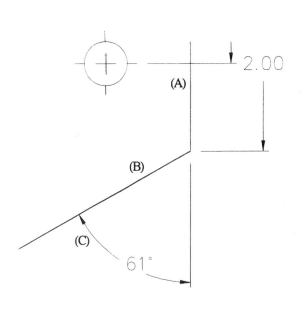

Step #13

The vertical dimension, .50, will be placed using the Dim-Vertical option. To prevent extension lines from being placed on top of object lines, the dimension variables Dimse1 and Dimse2 need to be turned On. This will turn off or suppress the extension lines for this dimension. Also, to place the dimension in between the arrowheads, set the dimension variable Dimtix from Off to On.

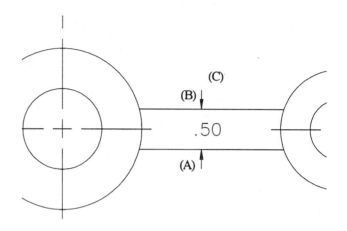

Dim: **Dimse1**
Current value <Off> New value: **On**
Dim: **Dimse2**
Current value <Off> New value: **On**
Dim: **Dimtix**
Current value <Off> New value: **On**
Dim: **Ver**
First extension line of RETURN to select: **Nea**
of *(Select the line at "A")*
Second extension line origin: **Per**
to *(Select the line at "B")*
Dimension line location: *(Select a point at "C")*
Dimension text <.50>: *(Strike Enter to accept this default value)*

Step #14

The Dim-Radius option will be used to place the .50 dimension illustrated at the right. Before performing this operation, reset the last dimension variables, Dimse1, Dimse2, and Dimtix back to their original values. When placing the radius dimension, notice that center lines are also placed; this is controlled by the dimension variable Dimcen which should already be set to a value of -.09. Since other corners of this drawing have the same radius value, the note "Typ." is typed in for new dimension text along with the .50 value. AutoCAD will first try to place the dimension in between the arc and center mark. If there is no room, the system prompts you to enter the leader length for the text. Follow the prompts below to place this dimension.

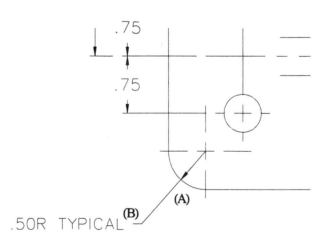

Dim: **Dimse1**
Current value <On> New value: **Off**
Dim: **Dimse2**
Current value <On> New value: **Off**
Dim: **Dimtix**
Current value <On> New value: **Off**
Dim: **Rad**
Select arc or circle: *(Select the arc at "A")*
Dimension text <.50>: **.50R TYPICAL**
Text does not fit.
Enter leader length for text: *(Select a point at "B")*

Step #15

Use the Dim-Diameter option to place the diameter dimension illustrated to the right. The term "Ream" is a precision drilling operation and needs to be called-out in note form. Reams usually carry decimal places of three or more places; therefore, when the dimension text is displayed in the diameter prompt, a new value of 1.000 REAM needs to be entered. Also, since center marks were placed in Step #1, and since the diameter mode automatically places a center mark, the dimension variable, Dimcen, needs to be changed from -.09 to 0. The zero will prevent a center mark from being placed. Follow the prompts below to place the diameter dimension.

Dim: **Dimcen**
Current value <-.09> New value: **0**
Dim: **Dia**
Select circle or arc: *(Select the circle at "A")*
Dimension text <1.00>: **1.000 REAM**
Text does not fit.
Enter leader length for text: *(Select near "B")*

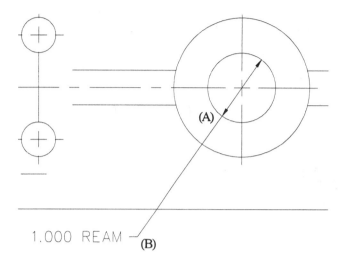

Step #16

Use the Dim-Diameter option to place the diameter dimension illustrated to the right. Again, as in the previous dimension, substitute the new dimension text .750 REAM for the default value <.75>. Leave the dimension variable Dimcen set to zero to prevent the placing of double center marks.

Dim: **Dia**
Select circle or arc: *(Select the circle at "A")*
Dimension text <.75>: **.750 REAM**
Text does not fit.
Enter leader length for text: *(Select near "B")*

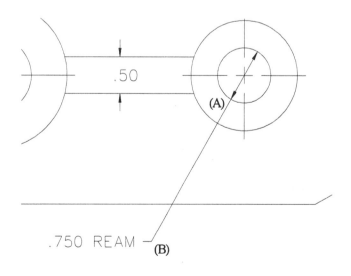

Step #17

The Dim-Leader option will now be used to place a value and note describing 3 holes that share the save diameter value of .50. When placing a leader dimension, use the Osnap-Nearest option when prompted for "Select circle or arc". This will select a point on the circle for the location of the tip of the arrowhead. The prompt continues with the familiar "To point" phrase. Follow the commands at the bottom to place this dimension. Once the dimension is placed, copy the text down a short distance and use the Change command to change the text from .50 DIA to 3 HOLES.

Dim: **Lea**
Leader start: **Nea**
of *(Select the circle at "A")*
To point: *(Select a point at "B")*
To point: *(Strike Enter to continue)*
Dimension text <.75>: **.50 DIA**
Dim: **Exit**

Command: **Copy**
Select objects: **L**
Select objects: *(Strike Enter to continue)*
<Base point or displacement>/Multiple: *(Select at "C")*
Second point of displacement: *(Select a point at "D")*

Command: **Change**
Select objects: *(Select the text at "D")*
Select objects: *(Strike Enter to continue)*
Properties/<Change point>: *(Strike Enter)*
Enter text insertion point: *(Strike Enter)*
New style or RETURN for none: *(Strike Enter)*
New Height <0.12>: *(Strike Enter)*
New rotation angle <0>: *(Strike Enter)*
New text <.50 DIA>: **3 HOLES**

The completed top view including dimensions is illustrated below. Use the example to check that all dimensions have been placed and all features (holes, fillets, etc.) have been properly identified.

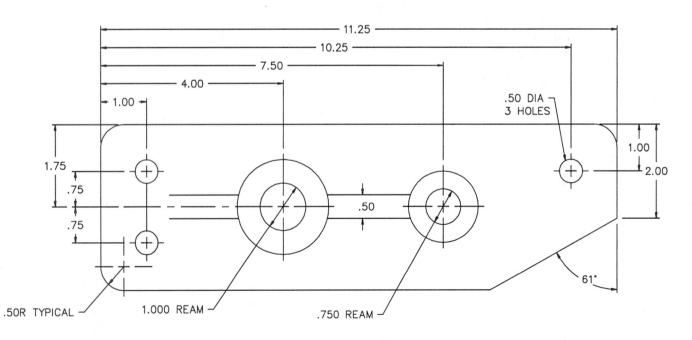

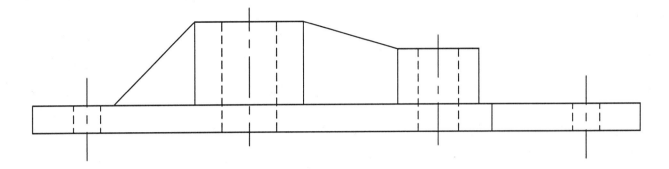

The front view will now be the focus of the next series of dimensioning steps. Again use the Zoom command whenever dimensioning to smaller surfaces.

Step #18

Use the Dim-Horizontal option to place the 2.00 dimension calling out the size of the cylinder as illustrated at the right.

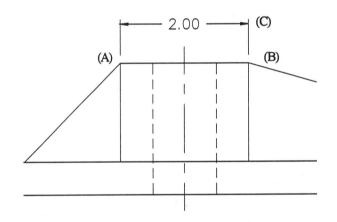

Dim: **Hor**
First extension line origin or RETURN to select: **Endp**
of *(Select the endpoint of the line at "A")*
Second extension line origin: **Endp**
of *(Select the endpoint of the line at "B")*
Dimension line location: *(Select a point at "C")*
Dimension text <2.00>: *(Strike Enter to accept this default value)*

Step #19

Use the Dim-Horizontal option to place the 1.50 dimension calling out the size of the smaller cylinder as illustrated at the right.

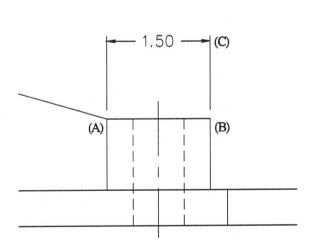

Dim: **Hor**
First extension line origin or RETURN to select: **Endp**
of *(Select the endpoint of the line at "A")*
Second extension line origin: **Endp**
of *(Select the endpoint of the line at "B")*
Dimension line location: *(Select a point at "C")*
Dimension text <1.50>: *(Strike Enter to accept this default value)*

Step #20

Use the Zoom command to magnify the right end of the front view as illustrated at the right. Next, use the Dim-Vertical option and place the 1.50 dimension. Follow the prompts below.

Dim: **Ver**
First extension line of RETURN to select: **Endp**
of *(Select the endpoint of the line at "A")*
Second extension line origin: **Endp**
of *(Select the endpoint of the line at "B")*
Dimension line location: *(Select a point at "C")*
Dimension text <1.50>: *(Strike Enter to accept this default value)*

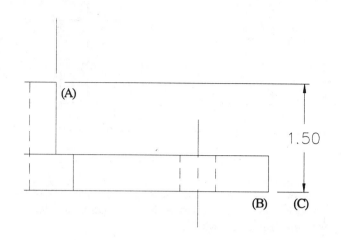

Step #21

Use the Zoom command and magnify the left end of the front view as illustrated to the right. Use the Dim-Vertical option and place the .50 dimension. Do not be concerned that the dimension is not in between the arrowheads. Remember, this is controlled by the dimension variable Dimtix. If you want the text to be inside the arrowheads, change the value of Dimtix from Off to On.

Dim: **Ver**
First extension line of RETURN to select: **Endp**
of *(Select the endpoint of the line at "A")*
Second extension line origin: **Endp**
of *(Select the endpoint of the line at "B")*
Dimension line location: *(Select a point at "C")*
Dimension text <.50>: *(Strike Enter to accept this default value)*

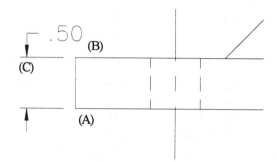

Step #22

Use the Dim-Baseline option to place the 2.00 dimension. Remember, when using the Baseline option, the first extension line location is remembered from the first or previous dimension placed.

Dim: **Bas**
Second extension line origin: **Endp**
of *(Select the endpoint of the line at "A")*
Dimension text <2.00>: *(Strike Enter to accept this default value)*

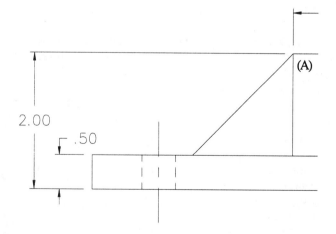

Step #23

Use the Dim-Angular option to place the 45-degree dimension illustrated to the right. Follow the system prompts carefully when performing angular dimensioning. If the results are not satisfactory, type "U" at the "Dim:" prompt to erase the last dimension and try again.

Dim: **Ang**
Select first line: *(Select the line at "A")*
Second line: *(Select the line at "B")*
Enter the dimension line arc location: *(Select near "C")*
Dimension text <45>: *(Enter)*
Enter text location: *(Select near "C")*

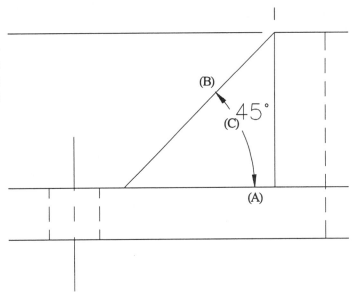

The front and top views, complete with dimensions, are shown below. Use the illustration below to check your final results.

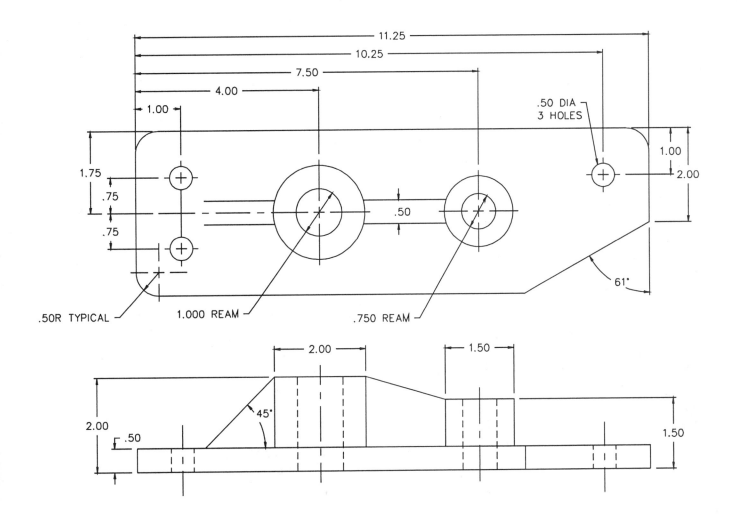

Tutorial Exercise #11
Tblk-iso.Dwg

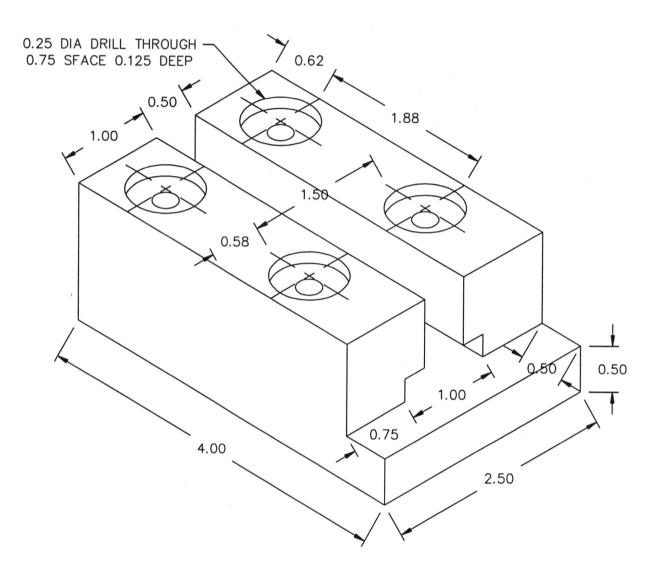

0.25 DIA DRILL THROUGH
0.75 SFACE 0.125 DEEP

0.62

0.50

1.00

1.88

1.50

0.58

0.50

0.50

1.00

0.75

4.00

2.50

PURPOSE:
The purpose of this tutorial is to convert aligned dimensions on an isometric drawing to oblique dimensions.

SYSTEM SETTINGS:
Since this drawing is provided on diskette completely drawn and dimensioned using aligned dimensions, edit an existing drawing called "Tblk-iso." Follow the steps in this tutorial for converting all dimensions to proper isometric format.

LAYERS:
All layers have already been created according to the following format:

Name-Color-Linetype
Dim - Yellow - Continuous
Center - Yellow - Center
Defpoints - White - Continuous

SUGGESTED COMMANDS:
All commands for this tutorial deal with the Oblique subcommand of Dim:.

DIMENSIONING:
Follow the tutorial for manipulating and setting dimension variables.

PLOTTING:
This tutorial exercise may be plotted on "A"-size paper (8.5" x 11"). Use a plotting scale of 1=1 to produce a scaled size plot.

VERSION OF AUTOCAD:
This tutorial exercise must be completed using AutoCAD Release 11.

Step #1

Before converting all dimensions to isometric form, create a new dimension style called "EXT-OFF". This style has the two variables Dimse1 and Dimse2 turned On, which will suppress or not show the extension lines of a dimension. This will be used later for one dimension. Then, use the Dim-Oblique option to rotate the dimension for isometric purposes.

Command: **Dim**
Dim: **Save**
?/Name for new dimension style: **Standard**

Dim: **Dimse1**
Current value <Off> New value: **On**

Dim: **Dimse2**
Current value <Off> New value: **On**

Dim: **Save**
?/Name for new dimension style: **Ext-off**

Dim: **Restore**
New dimension style to restore: **Standard**

Dim: **Obl**
Select objects: *(Select dimensions "A" through "G")*
Select objects: *(Strike Enter to continue)*
Enter obliquing angle (Return for none): **150**

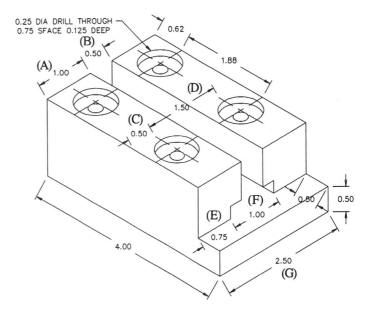

Step #2

Use the Dim-Oblique option to convert the three dimensions illustrated at the right to an isometric form at an obliquing angle of 30 degrees.

Dim: **Obl**
Select objects: *(Select dimensions "A" through "D")*
Select objects: *(Strike Enter to continue)*
Enter obliquing angle (Return for none): **210**

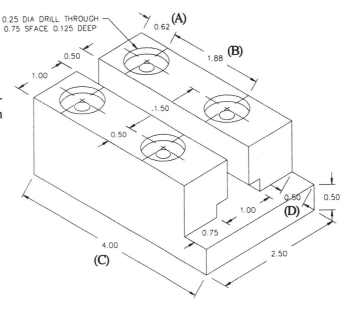

Step #3

Convert the dimension illustrated at the right to an isometric dimension and an obliquing angle of 30 degrees using the Dim-Oblique option.

Dim: **Obl**
Select objects: *(Select the dimension at "A")*
Select objects: *(Strike Enter to continue)*
Enter obliquing angle (Return for none): **30**

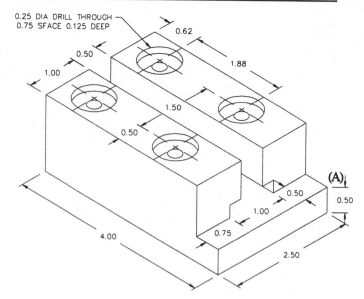

Step #4

Place the 1.00 dimension illustrated at the right using the Dim-Aligned option.

Dim: **Ali**
First extension line origin or Return to select: **Endp**
of *(Select the endpoint of the line at "A")*
Second extension line origin: **Endp**
of *(Select the endpoint of the line at "B")*
Dimension line location: *(Select a point at a convenient distance)*
Dimension text <1.00>: *(Strike Enter to accept this value)*

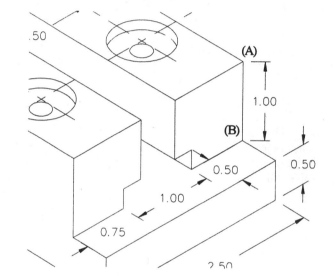

Step #5

Use the Dim-Oblique option to convert the dimension illustrated at the right to an isometric dimension by entering a value of 30 degrees for the obliquing angle. Then, place an aligned dimension using the Dim-Aligned option.

Dim: **Obl**
Select objects: *(Select the dimension at "A")*
Select objects: *(Strike Enter to continue)*
Enter obliquing angle (Return for none): **30**

Dim: **Ali**
First extension line origin or Return to select: **Endp**
of *(Select the endpoint of the line at "B")*
Second extension line origin: **Endp**
of *(Select the endpoint of the line at "C")*
Dimension line location: *(Select a point at a convenient distance)*
Dimension text <0.75>: *(Strike Enter to accept this value)*

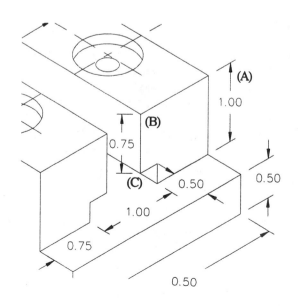

Step #6

In this final step, use the Dim-Oblique option to convert the dimension at "A" to an isometric form. Then, restore the dimension style, Ext-off, and use the Dim-Update option to update the dimension at "B". This dimension style has both Dimse1 and Dimse2 turned On which will turn off the extension lines of the 0.50 dimension and prevent lines from plotting over existing object lines.

Dim: **Obl**
Select objects: *(Select the dimension at "A")*
Select objects: *(Strike Enter to continue)*
Enter obliquing angle (Return for none): **30**

Dim: **Restore**
?/Name of dimension style to restore: **Ext-off**

Dim: **Upd**
Select objects: *(Select the dimension at "B" and "C")*
Select objects: *(Strike Enter to update the dimension to the new dimension style)*

Dim: **Redraw**

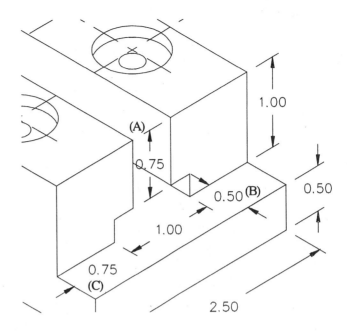

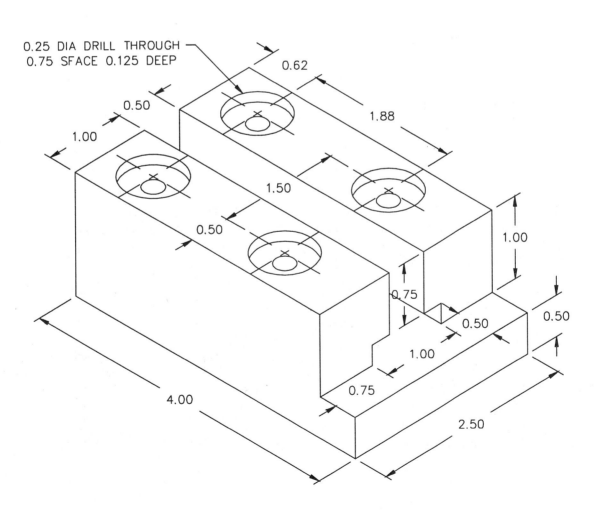

Tutorial Exercise #12
Bas-plat.Dwg

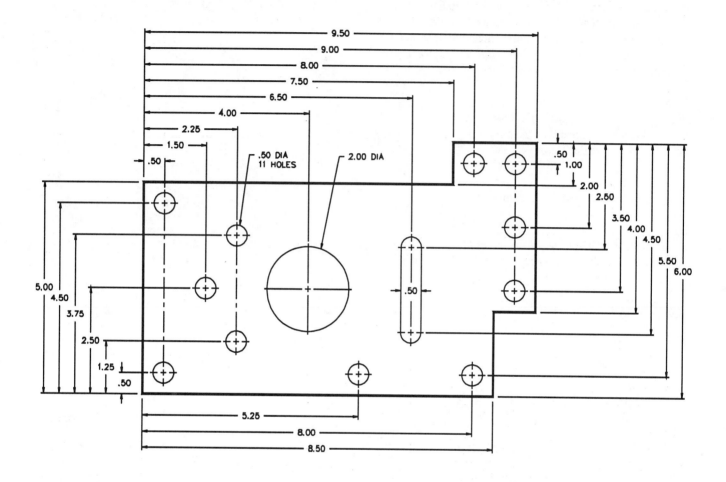

PURPOSE:

The purpose of this tutorial is to convert the drawing of the Bas-plat (Base plate) from the conventional dimensioning style to the ordinate dimensioning style.

SYSTEM SETTINGS:

Since this drawing is provided on diskette, edit an existing drawing called "Bas-plat". Follow the steps in this tutorial for creating ordinate dimensions.

LAYERS:

All layers have already been created:

Name-Color-Linetype
Object - White - Continuous
Dim - Yellow - Continuous

SUGGESTED COMMANDS:

All commands for this tutorial deal with the Ordinate subcommand of Dim:.

DIMENSIONING:

Follow the tutorial for manipulating and setting dimension variables.

PLOTTING:

This tutorial exercise may be plotted on "A"-size paper (8.5" x 11"). Use a plotting scale of 1=1 to produce a scaled size plot.

VERSION OF AUTOCAD:

This tutorial exercise must be completed using only AutoCAD Release 11.

Step #1

All ordinate dimensions make reference to the current 0,0 location identified by the position of the UCS. Since this icon is located in the lower left corner of the display screen by default, the coordinate system must be moved to a point on the object where all ordinate dimensions will be referenced from. First use the UCS command to define a new coordinate system with the origin at the lower left corner of the object. Then use the Ucsicon command to force the icon to display at the new origin.

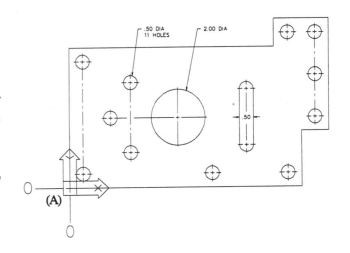

Command: **UCS**
Origin/ZAxis/3point/Entity/View/X/Y/Z/Prev/Restore/Save/
Del/?/<World>: **Origin**
Origin point <0,0,0>: **Int**
of *(Select the intersection at "A")*

Command: **Ucsicon**
ON/OFF/All/Noorigin/ORigin <ON>: **OR**

Step #2

Use the Zoom-Center option to magnify the screen similar to the illustration at the right.

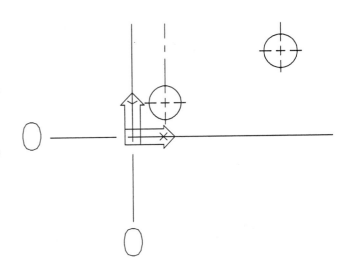

Command: **Zoom**
All/Center/Dynamic/Extents/Left/Previous/Vmax/Window/
<Scale(X/XP)>: **C**
Center point: **0,0**
Magnification or Height <11.23>: **4**

Step #3

Begin placing the first ordinate dimension using the Dim-Ordinate option. Use the Osnap-Quadrant option to select the circle as the feature. With the Snap and Ortho turned On, mode two grid dots to identify the leader endpoint. AutoCAD will determine if the dimension is Xdatum or Ydatum.

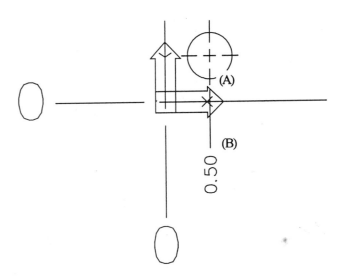

Command: **Dim**
Dim: **Ord**
Select feature: **Qua**
of *(Select the quadrant of the circle at "A")*
Leader endpoint (Xdatum/Ydatum): *(Pick a point two grid dots below the edge of the object at "B")*
Dimension text <0.50>: *(Strike Enter to accept this default value)*

Step #4

Use the Zoom-Previous transparent command to return to the previous screen display or exit completely out of dimensioning to use Zoom-Previous. Then, repeat the procedure in Step #3 to place ordinate dimensions at locations "A" through "G". Be sure Ortho mode is On and that the leader location is two grid dots below the bottom edge of the object. The Osnap-Quadrant option should be used on each circle and arc to satisfy the prompt "Select feature".

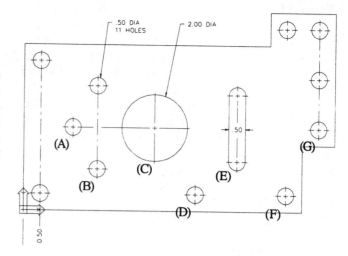

Step #5

Continue placing ordinate dimensions similar to the procedure used in Step #3. Use the Osnap-Endpoint option to select features at "A" and "B" in the illustration at the right. Have Ortho mode On and identify the leader endpoint two grid dots below the bottom edge of the object.

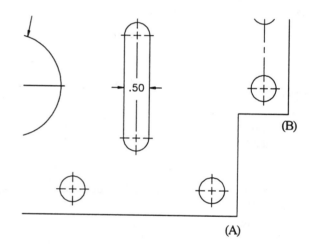

Step #6

Your display should appear similar to the illustration at the right. Notice the user coordinate system icon has disappeared. The Ucsicon command may be used to turn off the display of the icon while still keeping the 0,0 origin at the lower left corner of the object.

Dim: **Exit**

Command: **Ucsicon**
ON/OFF/All/Noorigin/ORigin <ON>: **OFF**

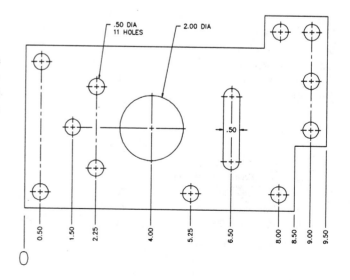

Step #7

With Snap On and Ortho On, begin placing the first vertical ordinate dimension; the procedure and prompts are identical to that of placing a horizontal ordinate dimension. Follow the example at the right and the prompts below for performing this operation.

Command: **Dim**
Dim: **Ord**
Select feature: **Qua**
of *(Select the quadrant of the circle at "A")*
Leader endpoint (Xdatum/Ydatum): *(Pick a point two grid dots to the left of the object at "B")*
Dimension text <0.50>: *(Strike Enter to accept this default value)*

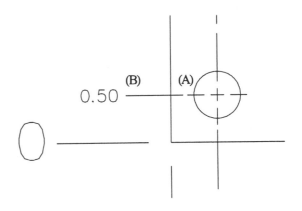

Step #8

Follow the procedure in the previous step to complete the vertical ordinate dimensions along this edge of the object. Use the Osnap-Quadrant option for "A" through "D" and Osnap-Endpoint for "E". Again have Ortho On and Snap On. For the leader endpoint count two grid dots to the left of the object and place the dimensions.

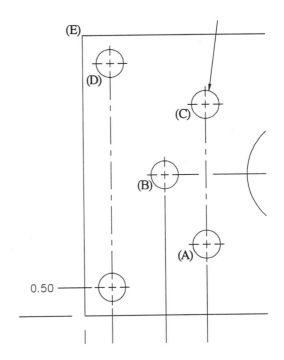

Step #9

Your display should appear similar to the illustration at the right.

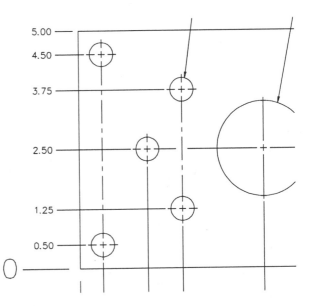

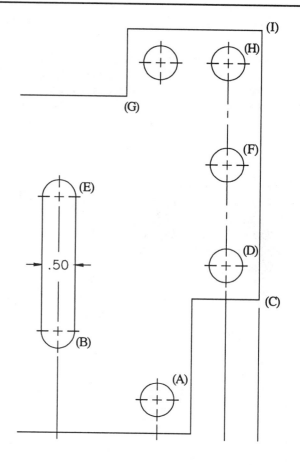

Step #10

Magnify the right portion of the object using the Zoom-Window option. Use ordinate dimensions and a combination of Osnap-Endpoint and Quadrant options to place vertical ordinate dimensions from "A" to "I".

Step #11

Your display should appear similar to the illustration at the right.

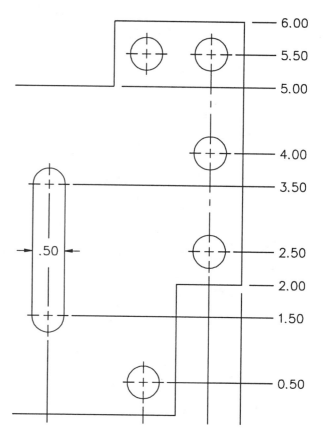

Step #12

Complete the dimensioning by placing two horizontal ordinate dimensions at the locations illustrated at the right. As in the past, have Ortho On, Snap On, and use a combination of the Osnap-Quadrant and Endpoint options to select the features. Place the leader endpoint two grid dots above the top line of the object. Use the Zoom-All option to return the entire object to your display.

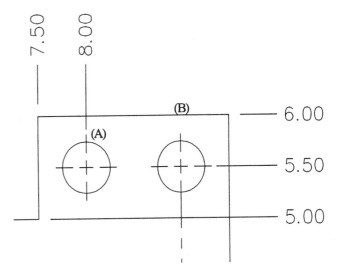

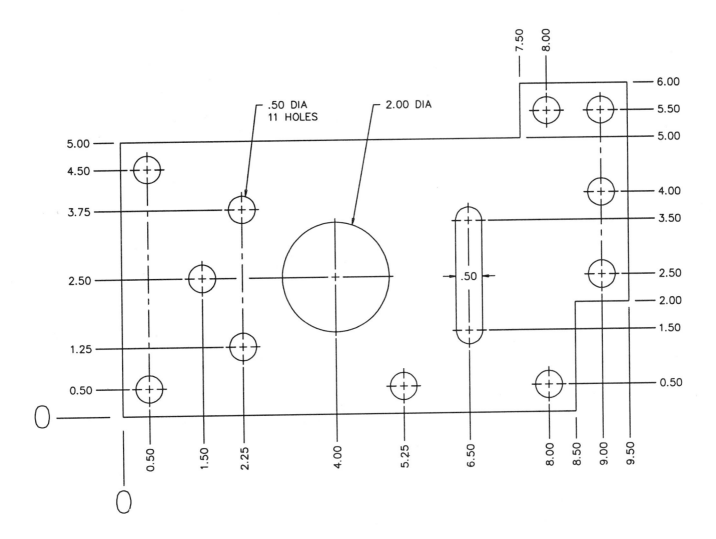

Problems for Unit 4

Problems 4–1 through 4–19

1. *Use the grid to determine all dimensions.*
2. *Redraw the views shown, and fully dimension your drawings.*

Problem 4–1

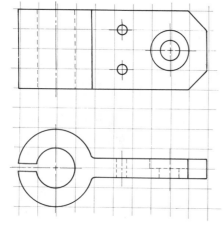

Problem 4–2

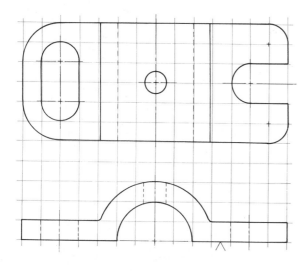

Problem 4–3

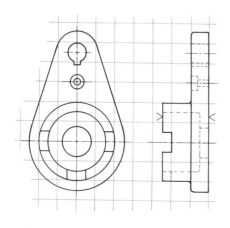

Problem 4–4

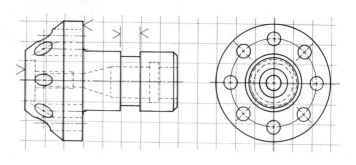

Problem 4–5

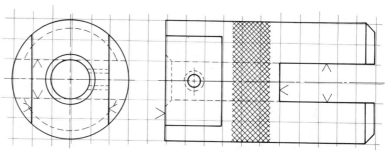

Problem 4–6

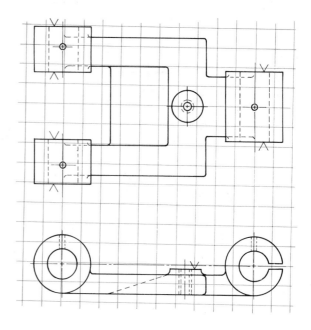

Problem 4-7

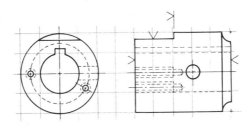

Problem 4-8

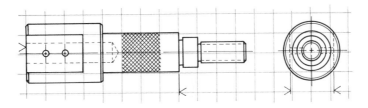

Problem 4-9

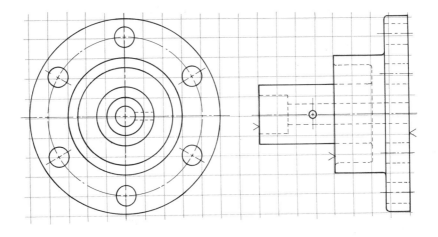

Problem 4-10

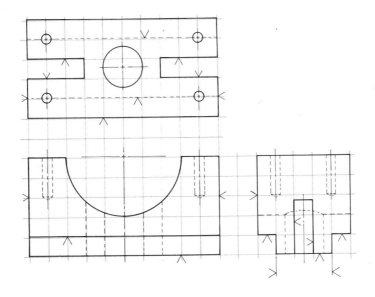

Problem 4-11

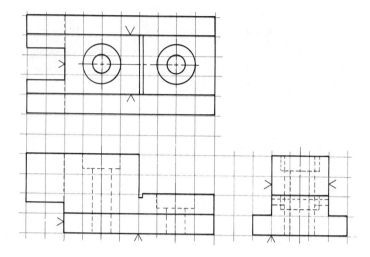

Problem 4-12

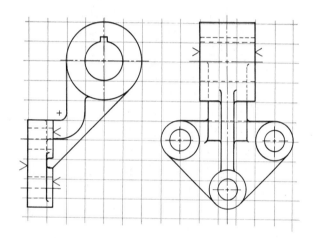

Problem 4-14

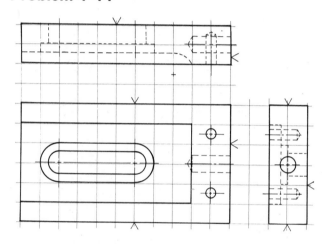

Problem 4-13

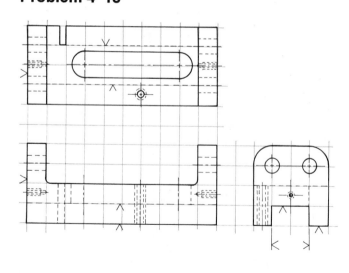

Problem 4-15

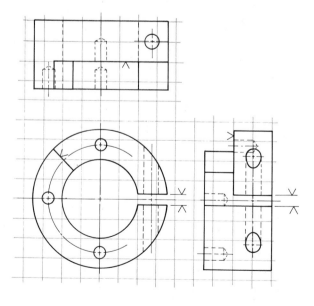

Problem 4-16

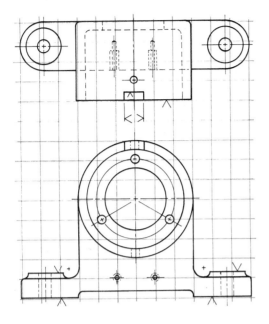

Problem 4-18

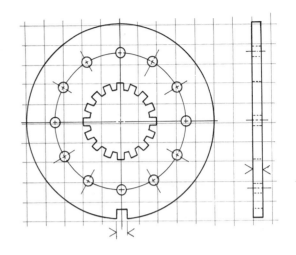

Problem 4-17

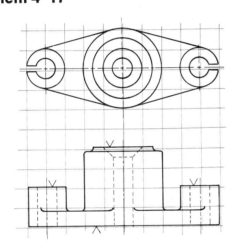

Problem 4-19

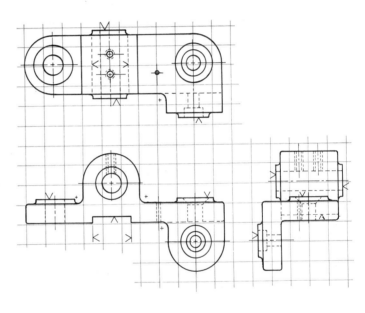

Problems 4-20 through 4-25

1. Convert the isometric drawings provided into orthogonal drawings, showing as many views as necessary to communi-
   cate the design.
2. Fully dimension your drawings.

Problem 4-20

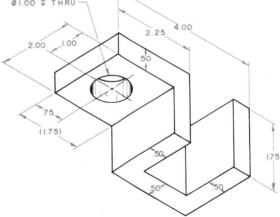

Ø1.00 ⊥ THRU

2.00   1.00   2.25   4.00
.50
.75
(1.75)
.50
.50   .50
1.75

Problem 4-21

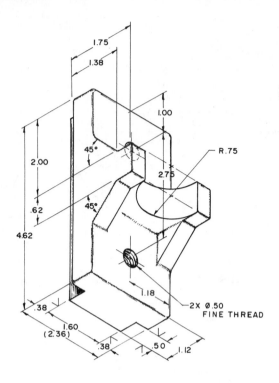

Problem 4-23

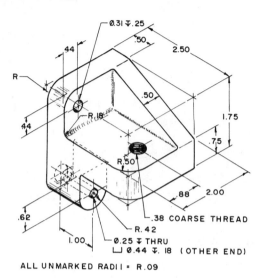

ALL UNMARKED RADII = R.09

Problem 4-22

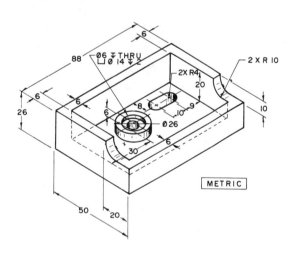

METRIC

Problem 4-24

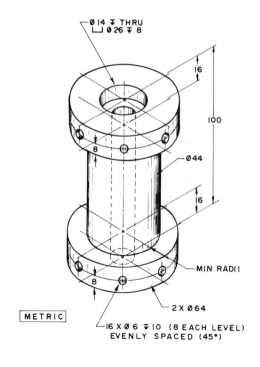

METRIC

Problem 4-25

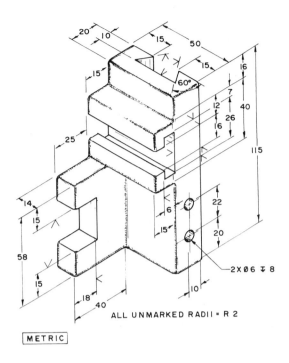

ALL UNMARKED RADII = R 2

2X Ø6 ∓ 8

METRIC

Directions for Problems 4-26 through 4-27
1. *Use the grid to determine all dimensions.*
2. *Redraw the views shown, and fully dimension your drawings.*

Problem 4-26

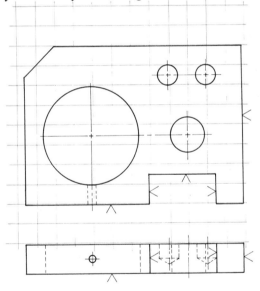

Problem 4-27

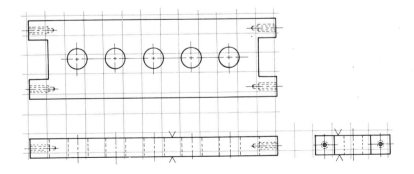

Directions for Problem 4–28
Convert Problem 4–28 into an ordinate drawing.

Problem 4–28

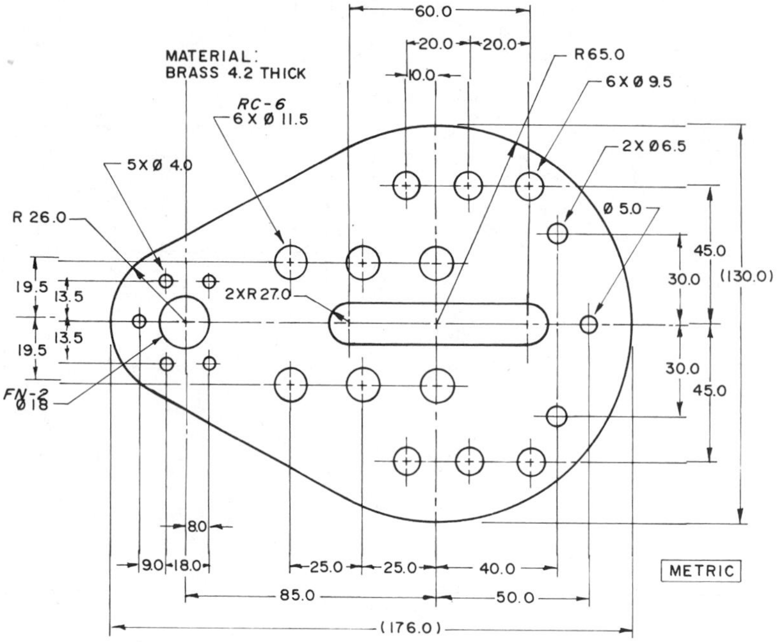

60.0
20.0
20.0
R65.0
6X Ø 9.5
MATERIAL:
BRASS 4.2 THICK
10.0
RC-6
6X Ø 11.5
2X Ø6.5
5X Ø 4.0
Ø 5.0
R 26.0
45.0
30.0
(130.0)
19.5
13.5
2XR27.0
13.5
19.5
30.0
45.0
FN-2
Ø 18
8.0
9.0
18.0
25.0
25.0
40.0
METRIC
85.0
50.0
(176.0)

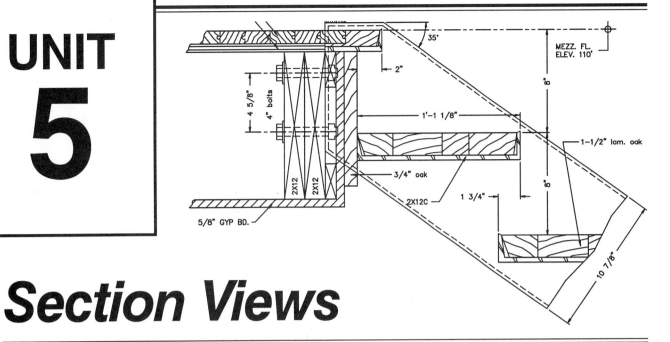

Section Views

Principles of orthographic projections remain the key method for the production of engineering drawings, whether using manual methods or CAD. As these drawings get more complicated in nature, the job of the operator or designer becomes more difficult in the interpretation of views, especially where hidden features are involved. The concept of slicing a view to expose these interior details is the purpose of performing a section. Section views then follow the same rules as orthographic or multi-view drawings except that the creation of a section makes the drawing easier to read since hidden features are converted to visible features. In this unit you will learn how sections are formed in addition to the many types of sections available to the designer. Two tutorial exercises at the end of the unit are designed to give you experience using two methods of cross-hatching when using AutoCAD as a drafting tool.

Section View Basics

The illustration at the right is a pictorial representation of a typical flange consisting of eight bolt holes and counterbore hole in the center.

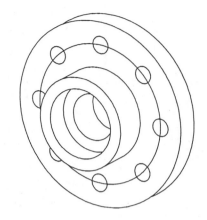

The drawings shown at the right show a typical solution to a multi-view problem complete with front and side views. The front view displaying the eight bolt holes is obvious to interpret; however, the numerous hidden lines in the side view make the drawing difficult to understand, and this is considered a relatively simple drawing. To relieve the confusion associated with a drawing too difficult to understand because of numerous hidden lines, a section is made of the part. Orthographic methods are followed up to the creation of the side view.

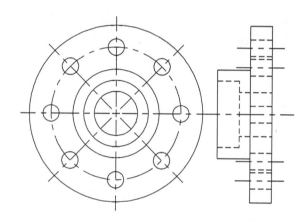

To understand section views more, see the illustration at the right. Creating a section view slices an object in such a way so as to expose what used to be hidden features and convert them into visible features. This slicing or cutting operation can be compared to that of using a glass plate or cutting plane to perform the section. In the object at the right, the glass plate cuts the object in half. It is the responsibility of the designer or CAD operator to convert one half of the object into a section and to discard the other half. Surfaces that come in contact with the glass plane are crosshatched to show where the actual cutting took place.

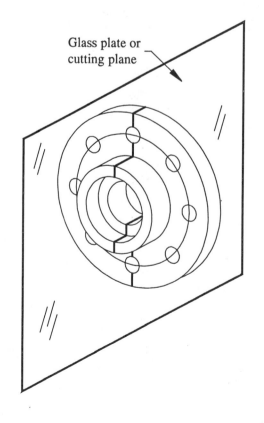

Glass plate or cutting plane

Section View Basics Continued

A completed section view drawing is shown at the right. Two new lines are also illustrated, a cutting plane line and section lines. The cutting plane line performs the cutting operation on the front view. In the side view, section lines show the surfaces that were cut. Notice that holes are not section lined since the cutting plane passes across the center of the hole. Notice also that hidden lines are not displayed in the side view. It is considered poor practice to merge hidden lines into a section view although there are always exceptions. The arrows of the cutting plane line tell the designer to view the section in the direction of the arrows and discard the other half.

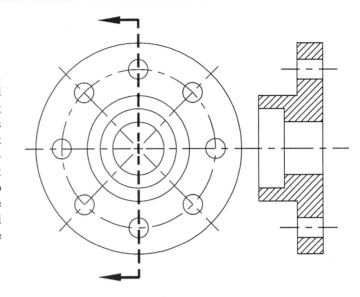

The cutting plane line consists of a very thick line at a series of dashes approximately 0.25″ in length. A polyline is used to create this line of 0.05 thickness. The arrows point in the direction of sight used to create the section with the other half generally discarded. Assign this line one of the dashed linetypes; the hidden linetype is reserved for detailing invisible features in views. The section line, by contrast with the cutting plane line, is a very thin line. This line identifies the surfaces being cut by the cutting plane line. The section line is usually drawn at an angle and at a specified spacing.

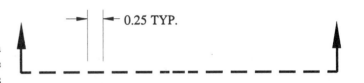

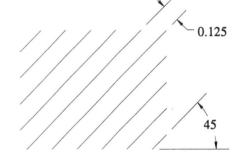

Depending on whether you are using AutoCAD Release 10 or 11, a wide variety of hatch patterns are already supplied with the software. One of these patterns, Ansi31, is displayed at the right. This is one of the more popular patterns with lines spaced 0.125 units apart from each other and at a 45-degree angle. AutoCAD Release 10 has 41 pattern styles; Release 11 has 56 pattern styles, including patterns used for architectural drawings.

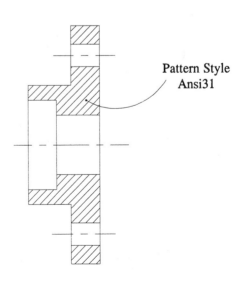

Pattern Style
Ansi31

Section View Basics Continued

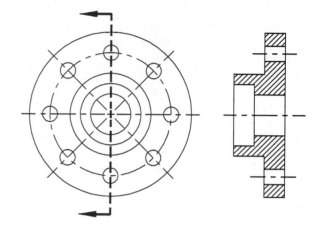

The object at the right illustrates proper section lining techniques. Much of the pain of spacing the section lines apart from each other and at angles has been eased considerably by using the computer as a tool. However, the designer must still practice proper section lining techniques at all times for clarity of the section.

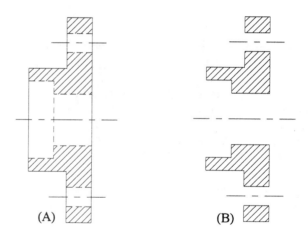

The next four examples illustrate common errors involved with section lines. At "A", section lines run the correct directions and at the same angle; however, the hidden lines have not been converted into object lines. This will confuse the more experienced designer since the presence of hidden lines in the section means more complicated invisible features. Example "B" is yet another error encountered when creating sections. Again, the section lines are properly placed; however all surfaces representing holes have been removed which displays the object as a series of sectioned blocks unconnected hinting four separate parts.

Example "C" appears to be a properly sectioned object; however upon closer inspection, we see the angle of the cross-hatch lines in the upper half different from the same lines in the lower left half. This suggests two different parts, when in actuality, it is the same part. In example "D" at the right, all section lines run the correct direction. The problem is the lines run through areas that were not sliced by the cutting plane line. These areas at "D" represent drill and counterbore holes and are left unsectioned.

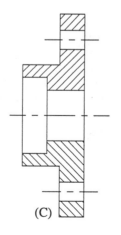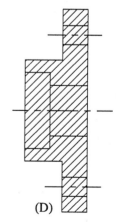

These have been identified as the most commonly made errors when cross-hatching an object. Remember just a few rules to follow: section lines are present only on surfaces that are cut by the cutting plane line; section lines are drawn in one direction when cross-hatching the same part; hidden lines are usually omitted when creating a section view; areas such as holes are not sectioned since the cutting line only passes across this feature.

Full Sections

When the cutting plane line passes through the entire object, a full section is formed. In the illustration at the right, a full section would be the same as taking an object and cutting it completely in half. Depending on the needs of the designer, one half is kept, the other half is discarded. The half that is kept is section lined.

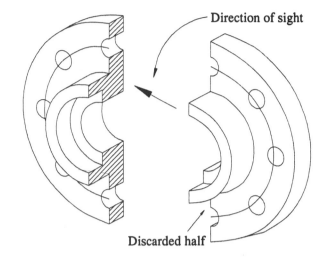

Direction of sight

Discarded half

At the right is the multi-view solution to the problem above. The front view is drawn with lines projected across to form the side view. To show the side view is in section, a cutting plane line is added to the front view. This line performs the physical cut. The designer has the option of keeping either half of the object. This is the purpose of adding arrowheads to the cutting plane line. The arrowheads define the direction of sight the designer views the object to form the section. The designer must then interpret what surfaces are being cut by the cutting plane line in order to properly add cross-hatching lines to the section which is located in the right side view. Hidden lines are not necessary once a section has been made.

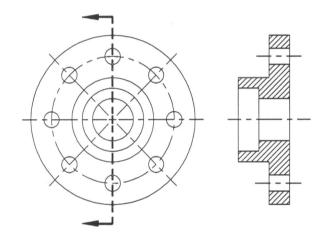

Numerous examples illustrate a cutting plane line with the direction of sight off to the the left. This does not mean that a cutting plane line cannot have a direction of sight going to the right as in the example. In this example, the section is formed from the left side view if the circular features are located in the front view.

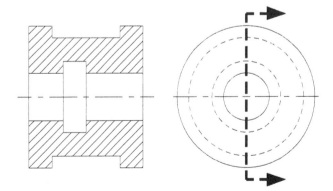

Half Sections

When symmetrical shaped objects are involved, sometimes it is not necessary to form a full section by cutting straight through the object. Instead, the cutting plane line passes only halfway through the object which makes the illustration at the right a half section. The rules for half sections are the same as for full sections; namely, a direction of sight is established, part of the object is kept, and part discarded.

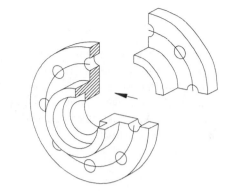

The views are laid out at the right in the usual multi-view format. To prepare the object as a half section, the cutting plane line passes halfway through the front view before being drawn off to the right. The right side view is converted into a half section by cross-hatching the upper half of the side view while leaving hidden lines in the lower half.

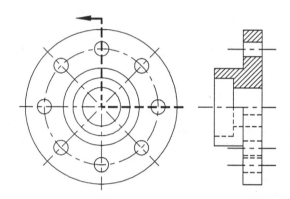

Depending on office practices, some designers prefer to omit hidden lines entirely from the side view similar to the illustration at the right. In this way, the lower half is drawn only what is visible.

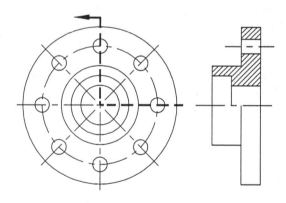

The illustration at the right shows another way of drawing the cutting plane line to conform to the right side view drawn in section. Hidden lines have been removed from the lower half; only those lines visible are displayed.

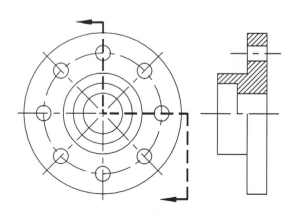

Assembly Sections

It would be unfair to give designers the impression that section views are only used for displaying internal features of individual parts. Yet another advantage of using section views is that it permits the designer to create numerous objects, assemble them, and then slice the assembly to expose internal details of all parts. This type of section is an assembly section similar to the illustration at the right. For all individual parts, notice the section lines running the same directions. This follows one of the basic rules of section views; keep section lines at the same angle for each individual part.

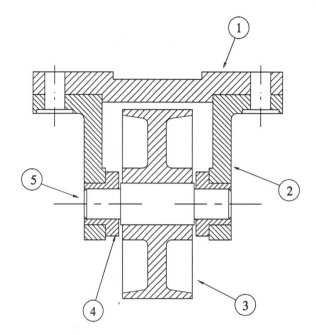

Illustrated at the right is the difference of assembly sections and individual parts that have been sectioned. For parts in an assembly that contact each other, it is considered good practice to alternate the directions of the section lines and make the assembly much more clear and distinguish the parts from each other. This can be accomplished by changing the angle of the hatch pattern or even the scale of the pattern.

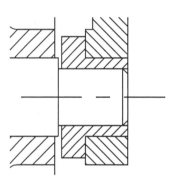

To identify parts in an assembly, an identifying part number along with a circle and arrowhead line are used. The line is very similar to a leader line used to call out notes for specific parts on a drawing. The addition of the circle highlights the part number. Sometimes this type of call out is referred to as a "bubble".

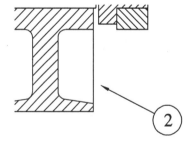

In the enlarged assembly illustrated at the right, the large area in the middle is actually a shaft used to support a pulley system. With the cutting plane passing through the assembly including the shaft, it is considered good practice to refrain from cross-hatching features such as shafts, screws, pins, or other types of fasteners. The overall appearance of the assembly is actually enhanced by not cross-hatching these items.

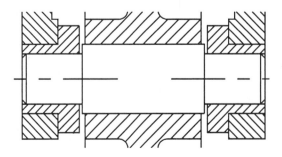

Aligned Sections

Aligned sections take into consideration the angular position of details or features of a drawing. Instead of drawing the cutting plane line vertically through the object at the right, the cutting plane is angled or aligned with the same angle the elements are at. Aligned sections are also made to produce better clarity of a drawing. At the right, with the cutting plane forming a full section of the object, it is difficult to obtain the true size of the angled elements. In the side view, the appear foreshortened or not to scale. Hidden lines were added as an attempt to better clarify the view.

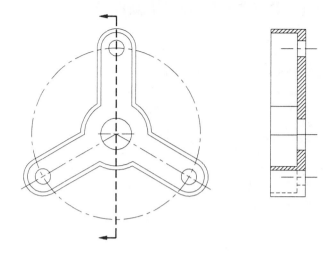

Instead of drawing the cutting plane line all the way through the object, the line is bent at the center of the object before being drawn through one of the angled legs. The direction of sight arrows on the cutting plane line not only determine which direction the view will be sectioned, but also shows another direction for rotating the angled elements so they line up with the upper elements. This rotation is usually never more than 90 degrees. As lines are projected across to form the side view, the section appears as if it were a full section. This is only because the features were rotated and projected in section for greater clarity of the drawing.

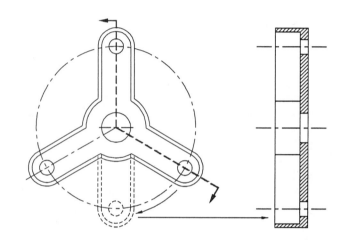

Offset Sections

Offset sections take their name from offsetting the cutting plane line to pass through certain details in a view. If the cutting plane line passes straight through any part of the object, details would be exposed while others would remain hidden. By offsetting the cutting plane line, the designer controls its direction and which features of a part it passes through. The view to section follows the basic section rules.

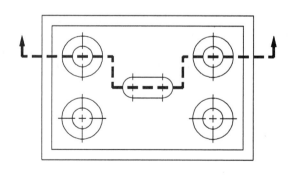

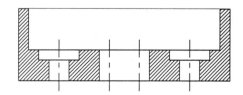

Sectioning Ribs

Contrary to section view principles, parts made out of cast iron with webs or ribs used for reinforcement do not follow basic rules of sections. In the example at the right, the front view has the cutting plane line passing through the entire view; the side view at "A" is cross-hatched according to section view basics. However, it is difficult to read the thickness of the base since the cross-hatching included the base along with the web. A more efficient method is to ignore cross-hatching webs as in "B". Therefore, not cross-hatching the web exposes other important details such as thicknesses of bases and walls.

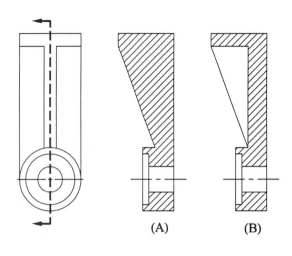

(A)   (B)

The object at the right is another example of performing a full section on an area consisting of webbed or ribbed features. By not cross-hatching the webbed areas, more information is available such as the thickness of the base and wall areas around the cylindrical hole. This may not be considered true projection, however it is considered good practice.

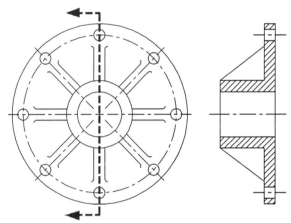

Broken Sections

At times only a partial section of an area needs to be created. For this reason, a broken section might be used. The object at "A" shows a combination of sectioned areas and conventional areas outlined by the hidden lines. When converting an area to a broken section, the designer creates a break line, cross-hatches one area and leaves the other area as in conventional drawing. Break lines may take the form of short freehanded line segments illustrated in example "B" or a series of long lines separated by break symbols as in example "C". Caution needs to be exercised when drawing the freehanded break line using the Sketch command. Be careful not to use a very small sketch increment since this will increase the size of the drawing file. The Line command can be used with Ortho-Off to produce the desired effect as in the examples at the right.

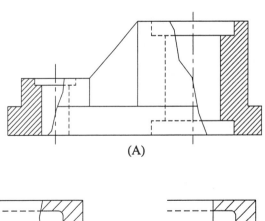

(A)

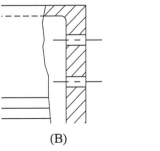
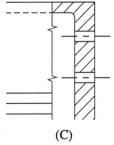

(B)   (C)

Revolved Sections

Section views may be constructed as part of a view by using revolved sections. In the example at the right, the elliptical shape is constructed while it is revolved into position and then crosshatched.

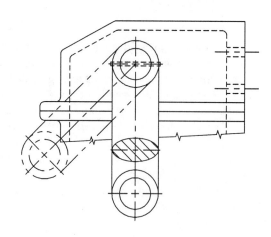

Illustrated at the right is another example of a revolved section where a cross-section of the C-clamp was cut away and revolved to display its shape.

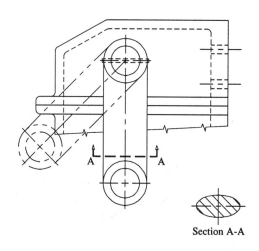

Removed Sections

Removed sections are very similar to revolved sections with the exception that instead of drawing the section somewhere inside of the object as is the case of a revolved section, the section is placed elsewhere or removed to a new location in the drawing. The cutting plane line is present with the arrows showing the direction of sight. Identifying letters are placed on the cutting plane and underneath the section to keep track of the removed sections especially when there are a number of them on the same drawing sheet.

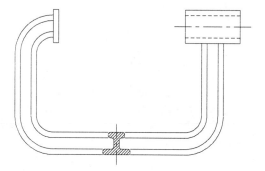

Section A-A

Another way of displaying removed sections is to use center lines as a substitute for the cutting plane line. In the example at the right, the center lines determine the three shapes of the chisel and display the basic shapes from circle to octagon to rectangle. Identification numbers are not required in this particular example.

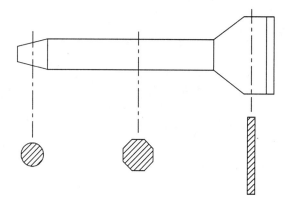

Isometric Sections

Section views may be incorporated into pictorial drawings that are illustrated at the right. The object at "A" is an example of a full isometric section with the cutting plane passing through the entire object. In keeping with basic section rules, only those surfaces sliced by the cutting plane line are cross-hatched. Isometric sections make it easy to view cut edges compared to holes or slots. Illustrated at "B" is an example of an isometric drawing converted into a half section.

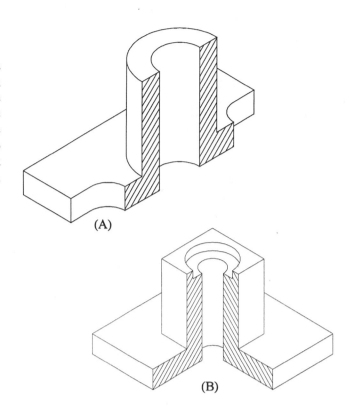

(A)

(B)

Architectural Sections

Mechanical representations of machine parts are not the only type of drawings where section views are used. Architectural drawings rely on sections to show the type of building materials that go into the construction of foundation plans, roof details, or wall sections in the example at the right. Here numerous types of cross-hatching symbols are used to call out the different types of building materials such as brick veneer at "A", insulation at "B", finished flooring at "C", floor joists at "D", concrete block at "E", earth at "F", and poured concrete at "G". Section symbols provided in AutoCAD were used to cross-hatch most of the building components with the exception of the floor joists and insulation.

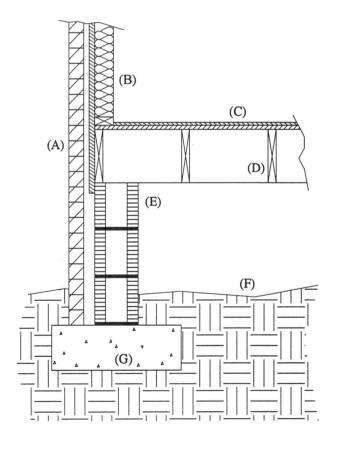

Hatching Techniques Using AutoCAD

The process of cross-hatching an object using AutoCAD is considered by some to be an art or even bordering on a science. This all revolves around boundary or area to be cross-hatched. Using manual methods, the designer followed a boundary with the aid of his eye along with the familiar T-square and triangle. Each line was stepped off and drawn individually. AutoCAD provides considerable help with this process since once a boundary is defined, the system cross-hatches the area automatically. The problem is defining the area to cross-hatch. The illustration at the right will be used to show two methods of cross-hatching available to the user.

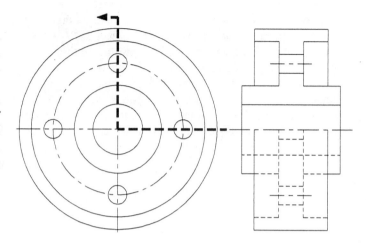

To perform a successful cross-hatching operation, a boundary must first be defined similar to the illustration at the right. This example shows the areas cut by the cutting plane line before cross-hatch lines are added. As this boundary is easily identified by eye, the actual construction of the view using AutoCAD may not yield the boundary at the right.

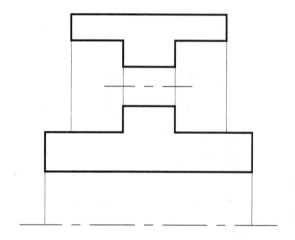

When constructing this view, horizontal lines were projected from the front view into the side view where they were sized using the Trim command. The dashed lines at the right merely represent the lines projected from the front view; in actuality, they are object lines. From this illustration, all horizontal lines are of the proper length to form a cross-hatching boundary.

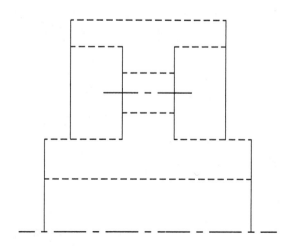

Hatching Techniques Using AutoCAD Continued

In this particular hatching example, problems will occur with the vertical being constructed as single entities labeled "A", "B", "C", "D", "E", and "F". These single entity lines will be difficult to define a boundary consisting of segments of these lines. Two methods may be used to define boundaries; using the Break command to split segments in two for boundary creation, or tracing a boundary using a polyline. Both methods are outlined in the text that follows.

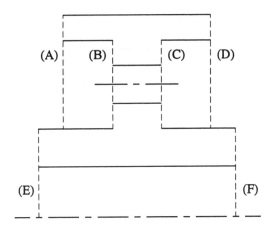

Lines may be formed into segments used to define boundaries using the Break command along with the @ option. In the illustration at the right, line "A" is the entity to be broken. Next, the first break point is selected at "B" followed by @ for the second break point. The significance of using @ is to select the previous point. The result is a break so small it is undetected by eye, yet breaks line "A" into two segments. Follow the prompt sequence below:

Command: **Break**
Select object: *(Select line segment "A")*
Enter second point (or F for first point): **F**
Enter first point: **Int**
of *(Select the intersection of the selected line at "B")*
Enter second point: **@**

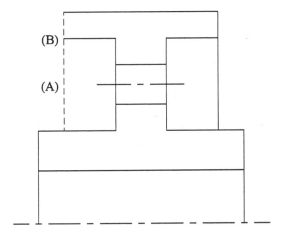

Repeat the procedure above for all points illustrated at the right. The result will be a series of broken vertical lines. When combined with the horizontal segments, they form the boundary needed to perform a proper cross-hatching operation using AutoCAD.

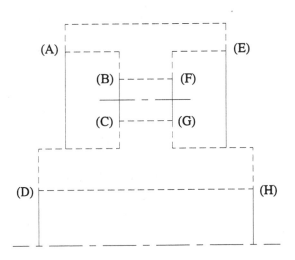

Hatching Techniques Using AutoCAD Continued

Once the boundaries are defined, use the Hatch command to place a certain cross-hatching pattern. Use the boxes illustrated at the right to form standard object selection set windows for performing the hatching operation. Follow the hatching prompts below:

Command: **Hatch**
Pattern (? or name/U,style): **Ansi31**
Scale for pattern <1.0000>: *(Strike Enter to accept this default)*
Angle for pattern <0>: *(Strike Enter to accept this default)*
Select objects: **W**
First corner: *(Select a point at "A")*
Other corner: *(Select a point at "B")*
Select objects: **W**
First corner: *(Select a point at "C")*
Other corner: *(Select a point at "D")*
Select objects: *(Strike Enter to perform the hatching operation)*

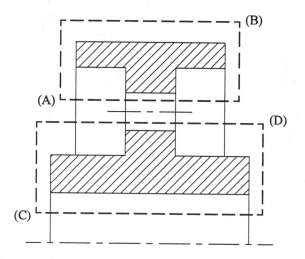

Another method of defining a cross-hatching boundary is to make a separate layer called "Pline", short for polyline. Use the Pline command along with Osnap-Intersection to trace the boundary over existing geometry similar to the illustration at the right.

Command: **Osnap**
Object snap mode(s): **Int**

Command: **Pline**
From point: *(Select the intersection at "A")*
Current line-width is 0.00000
Arc/Close/Halfwidth/Length/Undo/Width/<Endpoint of line>: *("B")*
Arc/Close/Halfwidth/Length/Undo/Width/<Endpoint of line>: *("C")*
Arc/Close/Halfwidth/Length/Undo/Width/<Endpoint of line>: *("D")*
Arc/Close/Halfwidth/Length/Undo/Width/<Endpoint of line>: *("E")*
Arc/Close/Halfwidth/Length/Undo/Width/<Endpoint of line>: *("F")*
Arc/Close/Halfwidth/Length/Undo/Width/<Endpoint of line>: *("G")*
Arc/Close/Halfwidth/Length/Undo/Width/<Endpoint of line>: *("H")*
Arc/Close/Halfwidth/Length/Undo/Width/<Endpoint of line>: **Cl**

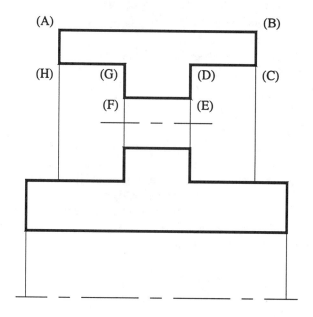

Hatching Techniques Using AutoCAD Continued

Use the Hatch command to cross-hatch the boundaries outlined by the polylines. Instead of using the Window option to group the selection set, simply select the polylines to perform the hatching.

Command: **Hatch**
Pattern (? or name/U,style): **Ansi31**
Scale for pattern <1.0000>: *(Strike Enter to accept this default)*
Angle for pattern <0>: *(Strike Enter to accept this default)*
Select objects: *(Select the polyline at "A")*
Select objects: *(Select the polyline at "B")*
Select objects: *(Strike Enter to perform the hatching operation)*

To prevent the original geometry lines from plotting underneath the polyline, use the Erase command to delete both polyline boundaries. A more efficient method would be to use the Layer command and turn off or freeze the layer named "Pline".

Hatching with AutoCAD is made easier only if a proper boundary to be hatched has been defined. The two methods of using the Break command or polyline to create a boundary will be illustrated in the tutorial exercises that follow.

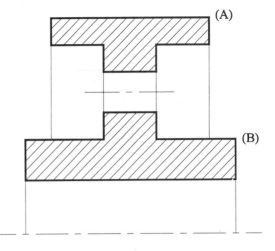

(A)

(B)

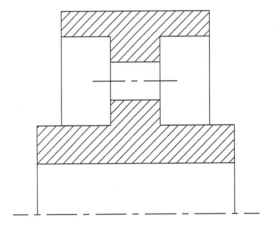

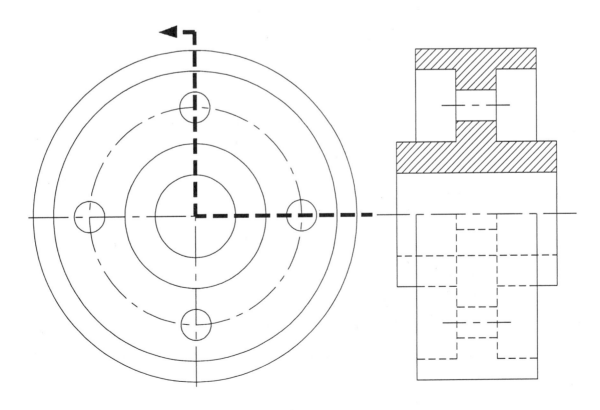

Section Views

215

Standard Hatch Patterns

AutoCAD has always been supplied with a generous amount of patterns to use when performing a cross-hatching operation. A total of 41 patterns are supplied with AutoCAD Release 10 to satisfy most hatching needs. It is also a fact that most users do not use all hatch patterns, however it is good to know that a pattern does exist for special needs. Illustrated at the right are eight patterns identified by the American National Standards Institute (ANSI) as being used to identify different material types. Below is a list of the patterns along with the material.

ANSI31 – Cast iron, Brick, Stone masonry
ANSI32 – Steel
ANSI33 – Bronze, Brass, Copper
ANSI34 – Plastic, Rubber
ANSI35 – Fire brick, any other refractory material
ANSI36 – Marble, Slate, Glass
ANSI37 – Lead, Zinc, Magnesium, Sound/Heat/Elec Insulation
ANSI38 – Aluminum

These hatch patterns may be called up by exact name, or from a graphical slide library located in the pull-down menu area. In order to see what material corresponds to a certain hatching pattern, follow the prompts below:

Command: **Hatch**
Pattern (? or name/U,style): **?** *(To list all patterns and materials)*

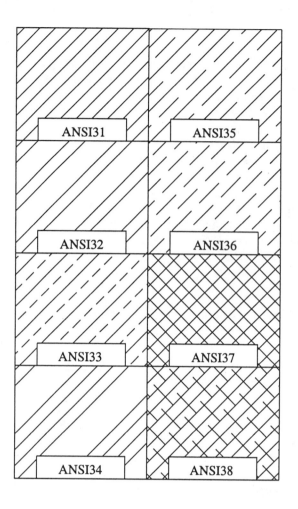

Illustrated at the right are six more patterns already defined by AutoCAD. All hatch patterns are located in a unique file called ACAD.Pat. This enables a user to make modifications to existing patterns or create entirely different patterns that are not supplied in AutoCAD. Below is a listing of the six additional patterns and the materials they support.

PLASTI – a modification of the ANSI34 pattern
EARTH – a pattern to symbolize ground, earth, or any other subterranean feature
GRASS – used to identify a grassy area
BRASS – a modification of the ANSI33 pattern
BRICK – used to identify any brick or masonry-type surface
SWAMP – used to identify a swampy area

Numerous architectural hatch patterns have been added to AutoCAD Release 11 making the total number of patterns closer to 56. For a listing of all patterns supported in AutoCAD Release 10 and Release 11, see the next page for a comparison.

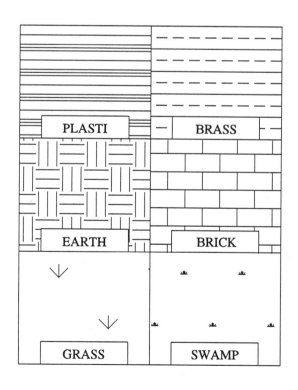

AutoCAD
Release 10 Patterns

Angle,Angle steel
Ansi31,ANSI Iron, Brick, Stone masonry
Ansi32,ANSI Steel
Ansi33,ANSI Bronze, Brass, Copper
Ansi34,ANSI Plastic, Rubber
Ansi35,ANSI Fire brick, Refractory material
Ansi36,ANSI Marble, Slate, Glass
Ansi37,ANSI Lead, Zinc, Magnesium, Sound/Heat/Elec Insulation
Ansi38,ANSI Aluminum
Box,Box steel
Brass,Brass material
Brick,Brick or masonry-type surface
Clay,Clay material
Cork,Cork material
Cross,A series of crosses
Dash,Dashed lines
Dolmit,Geological rock layering
Dots,A series of dots
Earth,Earth or ground (subterranean)
Escher,Escher pattern
Flex,Flexible material
Grass,Grass area
Grate,Grated area
Hex,Hexagons
Honey,Honeycomb pattern
Hound,Houndstooth check
Insul,Insulation material
Line,Parallel horizontal lines
Mudst,Mud and sand
Net,Horizontal / vertical grid
Net3,Network pattern 0-60-120
Plast,Plastic material
Plasti,Plastic material
Sacncr,Concrete
Square,Small aligned squares
Stars,Star of David
Steel,Steel material
Swamp,Swampy area
Trans,Heat transfer material
Triang,Equilateral triangles
Zigzag,Staircase effect

AutoCAD
Release 11 Patterns

angle,Angle steel
ansi31,ANSI Iron, Brick, Stone masonry
ansi32,ANSI Steel
ansi33,ANSI Bronze, Brass, Copper
ansi34,ANSI Plastic, Rubber
ansi35,ANSI Fire brick, Refractory material
ansi36,ANSI Marble, Slate, Glass
ansi37,ANSI Lead, Zinc, Magnesium, Sound/Heat/Elec Insulation
ansi38,ANSI Aluminum

The following hatch patterns AR-xxxxx
come from AEC/Architectural
AR-B816, 8x16 block elevation stretcher bond
AR-B816C, 8x16 block elevation stretcher bond with mortar joints
AR-B88, 8x8 block elevation stretcher bond
AR-BRELM, standard brick elevation english bond with mortar joints
AR-BRSTD, standard brick elevation stretcher bond
AR-CONC, random dot and stone pattern
AR-HBONE, standard brick herringbone pattern @ 45 degrees
AR-PARQ1, 2x12 parquet flooring: pattern of 12x12
AR-RROOF, roof shingle texture
AR-RSHKE, roof wood shake texture
AR-SAND, random dot pattern
box,Box steel
brass,Brass material
brick,Brick or masonry-type surface

brstone,Brick and stone
clay,Clay material
cork,Cork material
cross,A series of crosses
dash,Dashed lines
dolmit,Geological rock layering
dots,A series of dots
earth,Earth or ground (subterranean)
escher,Escher pattern
flex,Flexible material
grass,Grass area
grate,Grated area
hex,Hexagons
honey,Honeycomb pattern
hound,Houndstooth check
insul,Insulation material
line,Parallel horizontal lines
mudst,Mud and sand
net,Horizontal / vertical grid
net3,Network pattern 0-60-120
plast,Plastic material
plasti,Plastic material
sacncr,Concrete
square,Small aligned squares
stars,Star of David
steel,Steel material
swamp,Swampy area
trans,Heat transfer material
triang,Equilateral triangles
zigzag,Staircase effect

Note: Information on both sets of hatch patterns was obtained by viewing the ACAD.Pat file which holds the names and definitions of each pattern.

Hatch Pattern Scaling

AutoCAD has predefined hatching patterns already sized. When using the Hatch command, the pattern used is assigned a scale value of 1.00 which will draw the pattern exactly the way it was orginally defined. The example at the right shows the affects of the Hatch command when accepting the default value for the pattern scale.

Command: **Hatch**
Pattern (? or name/U,style): **Ansi31**
Scale for pattern <1.0000>: *(Strike Enter to accept this default)*
Angle for pattern <0>: *(Strike Enter to accept this default)*
Select objects: *(Select the desired areas at the right)*

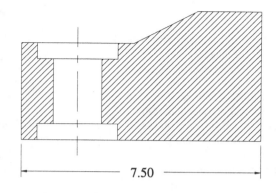

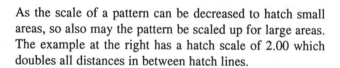

Entering a different scale value for the pattern will either increase or decrease the spacing in between cross-hatch lines. At the right is an example of the Ansi31 pattern with a new scale value of 0.50.

Command: **Hatch**
Pattern (? or name/U,style): **Ansi31**
Scale for pattern <1.0000>: **0.50**
Angle for pattern <0>: *(Strike Enter to accept this default)*
Select objects: *(Select the desired areas at the right)*

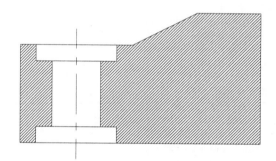

As the scale of a pattern can be decreased to hatch small areas, so also may the pattern be scaled up for large areas. The example at the right has a hatch scale of 2.00 which doubles all distances in between hatch lines.

Command: **Hatch**
Pattern (? or name/U,style): **Ansi31**
Scale for pattern <1.0000>: **2.00**
Angle for pattern <0>: *(Strike Enter to accept this default)*
Select objects: *(Select the desired areas at the right)*

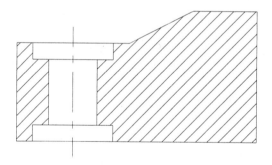

Care needs to be used when hatching large areas. In the example at the right, the distance measures 190.50 millimeters. If the hatch scale of 1.00 were used, the pattern would take on a filled appearance similar to the Solid command. The problem is that numerous lines are generated that increase the size of the drawing file. A value of 25.4 is used to scale hatch lines for metric drawings.

Command: **Hatch**
Pattern (? or name/U,style): **Ansi31**
Scale for pattern <1.0000>: **25.4**
Angle for pattern <0>: *(Strike Enter to accept this default)*
Select objects: *(Select the desired areas at the right)*

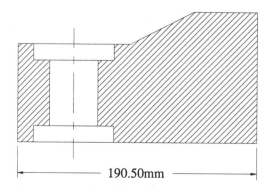

Hatch Pattern Angle Manipulation

As with the scale of the hatch pattern, the angle for the hatch pattern can be controlled by the designer depending on the affect the pattern has with the area being hatched. By default, the Hatch command displays a 0-degree angle for all patterns. In the example at the right, the angle for Ansi31 is 45 degrees. This is because the pattern was originally created at a 45-degree angle.

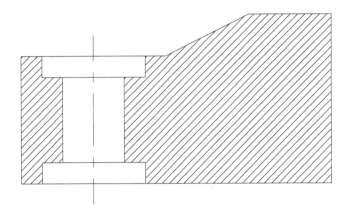

Command: **Hatch**
Pattern (? or name/U,style): **Ansi31**
Scale for pattern <1.0000>: *(Strike Enter to accept this default)*
Angle for pattern <0>: *(Strike Enter to accept this default)*
Select objects: *(Select the desired areas at the right)*

Entering any angle different from the default value of 0 will rotate the hatch pattern by that value. This means if a pattern was originally designed at a 45-degree angle like Ansi31, entering a new angle for the pattern would begin rotating the pattern starting at the 45 degree position. In the example at the right, a new angle of 45 degrees is entered. Since the original angle was already 45 degrees, this new angle value is added to the original to obtain a vertical cross-hatch pattern.

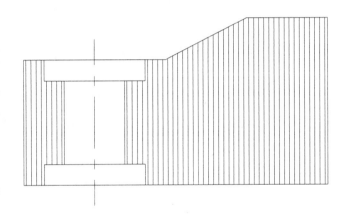

Command: **Hatch**
Pattern (? or name/U,style): **Ansi31**
Scale for pattern <1.0000>: *(Strike Enter to accept this default)*
Angle for pattern <0>: **45**
Select objects: *(Select the desired areas at the right)*

Again, entering an angle other than the default rotates the pattern from the original angle to a new angle. In the example at the right, the Ansi31 pattern was rotated by 90 degrees.

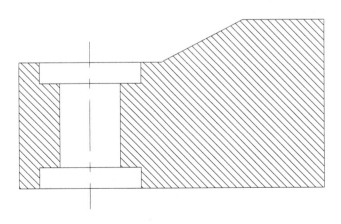

Command: **Hatch**
Pattern (? or name/U,style): **Ansi31**
Scale for pattern <1.0000>: *(Strike Enter to accept this default)*
Angle for pattern <0>: **90**
Select objects: *(Select the desired areas at the right)*

Providing different angles for patterns is useful when creating section assemblies where different parts are in contact with each other and patterns are placed at different angles making the parts easy to see.

Selecting Patterns from Icon Menus

Entering the Hatch command from the keyboard, the designer or CAD operator must know the exact name of the pattern to be used. If the pattern is misspelled, the pattern is rejected by AutoCAD. There is a more efficient way of making a pattern current so it already appears as the default value in the Hatch command. AutoCAD Release 11 has enhanced the pull-down menu areas to include a separate hatching options area located in the Options pull-down menu. Through this area, the operator has the option of setting a new current hatch pattern, hatch style, hatch scale factor, and hatch rotation angle. The hatching operation still needs to be performed using the Hatch command. This area in Options merely sets new default when using the hatch command. One of these options has to do with the hatch pattern itself. Instead of remembering names of patterns, an icon menu appears when selecting "Hatch Patterns..." from the pull-down menu. The icon menu displays a graphical representation of the desired pattern. Select the box to the left of the pattern and the pattern becomes the new default value for the Hatch command. Selecting Hatch Scale and Hatch Angle from the pull-down menu will make these values default for the Hatch command.

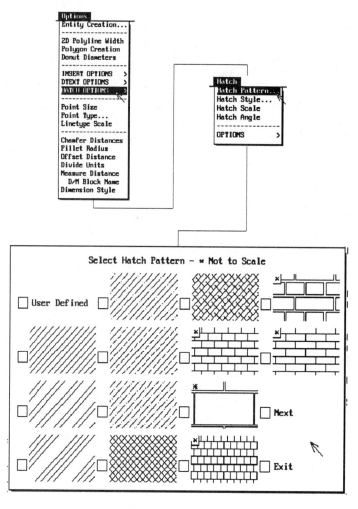

Another check on hatching exists to better control cross-hatching where multiple objects are involved. This is the purpose of "Hatch Style...", also located in the Options pull-down menu area. By default, AutoCAD hatches groups of entities illustrated in "A" at the right. Here, the outermost area is hatched, the next innermost area is ignored, followed by hatching the next innermost area and so on. If text is present in the selection set and a window is used to select the objects to hatch, the cross-hatch lines will not be placed through the text. In "B" at the right, only the outermost area is hatched with all innermost areas ignored. This style is also achieved by entering "Pattern Name, O" when prompted for the name of the pattern in the Hatch command. In "C" at the right, the default hatching style is ignored and all objects have cross-hatch lines added.

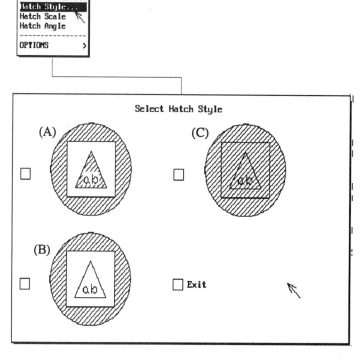

Tutorial Exercise #13
Dflange.Dwg

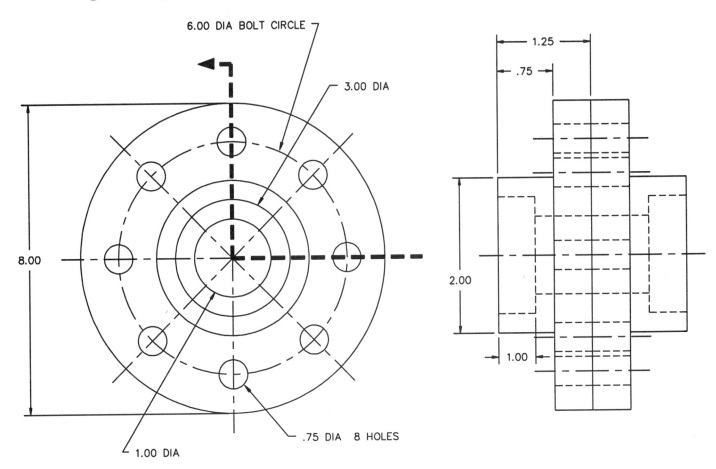

6.00 DIA BOLT CIRCLE

3.00 DIA

8.00

2.00

1.25

.75

1.00

.75 DIA   8 HOLES

1.00 DIA

PURPOSE:
This tutorial is designed to use the Mirror, Change, and Hatch commands to convert the Dflange (Double Flange) into a half section assembly.

SYSTEM SETTINGS:
Since this drawing is provided on diskette, edit an existing drawing called "Dflange." Follow the steps in this tutorial for creating the upper half of the side view into a half section.

LAYERS:
The following layers have already been created:

Name-Color-Linetype
Object - White - Continuous
Cen - Yellow - Center
Hid - Red - Hidden
Xhatch - Magenta - Continuous
Cpl - Yellow - Dashed
Pline - Green - Continuous
Dim - Yellow - Continuous

SUGGESTED COMMANDS:
Begin this tutorial by converting one-half of the object into a section by erasing unnecessary hidden lines. Next, use the Change command and change the remaining hidden lines to the Object layer. Use the Break command to define boundaries to hatch. The selected boundaries are section lined using the Hatch command and the entire object is duplicated and copied to form a matching flange using the Mirror command. Use the Ansi31 hatching pattern for this exercise.

DIMENSIONING:
Dimensions may be added to this tutorial at a later time. Use the Dim layer already created. Consult your instructor.

PLOTTING:
This tutorial exercise may be plotted on "C"-size paper (18" x 24"). Use a plotting scale of 1=1 to produce a full size plot.

VERSION OF AUTOCAD:
This tutorial exercise may be completed using either AutoCAD Release 10 or Release 11.

Step #1

Prepare the object to be mirrored by creating a selection set of the side view. Be sure snap is turned off by striking the F9 key. Use the Select command to accomplish this by windowing all entities from "A" to "B".

Command: **Select**
Select objects: **Window**
First corner: *(Mark a point at "A")*
Other corner: *(Mark a point at "B")*
Select objects: *(Strike Enter to exit this command and create the selection set)*

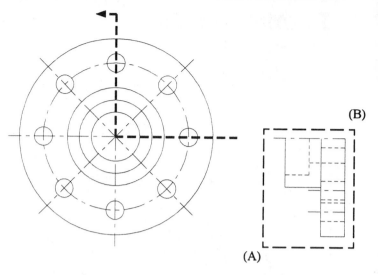

(B)

(A)

Step #2

Use the Mirror command to create a duplicate copy of the side view of the Dflange. Use the vertical line from "A" to "B" as the mirror line.

Command: **Mirror**
Select objects: **Previous**
Select objects: *(Strike Enter to continue)*
First point of mirror line: **Endp**
of *(Select the endpoint of the line at "A")*
Second point: **Endp**
of *(Select the other endpoint of the line at "B")*
Select old objects? <N>: *(Strike Enter to exit this command)*

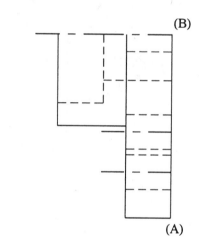

(B)

(A)

Step #3

Again use the Mirror command to copy and duplicate one-half of the bottom of the Dflange. It is this half that will be converted into a half section.

Command: **Mirror**
Select objects: **Previous**
Select objects: *(Strike Enter to continue)*
First point of mirror line: **Endp**
of *(Select the endpoint of the center line at "A")*
Second point: **Endp**
of *(Select the other endpoint of the center line at "B")*
Select old objects? <N>: *(Strike Enter to exit this command)*

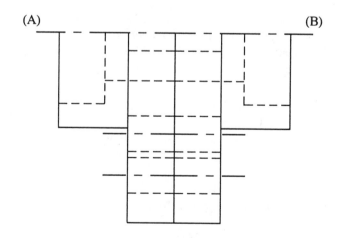

(A) (B)

Step #4

Begin the preparation of the section by deleting any unnecessary entities such as the hidden and center lines located at "A", "B", "C", and "D" shown in the illustration at the right. Line "E" was copied during the last mirror operation and must also be deleted. Line "F" also represents a duplicate entity and must be deleted.

Command: **Erase**
Select objects: *(Select the lines at "A", "B", "C", "D" "E" and "F")*
Select objects: *(Strike Enter to execute this command)*

Command: **Redraw**

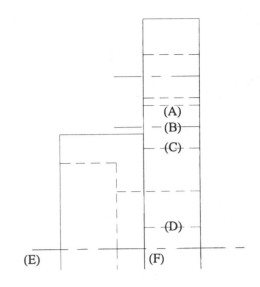

Step #5

Use the Chprop command to convert the five hidden lines illustrated at the right from hidden lines to object lines. Perform the change from the layer Hid to the layer Object.

Command: **Chprop**
Select objects: *(Select the five lines labeled "A" to "E")*
Select objects: *(Strike Enter to continue)*
Change what property(Color/LAyer/LType/Thickness)? **LA**
New layer <HID>: **Object**
Change what property(Color/LAyer/LType/Thickness)?
    (Strike Enter to exit this command)

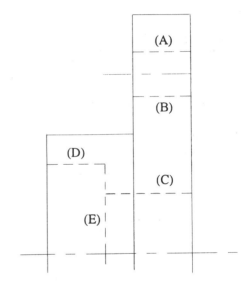

Step #6

Use the Trim command, select the horizontal line at "A" as the cutting edge, and select the vertical line at "B" as the entity to trim.

Command: **Trim**
Select cutting edge(s)...
Select objects: *(Select the horizontal line at "A")*
Select objects: *(Strike Enter to continue)*
Select object to trim: *(Select the vertical line at "B")*
Select object to trim: *(Strike Enter to exit this command)*

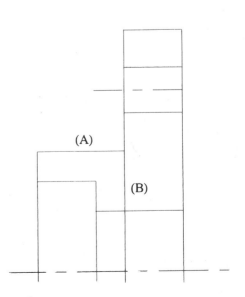

Step #7

The side view now needs to have the surfaces sliced by the cutting plane line cross-hatched. Two areas labeled "A" and "B" need to be prepared for cross-hatching. However, since all vertical lines are continuous and not broken into individual segments, polylines will be used to trace a boundary to be accepted by the Hatch command.

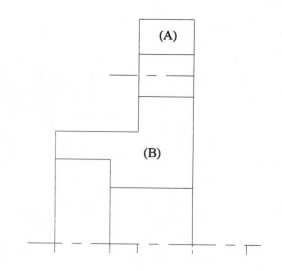

Step #8

Use the Layer command and make the layer, Pline, the new current layer. Then use the Pline command to trace a boundary using the illustration at the right as a guide. Because of the many intersections selected, use the Osnap command to lock into the intersection mode.

Command: **Layer**
?/Make/Set/New/ON/OFF/Color/Ltype/Freeze/Thaw: **Set**
New current layer <0>: **Pline**
?/Make/Set/New/ON/OFF/Color/Ltype/Freeze/Thaw:
    (Strike Enter to exit this command)

Command: **Osnap**
Object snap mode(s): **Int**

Command: **Pline**
From point: *(Select the intersection of the lines at "A")*
Arc/Close/Halfwidth/Length/Undo/Width/<Endpoint of
    line>: *(Select the intersection at "B")*
Arc/Close/Halfwidth/Length/Undo/Width/<Endpoint of
    line>: *(Select the intersection at "C")*
Arc/Close/Halfwidth/Length/Undo/Width/<Endpoint of
    line>: *(Select the intersection at "D")*
Arc/Close/Halfwidth/Length/Undo/Width/<Endpoint of
    line>: **C**

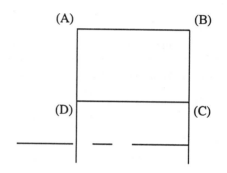

Step #9

Repeat the procedure above for tracing a polyline using the illustration at the right as a guide. Draw the polyline from the intersection of "A" to "B", "C", "D", "E", "F", "G", "H", and complete the polyline by using the Close option. Since you are still locked into the Osnap-Intersection option, use the Osnap command again to free up any Osnap options previously in use by entering the "None" option.

Command: **Osnap**
Object selection mode(s): **None**

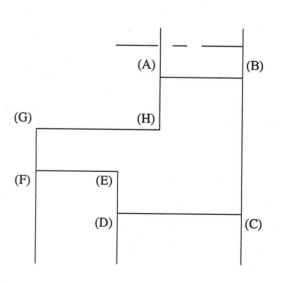

Step #10

Using the polylines constructed in the previous step as boundaries, use the Hatch command and the pattern Ansi31 to cross-hatch the two areas at the right. Before performing this step, change layers from Pline to Xhatch.

Command: **Layer**
?/Make/Set/New/ON/OFF/Color/Ltype/Freeze/Thaw: **Set**
New current layer<Pline>: **Xhatch**
?/Make/Set/New/ON/OFF/Color/Ltype/Freeze/Thaw:
　　　(Strike Enter to exit this command)

Command: **Hatch**
Pattern (? or name/U,style): **Ansi31**
Scale for pattern <1.0000>: *(Strike Enter to accept this default)*
Angle for pattern <0>: *(Strike Enter to accept this default)*
Select objects: *(Select the polyline at "A")*
Select objects: *(Select the polyline at "B")*
Select objects: *(Strike Enter to execute this command)*

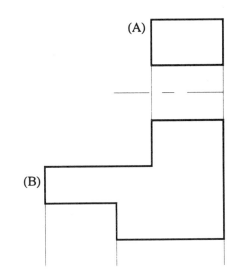

Step #11

The complete cross-hatching of the upper half of the side view is illustrated at the right. The pattern Ansi31 is drawn at a 45-degree angle with a spacing of approximately .125" in between lines. Because the polyline used to define the boundary is on top of existing lines, use the Layer command to turn the "Pline" layer off.

Command: **Layer**
?/Make/Set/New/ON/OFF/Color/Ltype/Freeze/Thaw: **OFF**
New current layer <0>: **Pline**

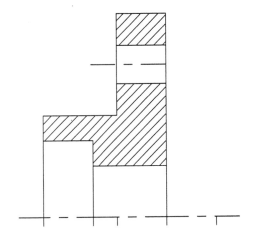

Step #12

Duplicate and flip the upper half of the side view to form the opposite half of the flange assembly using the Mirror command. Use the crossing option to select the objects illustrated at the right. Select "C" and "D" as the first and second points of the mirror lines.

Command: **Mirror**
Select objects: **C**
First corner: *(Select at "A")*
Other corner: *(Select at "B")*
Select objects: *(Strike Enter to continue)*
First point of mirror line: **Endp**
of *(Select the endpoint of the vertical line at "C")*
Second point: **Endp**
of *(Select the endpoint of the vertical line at "D")*
Delete old objects? <N> *(Strike Enter to accept the default)*

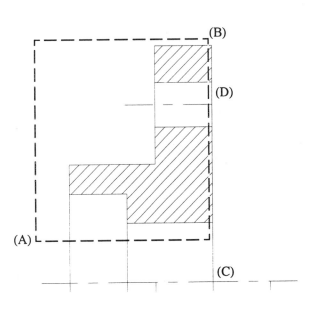

Step #13

The completed assembly consisting of both sectioned halves
is illustrated at the right. It is acceptable for cross-hatch
lines to run in opposite directions whenever two different
parts attach to each other.

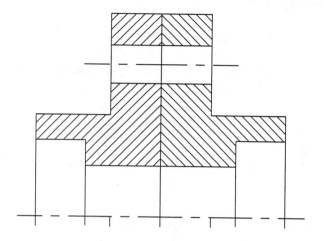

Step #14

Dimensions may be added at a later time, similar to the
illustration at the right. Notice the bottom halves of the
side views show conventional hidden lines. This is one
method of representing half sections.

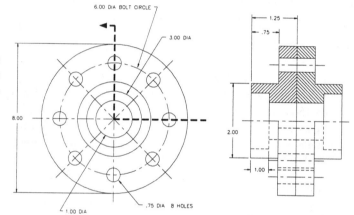

Step #15

Yet another method of displaying half sections is to elimi-
nate all hidden and center lines in the view that is not cross-
hatched. In this way, the sectioned half is interpreted to be
of the same makeup as the lower half that is not sectioned.

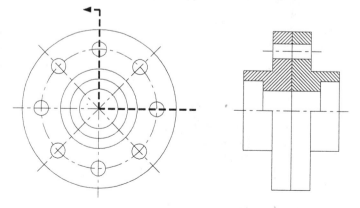

Tutorial Exercise #14
Coupler.Dwg

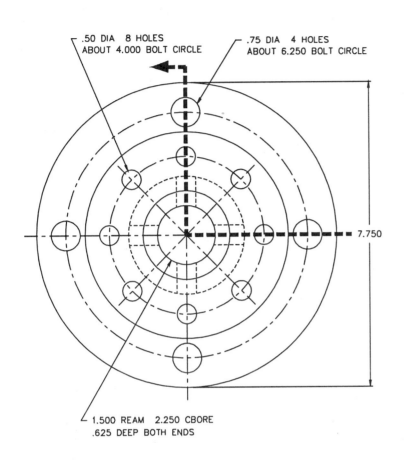

.50 DIA 8 HOLES
ABOUT 4.000 BOLT CIRCLE

.75 DIA 4 HOLES
ABOUT 6.250 BOLT CIRCLE

7.750

1.500 REAM 2.250 CBORE
.625 DEEP BOTH ENDS

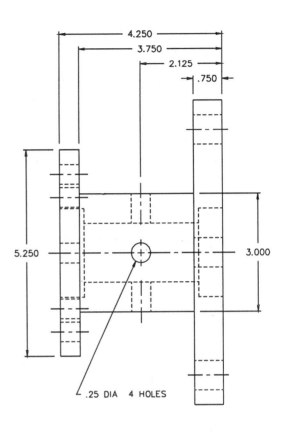

4.250

3.750

2.125

.750

5.250

3.000

.25 DIA 4 HOLES

PURPOSE:
This tutorial is designed to use the Mirror, Change, and Hatch commands to convert the Coupler into a half section. The methods will be similar to the previous tutorial, with the exception of the cross-hatching segment being different.

SYSTEM SETTINGS:
Since this drawing is provided on diskette, edit an existing drawing called "Coupler." Follow the steps in this tutorial for creating the upper half of the side view into a half section.

LAYERS:
Create the following layers with the format:
Name-Color-Linetype
Object - White - Continuous
Cen - Yellow - Center
Hid - Red - Hidden
Xhatch - Magenta - Continuous
Cpl - Yellow - Dashed
Dim - Yellow - Continuous

SUGGESTED COMMANDS:
This tutorial begins similar to the previous exercise, Dflange.Dwg, with the exception of the cross-hatching segment. Instead of tracing a boundary using a polyline, the Break command will be used along with the @ option. Boundaries will be formed when two lines are broken at an exact intersection so that the break will not be seen. This will also define the boundary to be cross-hatched with the Ansi31 pattern.

DIMENSIONING:
Dimensions may be added to this tutorial at a later time. Consult your instructor.

PLOTTING:
This tutorial exercise may be plotted on "C"-size paper (18" x 24"). Use a plotting scale of 1=1 to produce a full-size plot.

VERSION OF AUTOCAD:
This tutorial exercise may be completed using either AutoCAD Release 10 or Release 11.

Step #1

Magnify the side view using the Zoom-Window command. Turn snap off by striking the F9 function key. Then, use the Select command to group all entities illustrated at the right into one selection set. The Remove option of the Select command will be used to remove the center line at "A" and "C" and the circle at "B" from the selection set.

Command: **Select**
Select objects: **Window**
First corner: *(Mark a point at "X")*
Other corner: *(Mark a point at "Y")*
Select objects: **Remove**
Remove objects: *(Select the center line at "A" and "C")*
Remove objects: *(Select the circle at "B")*
Remove objects: *(Strike Enter to exit this command and create the selection set)*

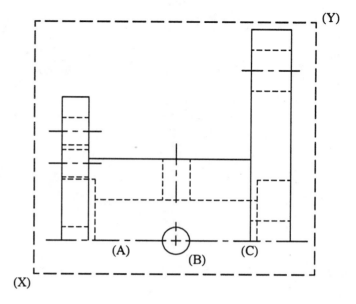

Step #2

Use the Mirror command to copy and flip the upper half of the side view and form the lower half.

Command: **Mirror**
Select objects: **Previous**
Select objects: *(Strike Enter to continue)*
First point of mirror line: **Endp**
of *(Select the endpoint of the center line at "A")*
Second point: **Endp**
of *(Select the other endpoint of the center line at "B")*
Select old objects? <N>: *(Strike Enter to exit this command)*

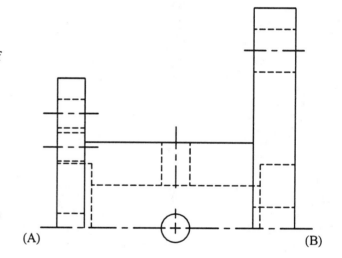

Step #3

Begin converting the upper half of the side view to a half section by using the Erase command to remove any unnecessary hidden and center lines from the view.

Command: **Erase**
Select objects: *(Carefully select the hidden lines labeled "A", "B", "C", and "D")*
Select objects: *(Select the center line at "E")*
Select objects: *(Strike Enter to execute this command)*

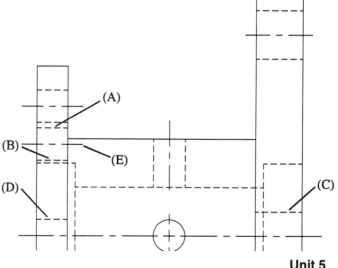

Step #4

Since the remaining hidden lines actually represent object lines when shown in section, use the Chprop command to convert all hidden lines labeled at the right from the Hid layer to the Object layer.

Command: **Chprop**
Select objects: *(Select all hidden lines labeled "A" to "K")*
Select objects: *(Strike Enter to continue)*
Change what property(Color/LAyer/LType/Thickness)? **LA**
New layer <HID>: **Object**
Change what property (Color/LAyer/LType/Thickness)?
    (Strike Enter to exit this command)

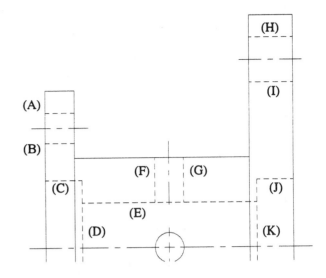

Step #5

Remove unnecessary line segments from the upper half of the converted section using the Trim command. Use the horizontal line at "A" as the cutting edge, and select the two vertical segments at "B" and "C" as the entities to trim.

Command: **Trim**
Select cutting edges...
Select objects: *(Select the horizontal line at "A")*
Select objects: *(Strike Enter to continue)*
Select object to trim: *(Select the vertical line at "B")*
Select object to trim: *(Select the vertical line at "C")*
Select object to trim: *(Strike Enter to exit this command)*

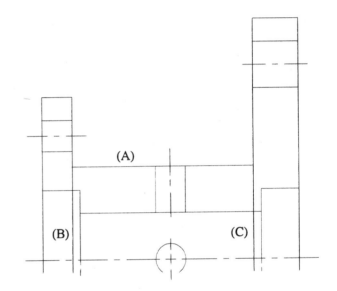

Step #6

Before beginning the cross-hatching segment, four boundaries need to be defined in order to succeed with the Hatch command. The four areas to hatch are labeled at the right. One previous method used was to trace the profile of the boundary with a polyline, hatch the area outlined by the polyline, and finally delete the polyline. For this tutorial, the Break command will be featured along with the @ option. The task will be to perform breaks at key intersections to define boundaries. The breaks, however, are to be small enough so as not to be noticed when performing a Zoom of the area or a plot.

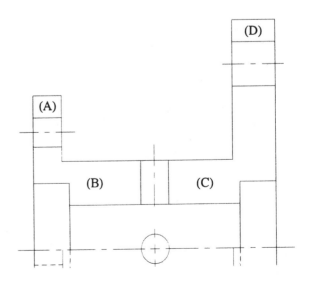

Step #7

Since all vertical and horizontal lines consist of continuous segments, the Break-@ option will split lines into numerous segments. Follow the prompts below to understand this operation:

Command: **Break**
Select object: *(Select the vertical line at "A")*
Enter second point (or F for first point): **F**
Enter first point: **Int**
of *(Select the intersection of the two lines at "A")*
Enter second point: **@**

The @ means "last point" and performs the break at the exact same location as the first point selected. In this way, the line is separated into two segments without the separation being noticed. Repeat the procedure above for intersections "B" through "H".

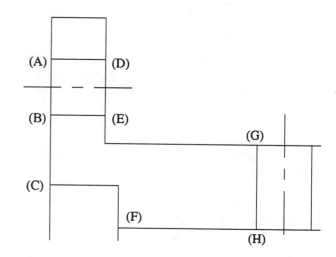

Step #8

Repeat the Break-@ option on the shape at the right to lay out the necessary boundaries to perform the hatching operation.

Command: **Break**
Select object: *(Select the horizontal line at "A")*
Enter second point (or F for first point): **F**
Enter first point: **Int**
of *(Select the intersection of the two lines at "A")*
Enter second point: **@**

Repeat the procedure above for intersections "B" through "H".

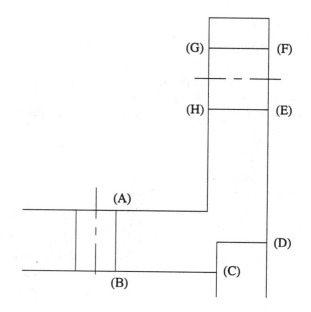

Step #9

Use the Layer command to set the new current layer to "Xhatch." Once all areas have been broken identifying boundaries, use the Hatch command and select each boundary to hatch.

Command: **Hatch**
Pattern (? or name/U,style): **Ansi31**
Scale for pattern <1.0000>: *(Strike Enter to accept this default)*
Angle for pattern <0>: *(Strike Enter to accept this default)*
Select objects: **W**
First corner: *(Select a point at "A")*
Other corner: *(Select a point at "B")*
Select objects: *(Use another window to select Area "C")*
Select objects: *(Use another window to select Area "D")*
Select objects: *(Use another window to select Area "E")*
Select objects: *(Strike Enter to execute this command)*

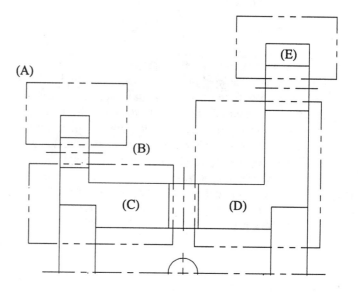

Step #10

As a type of check on progress, using the Window option of the Hatch command and selecting the boundaries in the previous illustration should highlight the boundaries similar to the illustration at the right. Here, the dashed lines signify the selected boundaries for the Hatch command to properly cross-hatch the area. If your display is not similar to the illustration at the right, use the Break-@ option to complete all breaks before using the Hatch command.

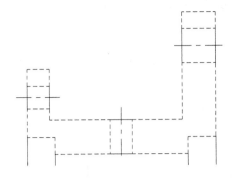

Step #11

Selecting multiple Window options at the "Select objects" prompt of the Hatch command will cross-hatch all defined areas as in the illustration at the right. Using multiple selections for the same command will also prevent having to use the command four separate times. The completed drawing is also illustrated below, including dimensions. As with other half section examples, it is the option of the operator or designer to show all hidden lines in the lower half or delete all hidden and center lines from the lower half and simply interpret the section in the upper half.

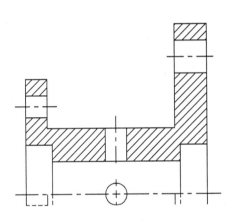

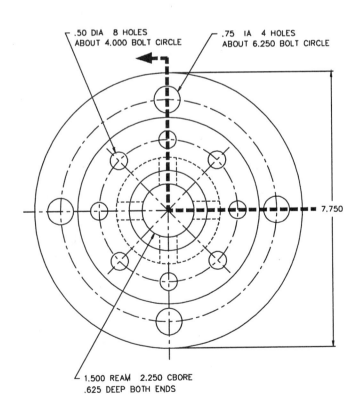

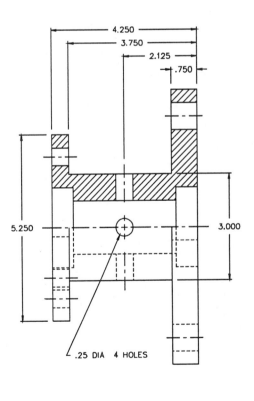

Problems for Unit 5

Problem 5-1
Center a three-view drawing and make the front view a full section.

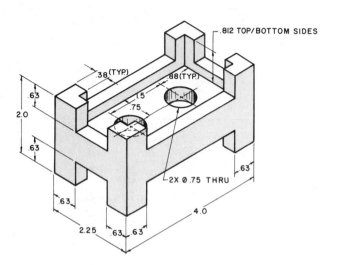

.812 TOP/BOTTOM SIDES
.38 (TYP.)
.88 (TYP.)
1.5
.75
.63
2.0
.63
.63
.63
2.25
.63 .63
4.0
.63
2X Ø.75 THRU

Problem 5-3
Center two views within the work area, and make one view a full section.

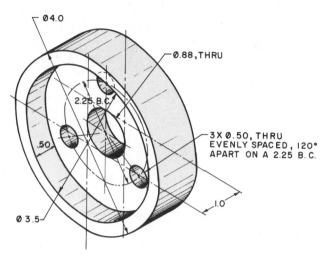

Ø4.0
Ø.88, THRU
2.25 B.C.
.50
3X Ø.50, THRU
EVENLY SPACED, 120°
APART ON A 2.25 B.C.
1.0
Ø3.5

Problem 5-2
Center two views within the work area, and make one view a full section. Use correct drafting practices for the ribs.

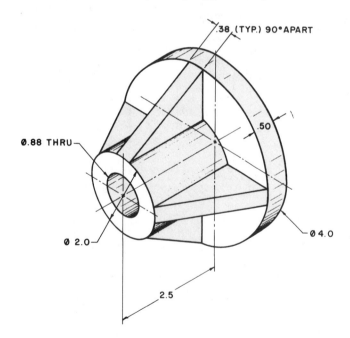

.38 (TYP.) 90°APART
.50
Ø.88 THRU
Ø 2.0
Ø 4.0
2.5

Problem 5-4
Center the front view and top view within the work area. Make one view a full section.

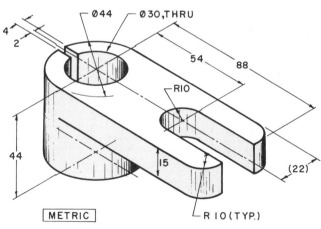

Ø44   Ø30, THRU
4
2
54   88
R10
44
15
(22)
METRIC
R 10 (TYP.)

Problem 5-5

Center two views within the work area, and make one view a full section. Use correct drafting practices for the keyway, ribs, and holes.

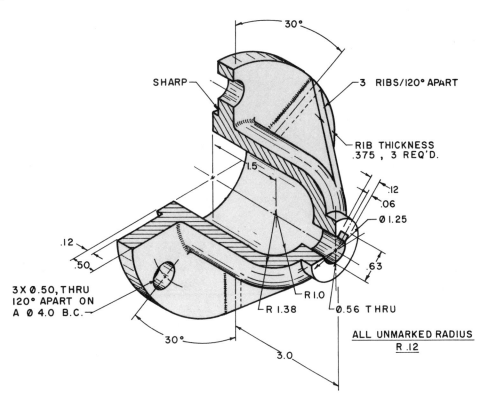

Problem 5-6

Center three views within the work area, and make one view an offset section.

Problem 5-7

Center three views within the work area, and make one view an offset section.

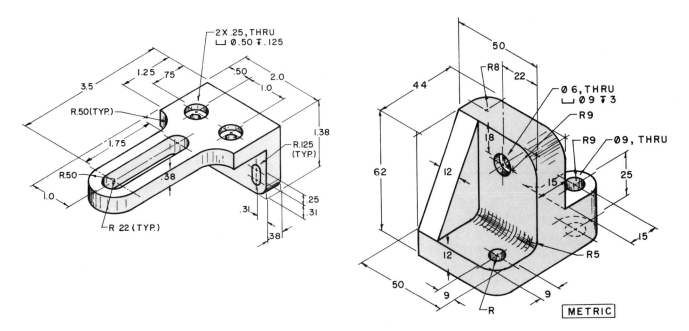

Section Views

Problem 5-8
Center three views within the work area, and make one view an offset section.

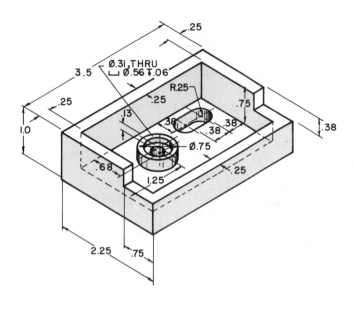

Problem 5-10
Center two views within the work area, and make one view an offset section.

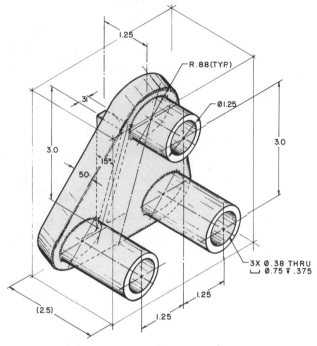

Problem 5-9
Center three views within the work area, and make one view an offset section.

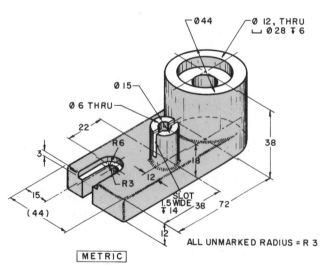

Problem 5-11
Center the front view and top view within the work area. Make one view a half section.

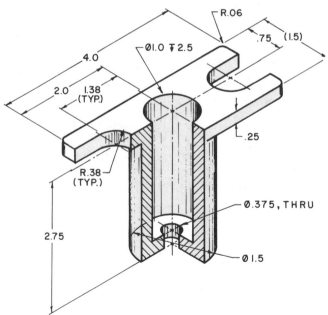

Problem 5-12
Center two views within the work area, and make one view a half section.

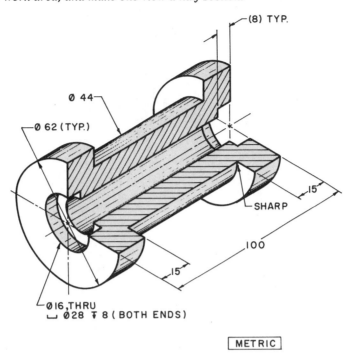

(8) TYP.

Ø 44

Ø 62 (TYP.)

SHARP

15

100

15

Ø16,THRU
⌴ Ø28 ⊤ 8 (BOTH ENDS)

METRIC

Problem 5-13
Center the two views within the work area, and make one view a half section.

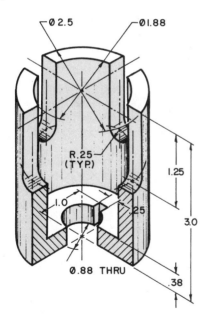

Ø 2.5   Ø1.88

R.25
(TYP.)

1.25

1.0   .25

3.0

Ø0.88 THRU

.38

Problem 5-14
Center two views within the work area, and make one view a half section.

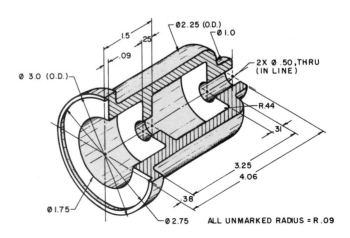

Ø2.25 (O.D.)   Ø1.0

1.5   .25

.09

Ø 3.0 (O.D.)

2X Ø .50,THRU
(IN LINE)

R.44

.31

3.25

4.06

.38

Ø1.75

Ø2.75   ALL UNMARKED RADIUS = R.09

Problem 5-15
Center two views within the work area, and make one view a half section.

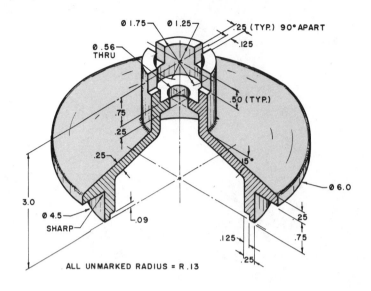

Ø1.75   Ø1.25   .25 (TYP.) 90° APART
Ø.56
THRU   .125

.50 (TYP.)

.75
.25

.25   15°

Ø6.0

3.0

Ø4.5   .25
SHARP   .09   .75

.125

.25

ALL UNMARKED RADIUS = R.13

Problem 5-16
Center the required views within the work area, and make one view a broken-out section to illustrate the complicated interior area.

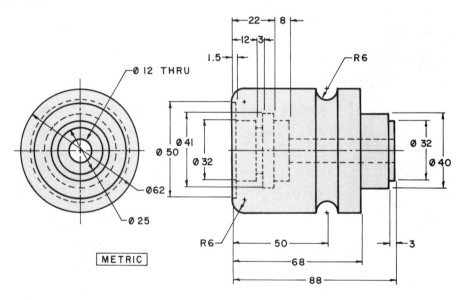

22   8
12   3
1.5
Ø 12 THRU   R6

Ø 41
Ø 50   Ø 32   Ø 32
Ø 40
Ø62
Ø 25   R6   50   3

METRIC   R6   68

88

Problem 5-17
Center three views within the work area, and add removed sections A-A and B-B.

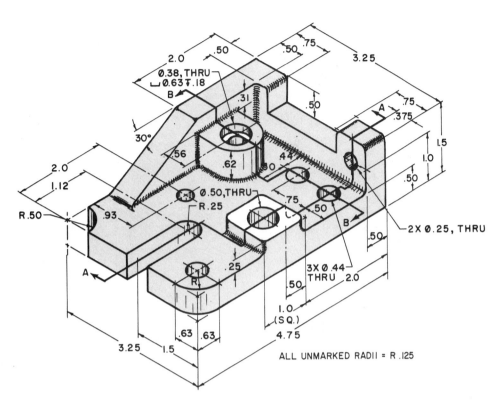

2.0
.50
.50
.75
3.25
Ø.38, THRU
⊔ Ø.63 ∓.18
B
.31
.50
A
.75
.375
30°
.56
.62
.44
.80
1.5
1.0
2.0
1.12
.50
R.50
.93
Ø.50, THRU
R.25
.75
.50
B
.50
2X Ø.25, THRU
A
.25
3X Ø.44
THRU 2.0
.50
.63 .63
1.0
(SQ.)
4.75
3.25
1.5

ALL UNMARKED RADII = R .125

Problem 5-18
Center the required views within the work area, and add removed section A-A.

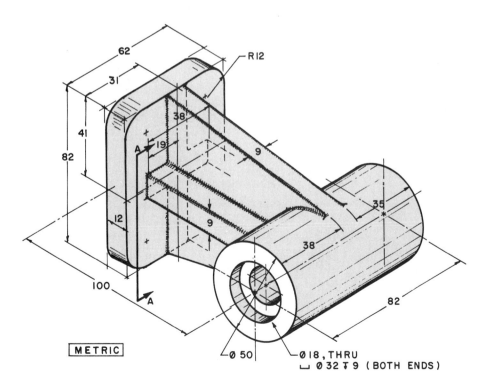

62
R12
31
38
41
A
19
9
82
35
12
9
100
38
A
82
METRIC
Ø 50
Ø18, THRU
⊔ Ø 32 ∓ 9 (BOTH ENDS)

Problem 5-19
Center the required views within the work area, and add removed section A-A.

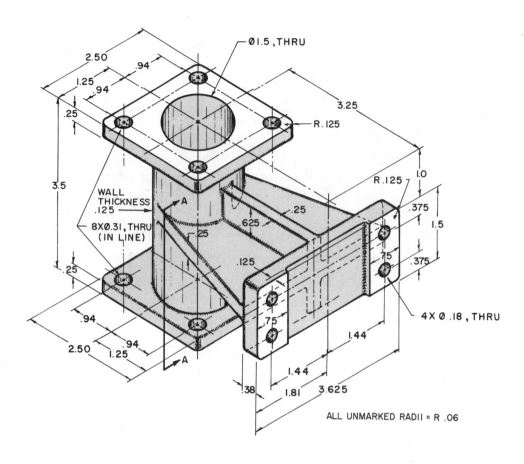

Problem 5-20
Center the required views within the work area, and add removed section A-A.

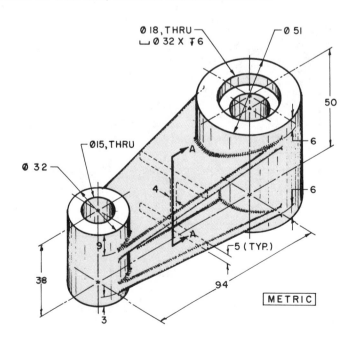

Directions for Problems 5-21 through 5-25
Center required views within the work area. Leave a 1-inch or 25-mm space between views. Make one view into a section view to fully illustrate the object. Use either a full half, offset, broken-out, revolved, or removed section. Consult your instructor if dimensions are to be added.

Problem 5-21

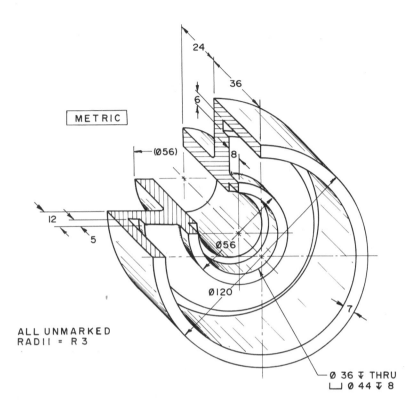

METRIC

(Ø56)

Ø56

Ø120

ALL UNMARKED
RADII = R 3

Ø 36 ⊥ THRU
⌴ Ø 44 ⊥ 8

Problem 5-22

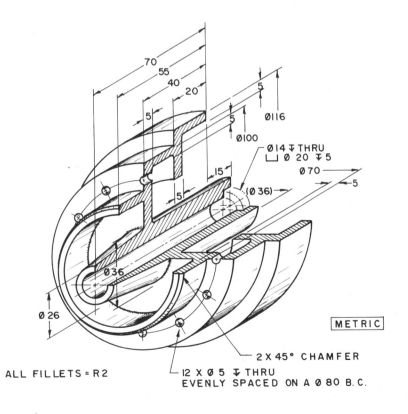

Ø116

Ø100

Ø14 ⊥ THRU
⌴ Ø 20 ⊥ 5

Ø70

(Ø 36)

METRIC

Ø36

Ø 26

2 X 45° CHAMFER

ALL FILLETS = R2

12 X Ø 5 ⊥ THRU
EVENLY SPACED ON A Ø 80 B.C.

Problem 5-23

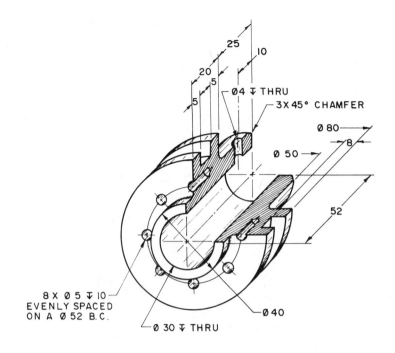

25
10
20
5
5

Ø 4 ⊽ THRU
3 X 45° CHAMFER
Ø 80
8
Ø 50
52

8 X Ø 5 ⊽ 10
EVENLY SPACED
ON A Ø 52 B.C.

Ø 40

Ø 30 ⊽ THRU

Problem 5-24

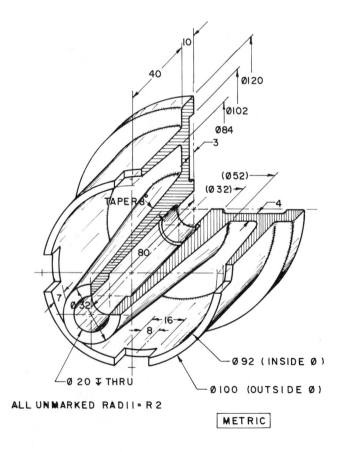

10
40
Ø120
Ø102
Ø84
3
(Ø52)
(Ø32)
4
TAPER 8°
80
7
Ø 32
16
8
Ø 92 (INSIDE Ø)
Ø 20 ⊽ THRU
Ø 100 (OUTSIDE Ø)

ALL UNMARKED RADII = R 2

METRIC

Problem 5-25

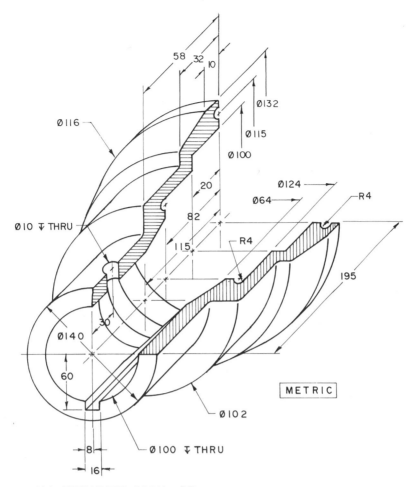

Ø116

Ø10 ↧ THRU

58 32 10

Ø132

Ø115

Ø100

20

82

115

30

Ø140

60

8

16

Ø100 ↧ THRU

Ø102

Ø124

Ø64

R4

R4

195

METRIC

ALL UNMARKED RADII = R5

Directions for Problems 5-26 through 5-27
Using the background grid as a guide, reproduce each problem on a CAD system. Add all dimensions.

Problem 5-26

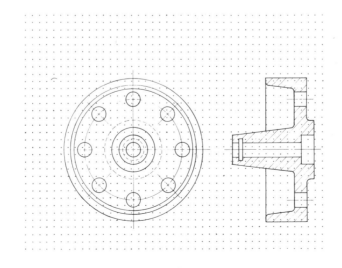

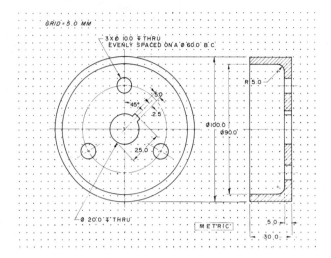

UNIT
6

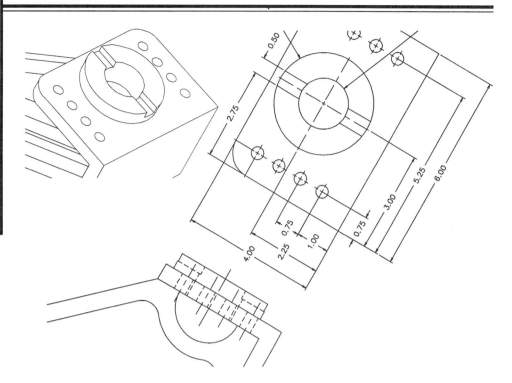

Auxiliary Views

Contents

During the discussion of multi-view drawings it was pointed out the need to draw enough views of an object in order to accurately describe it. This requires a front, top, and right side view in most cases. Sometimes additional views are required such as left side, bottom, and back views to show features not visible in the three primary views. Other special views like sections are taken to expose interior details for better clarity. Sometimes all of these views are still not enough to describe the object, especially when

features are located on an inclined surface. To produce a view perpendicular to this inclined surface, an auxiliary view is drawn. This unit will describe where auxiliary views are used and how they are projected from one view to another. A tutorial exercise is presented to go through the steps in the construction of an auxiliary view. Additional problems are provided at the end of this unit for further study of auxiliary views.

Auxiliary View Basics

The illustration at the right presents interesting results if constructed as a multi-view drawing or orthographic projection. Let us see how this object differs from others previously discussed in Unit 3.

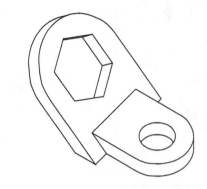

The illustration at the right should be quite familiar; it represents the standard glass box with object located in the center. The purpose again of this box is to prove how orthographic views are organized and laid out. The example at the right is no different. First the primary views, front, top, and right side views are projected from the object to intersect perpendicular with the glass plane. Under normal circumstances, this procedure would satisfy most multi-view drawing cases. Remember, only those views necessary to describe the object are drawn. However, under closer inspection, we notice the object in the front view consists of an angle forming an inclined surface.

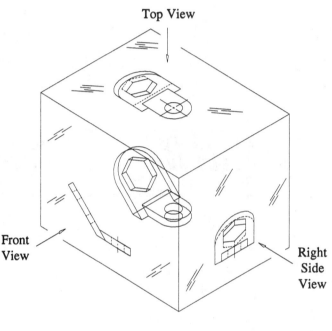

When laying out the front, top, and right side views, a problem occurs. The front view shows the basic shape of the object; the angle of the inclined surface. The top view shows true size and shape of the surface formed by the circle and arc. The right side view shows the true thickness of the hole from information found in the top view. However, there does not exist a true size and shape of the features found in the inclined surface at "A". We see the hexagonal hole going through the object in the top and right side views. These views, however, show the detail not to scale or foreshortened. For that matter, the entire inclined surface is foreshortened in all views. This is one case where the front, top, and right side views are not enough to describe the object. An additional view, or auxiliary view, is used to display the true shape of surfaces along an incline.

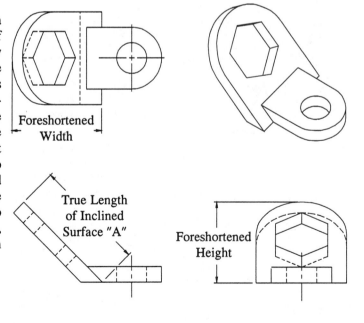

Auxiliary View Basics Continued

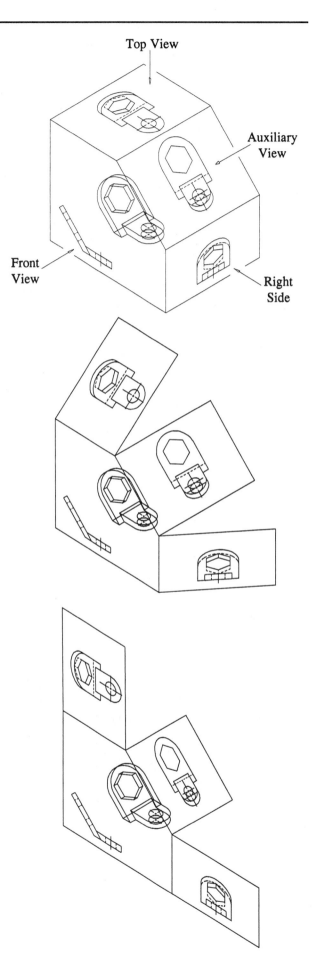

To prove the formation of an auxiliary view, lets create another glass box; this time an inclined plane is formed. This plane is always parallel to the inclined surface of the object. Instead of just projecting the front, top, and right side views, the geometry describing the features along the inclined surface is projected to the auxiliary plane similar to the illustration at the right.

As in multi-view projection, the edges of the glass box are unfolded out using the edges of the front view plane as the pivot.

All planes are extended perpendicular to the front view where the rotation stops. The result is the organization of the multi-view drawing complete with an auxiliary viewing plane.

Auxiliary View Basics Continued

Illustrated at the right is the final layout complete with auxiliary view. This example shows the auxiliary being formed as a result of the inclined surface being in the front view. An auxiliary view may be made in relation to any inclined surface located in any view. Also, the illustration displays circles and arcs in the top view which appear as ellipses in the auxiliary view. It is usually not required to draw elliptical shapes in one view where the feature is shown true size and shape in another. The resulting view minus these elliptical features is called a partial view which is used extensively in auxiliary views. An example of the top view converted into a partial view is displayed at "A".

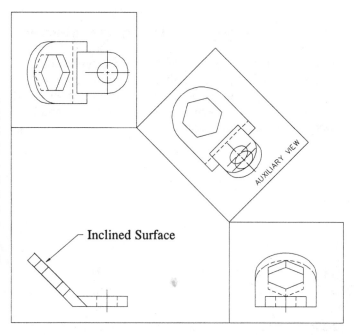

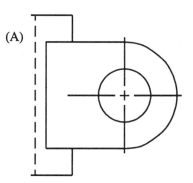

(A)

A few rules to follow when constructing auxiliary views are displayed pictorially at the right. First, the auxiliary plane is always constructed parallel to the inclined surface. Once this is established, visible as well as hidden features are projected from the incline to the auxiliary view. These projection lines are always drawn perpendicular to the inclined surface and the auxiliary view.

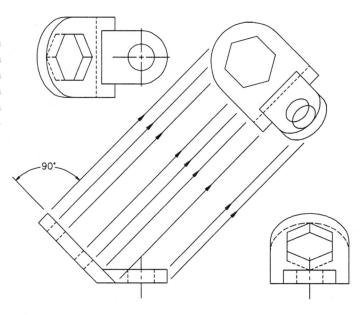

Constructing an Auxiliary View

Given at the right is a multi-view drawing consisting of front, top and right side views. The inclined surface in the front view is displayed in the top and right side views; however, the surface appears foreshortened in both adjacent views. An auxiliary view of the incline needs to be made to show its true size and shape. Currently the display screen has Grid On in addition to the position of the typical AutoCAD cursor. Follow the next series of illustrations for one suggested method for projecting to find auxiliary views.

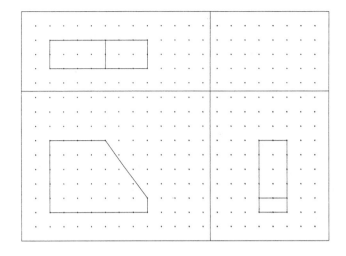

To assist with the projection process, it would help if the current grid display could be rotated parallel and perpendicular to the inclined surface. In fact, this can be accomplished using the Snap command and the prompts below:

Command: **Snap**
Snap spacing or ON/OFF/Aspect/Rotate/Style <0.50>: **Rotate**
Base point <0,0,0>: **Endp**
of *(Select the endpoint of the line at "A" as the new base point)*
Rotation angle <0>: **Endp**
of *(Select the endpoint of the line at "B" to define the rotation angle by pointing)*

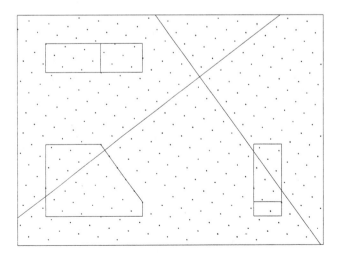

The results of rotating the grid through the Snap command are illustrated at the right. This operation has had no affect on the already existing views; however, the grid is now placed rotated in relation to the incline located in the front view. Notice the appearance of the standard AutoCAD cursor has also changed to conform to the new grid orientation. In addition to snaping to these new grid dots, lines are easily drawn perpendicular to the incline using Ortho On, which will draw lines in relation to the current cursor.

Constructing an Auxiliary View Continued

Use the Offset command to construct a reference line at a specified distance away from the incline in the front view. This reference line becomes the start for the auxiliary view.

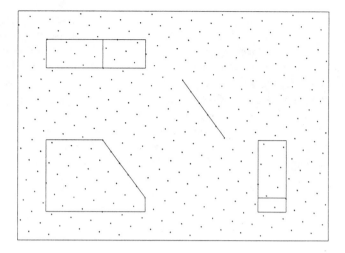

Use the Change command to extend the two endpoints of the previous line. The exact distances are not critical; however, the line should be long enough to accept projector lines from the front view.

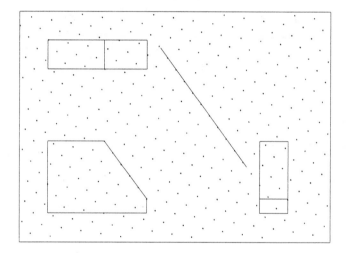

Use the Offset command to copy the auxiliary reference line the thickness of the object. This distance may be retrieved from the depth of the top or right side views since they both contain the depth measurement of the object.

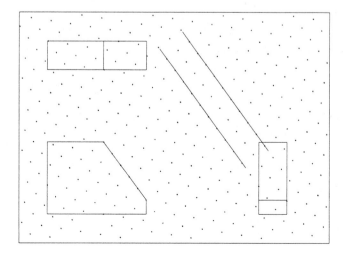

Constructing an Auxiliary View Continued

Use the Line command and connect each intersection on the front view perpendicular to the outer line on the auxiliary. Draw the lines starting with the Osnap-Intersec option and ending with the Osnap-Perpend option.

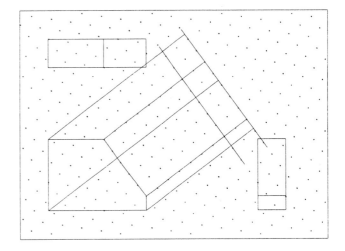

Before editing the auxiliary view, analyze the drawing to see if any corners in the front view are to be represented as hidden lines in the auxiliary view. It turns out that the lower left corner of the front view is hidden in the auxiliary view. Use the Chprop command to convert this projection line from a linetype of continuous to hidden. This is best accomplished by assigning the hidden linetype to a layer and then using Chprop to convert the line to the different layer.

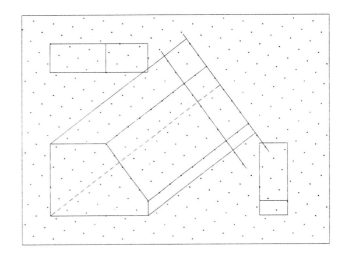

Use the Trim command to partially delete all projection lines using the inside auxiliary view line as the cutting edge.

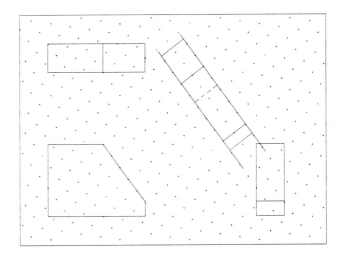

Constructing an Auxiliary View Continued

Use the Trim command to trim the corners of the auxiliary view. A better solution would be to use the Fillet command set to a radius value of 0. Selecting two lines would automatically corner the view.

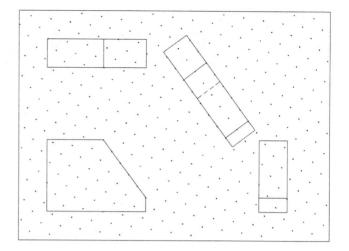

The result is a multi-view drawing complete with auxiliary view displaying the true size and shape of the inclined surface.

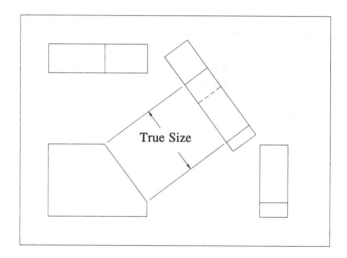

True Size

For dimensioning purposes, use the Snap command and set the grid and cursor appearance back to normal.

Command: **Snap**
Snap spacing or ON/OFF/Aspect/Rotate/Style <0.50>:
      Rotate
Base point <Current default>: **0,0,0**
Rotation angle <Current default>: **0**

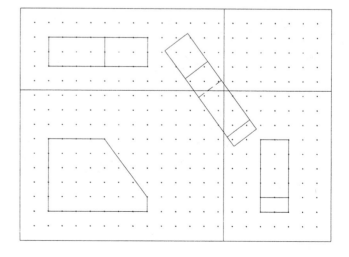

Tutorial Exercise #15
Bracket.Dwg

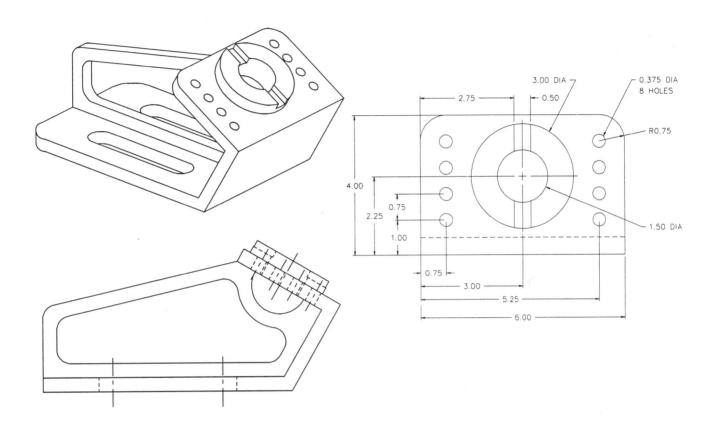

PURPOSE:
This tutorial is designed to allow the user to construct an auxiliary view of the inclined surface above in the Bracket.

SYSTEM SETTINGS:
Since this drawing is provided on diskette, edit an existing drawing called "Bracket." Follow the steps in this tutorial for the creation of an auxiliary view. All units, limits, grid, and snap values have been previously set.

LAYERS:
The following layers have already been created with the following format:

Name-Color-Linetype
Cen - Yellow - Center
Dim - Yellow - Continuous
Hid - Red - Hidden
Obj - Cyan - Continuous

SUGGESTED COMMANDS:
Begin this tutorial by using the Offset command to construct a series of lines parallel to the inclined surface containing the auxiliary view. Next construct lines perpendicular to the inclined surface. Use the circle command to begin laying out features that lie in the auxiliary view. Use Array to copy the circle in a rectangular pattern. Add center lines using the Dim-Center command. Insert a predefined view called "Top." A three-view drawing consisting of Front, Top, and Auxiliary views is completed.

DIMENSIONING:
This drawing may be dimensioned at a later date. Consult your instructor before continuing.

PLOTTING:
This tutorial exercise may be plotted on "D"-size paper (24" x 36"). Use a plotting scale of 1=1 to produce a full size plot.

VERSION OF AUTOCAD:
This tutorial exercise may be completed using either AutoCAD Release 10 or Release 11.

Step #1

Before beginning, understand that an auxiliary view will be taken from a point of view illustrated at the right. This direction of sight is always perpendicular to the inclined surface. This perpendicular direction ensures the auxiliary view of the inclined surface will be of true size and shape. Restore a previously saved view called "Front."

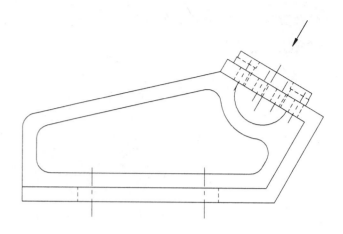

Command: **View**
?/Delete/Restore/Save/Window: **R**
View name to restore: **Front**

Step #2

Use the Snap command to rotate the grid perpendicular to the inclined surface. For the base point, identify the endpoint of the line at "A". For the rotation angle, use the "rubberband" cursor and mark a point at the endpoint of the line at "B". The grid should change along with the standard AutoCAD cursor.

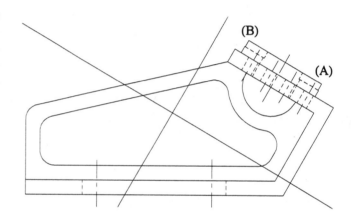

Command: **Snap**
Snap spacing or ON/OFF/Aspect/Rotate/Style <0.50>:
    Rotate
Base point <0,0>: **Endp**
of *(Select the endpoint of the line at "A")*
Rotation angle <0>: **Endp**
of *(Select the endpoint of the line at "B")*

Command: **View**
?/Delete/Restore/Save/Window: **R**
View name to restore: **Overall**

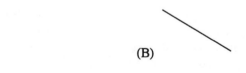

Step #3

Turn snap Off by striking the F9 function key. Begin the construction of the auxiliary view by using the Offset command to copy a line parallel to the inclined line. Use an offset distance of 8.50 as the distance between the front and auxiliary view.

Command: **Offset**
Offset distance or Through <Through>: **8.50**
Select object to offset: *(Select the inclined line at "A")*
Side to offset? *(Select a point anywhere near "B")*
Select object to offset: *(Strike Enter to exit this command)*

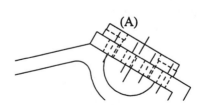

Step #4

Refer to the working drawing on page 251 for the necessary dimensions required to construct the auxiliary view. Use the Offset command again to add the depth of the auxiliary. Remember, the depth of the auxiliary view is the same dimension found in the top and right side views. Set the offset distance to 6.00. Set the new current layer to "Obj."

Command: **Offset**
Offset distance or Through <8.50>: **6.00**
Select object to offset: *(Select the inclined line at "A")*
Side to offset? *(Select a point anywhere near "B")*
Select object to offset: *(Strike Enter to exit this command)*

Command: **Layer**
?/Make/Set/New/ON/OFF/Color/Ltype/Freeze/Thaw: **Set**
New current layer <0>: **Obj**
?/Make/Set/New/ON/OFF/Color/Ltype/Freeze/Thaw:
    (Strike Enter to exit this command)

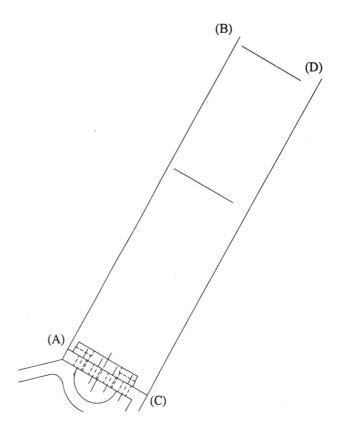

Step #5

Project two lines from the endpoints of the front view at "A" and "B". These lines should extend past the outer line of the auxiliary view. Turn the Snap Off and Ortho On. This should aid in this operation.

Command: **Ortho**
ON/OFF <off>: **On**

Command: **Line**
From point: **Endp**
of *(Pick the endpoint of the line at "A")*
To point: *(Pick a point anywhere near "B")*
To point: *(Strike Enter to exit this command)*

Command: **Line**
From point: **Endp**
of *(Pick the endpoint of the line at "C")*
To point: *(Pick a point anywhere near "D")*
To point: *(Strike Enter to exit this command)*

Step #6

Use the Zoom-Window option to magnify the display of the auxiliary view similar to the illustration at the right. Then use the Extend command, select the boundary edges at "A" and "B", and extend the four endpoints of the lines.

Command: **Extend**
Select boundary edge(s)...
Select objects: *(Select the two lines at "A" and "B")*
Select objects: *(Strike Enter to continue)*
Select object to extend: *(Select the end of the line at "C")*
Select object to extend: *(Select the end of the line at "D")*
Select object to extend: *(Select the end of the line at "E")*
Select object to extend: *(Select the end of the line at "F")*
Select object to extend: *(Strike Enter to exit this command)*

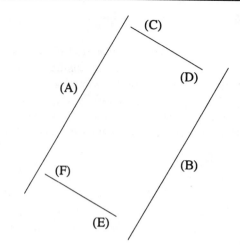

Step #7

Use the Trim command, select the lines at "A" and "B" as cutting edges, and trim away the ends of the four lines labeled at the right. While in this display, use the View command to save as "Aux" for future reference.

Command: **Trim**
Select cutting edge(s)...
Select objects: *(Select the two lines at "A" and "B")*
Select objects: *(Strike Enter to continue)*
Select object to trim: *(Select the line at "C")*
Select object to trim: *(Select the line at "D")*
Select object to trim: *(Select the line at "E")*
Select object to trim: *(Select the line at "F")*
Select object to trim: *(Strike Enter to exit this command)*

Command: **View**
?/Delete/Restore/Save/Window: **Save**
View name to save: **Aux**

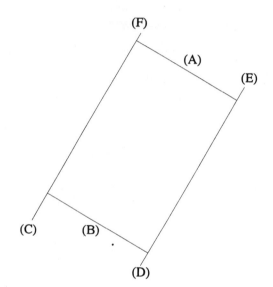

Step #8

Use the Zoom-Previous option to demagnify the screen back to the original display. Once here, draw a line from the endpoint of the center line in the front view to a point past the auxiliary view. Check to see that Ortho mode is On. This line will assist in constructing circles in the auxiliary view.

Command: **Line**
From point: **Endp**
of *(Select the endpoint of the center line at "A")*
To point: *(Select a point anywhere near "B")*
To point: *(Strike Enter to exit this command)*

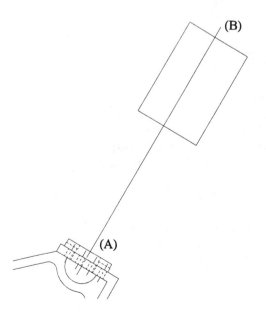

Step #9

Use the Offset command and offset the line at "A" a distance of 3.00 units. The intersection of this line and the previous line form the center for placing two circles.

Command: **Offset**
Offset distance or Through <6.00>: **3.00**
Select object to offset: *(Select the line at "A")*
Side to offset? *(Select a point anywhere near "B")*
Select object to offset: *(Strike Enter to exit this command)*

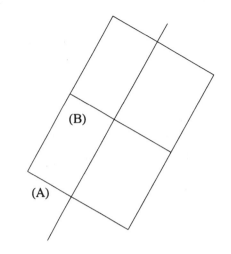

Step #10

Use the View command and restore the view "Aux". Then, draw two circles of diameters 3.00 and 1.50 from the center at "A" using the Circle command. For the center of the second circle, the @ option may be used to pick up the previous point that was the center of the 3.00 diameter circle.

Command: **View**
?/Delete/Restore/Save/Window: **Restore**
View name: **Aux**

Command: **Circle**
3P/2P/TTR/<Center point>: **Int**
of *(Select the intersection of the two lines at "A")*
Diameter/<Radius>: **D**
Diameter: **3.00**

Command: **Circle**
3P/2P/TTR/<Center point>: **@** *(To reference the last point)*
Diameter/<Radius>: **D**
Diameter: **1.50**

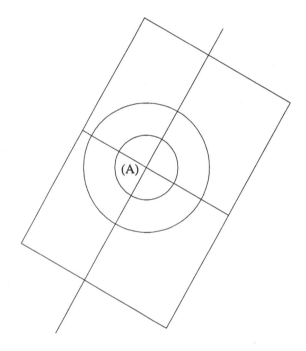

Step #11

Use the Offset command to offset the center line the distance of 0.25. Perform this operation on both sides of the center line. Both offset lines form the width of the 0.50 slot.

Command: **Offset**
Offset distance or Through <3.00>: **0.25**
Select object to offset: *(Select the middle line at "A")*
Side to offset? *(Select a point anywhere near "B")*
Select object to offset: *(Select the middle line at "A" again)*
Side to offset? *(Select a point anywhere near "C")*
Select object to offset: *(Strike Enter to exit this command)*

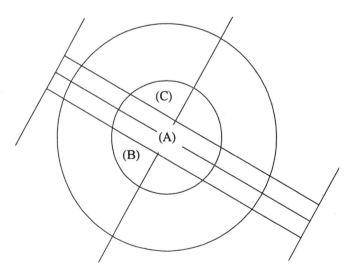

Step #12

Use the Trim command to trim away portions of the lines illustrated at the right.

Command: **Trim**
Select cutting edge(s)...
Select objects: *(Select both circles as cutting edges)*
Select objects: *(Strike Enter to continue)*
Select object to trim: *(Select the line at "A")*
Select object to trim: *(Select the line at "B")*
Select object to trim: *(Select the line at "C")*
Select object to trim: *(Select the line at "D")*
Select object to trim: *(Select the line at "E")*
Select object to trim: *(Select the line at "F")*
Select object to trim: *(Strike Enter to exit this command)*

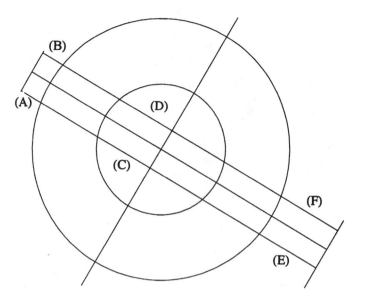

Step #13

Use the Erase command to delete the two lines at "A" and "B". Standard center lines will be placed here later marking the center of both circles.

Command: **Erase**
Select objects: *(Select the lines at "A" and "B")*
Select objects: *(Strike Enter to execute this command)*

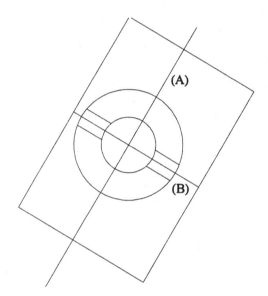

Step #14

To identify the center of the small 0.375 diameter circle, use the Offset command to copy parallel the line at "A" a distance of 0.75 units and the line at "C" the distance of 1.00 units.

Command: **Offset**
Offset distance or Through <0.25>: **0.75**
Select object to offset: *(Select the line at "A")*
Side to offset? *(Select a point anywhere near "B")*
Select object to offset: *(Strike Enter to exit this command)*

Command: **Offset**
Offset distance or Through <0.75>: **1.00**
Select object to offset: *(Select the line at "C")*
Side to offset? *(Select a point anywhere near "D")*
Select object to offset: *(Strike Enter to exit this command)*

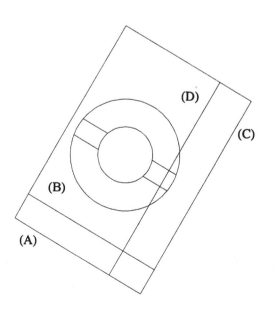

Step #15

Draw a circle of 0.375 diameter from the intersection of the two lines created in the last offset command. Use the Erase command to delete the two lines at "A" and "B". A standard center marker will be placed at the center of this circle.

Command: **Circle**
3P/2P/TTR/<Center point>: **Int**
of *(Select the intersection of the two lines at "A")*
Diameter/<Radius>: **D**
Diameter: **0.375**

Command: **Erase**
Select objects: *(Select the two lines at "B" and "C")*
Select objects: *(Strike Enter to execute this command)*

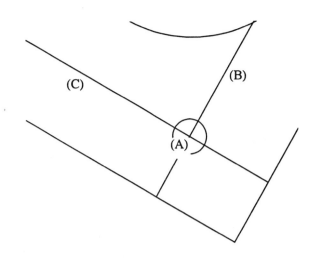

Step #16

Set the new current layer to "Cen." Prepare the following parameters before placing a center marker at the center of the 0.375 diameter circle. Set the dimension variable Dimcen to a value of -0.07. The negative value will construct lines that are drawn outside of the circle. Use the Dim-Cen command to place the center marker.

Command. **Layer**
?/Make/Set/New/ON/OFF/Color/Ltype/Freeze/Thaw: **Set**
New current layer <OBJ>: **Cen**
?/Make/Set/New/ON/OFF/Color/Ltype/Freeze/Thaw:
    (Strike Enter to exit this command)

Command: **Dim**
Dim: **Dimcen**
Current value <0.09> New value: **-0.07**

Dim: **Cen**
Select arc or circle: *(Select the small circle at "A")*

Dim: **Exit** *(To exit dimensioning and return to the Command prompt)*

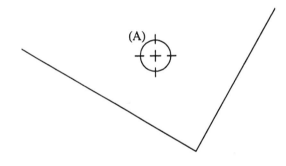

Step #17

Use the Rotate command to rotate the center marker parallel to the edges of the auxiliary view. Select the center marker and circle as the entities to rotate. Check to see that Ortho is On.

Command: **Rotate**
Select objects: *(Select the small circle and all entities that make up the center marker; the Window option is recommended here)*
Select objects: *(Strike Enter to continue)*
Base point: **Cen**
of *(Select the edge of the small circle at "A")*
<Rotation angle>/Reference: *(Pick a point anywhere at "B")*

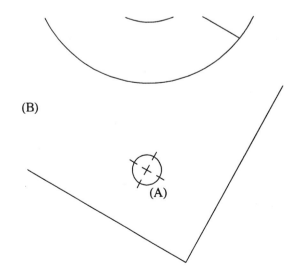

Step #18

Since the remaining seven holes are along a set pattern, use the Array command and perform a rectangular array. The number of rows are 2 and number of columns 4. Distance between rows is 4.50 units and between columns is -0.75 units; this will force the circles to be patterned to the left which is where we want them to go.

Command: **Array**
Select objects: *(Select the small circle and center marker)*
Select objects: *(Strike Enter to continue)*
Rectangular or Polar array (R/P): **R**
Number of rows(---) <1>: **2**
Number of columns(|||) <1>: **4**
Unit cell or distance between rows (---): **4.50**
Distance between columns (|||): **-0.75**

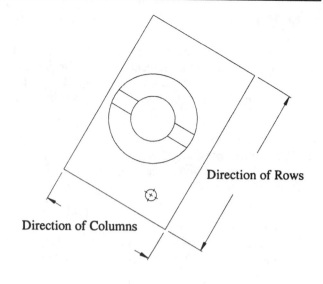

Direction of Rows

Direction of Columns

Step #19

Use the Fillet command set to a radius of 0.75 to place a radius along the two corners of the auxiliary following the prompts and the illustration below.

Command: **Fillet**
Polyline/Radius/<Select two objects>: **R**
Enter fillet radius <0.00>: **0.75**

Command: **Fillet**
Polyline/Radius/<Select two objects>: *(Select lines "A" and "B")*

Command: **Fillet**
Polyline/Radius/<Select two objects>: *(Select lines "B" and "C")*

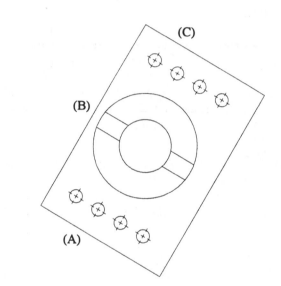

(C)

(B)

(A)

Step #20

Place a center marker in the center of the two large circles using the existing value of the dimension variable Dimcen. Since the center marker is placed in relation to the World coordinate system, use the Rotate command to rotate it parallel to the auxiliary view. Ortho should be on.

Command: **Dim**
Dim: **Cen**
Select arc or circle: *(Select the large circle at "A")*
Dim: **Exit**

Command: **Rotate**
Select objects: *(Select all lines that make up the large center marker)*
Select objects: *(Strike Enter to continue)*
Base point: **Cen**
of *(Select the edge of the large circle at "A")*
<Rotation angle>/Reference: *(Pick a point anywhere at "B")*

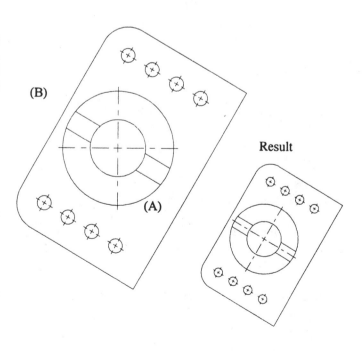

(B)

(A)

Result

Step #21

Restore the view named "Overall." Complete the multi-view drawing of the Bracket by inserting an existing block called "Top" into this drawing. This block represents the complete top view of the drawing. Use an insertion point of −3.0325,0 for this insertion.

Command: **View**
?/Delete/Restore/Save/Window: **R**
View name to restore: **Overall**

Command: **Insert**
Block name (or ?): **Top**
Insertion point: **−3.0325,0**
X scale factor <1>/Corner/XYZ: *(Strike Enter to accept default)*
Y scale factor (default=X): *(Strike Enter to accept default)*
Rotation angle <0>: *(Strike Enter to accept this default)*

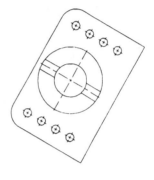

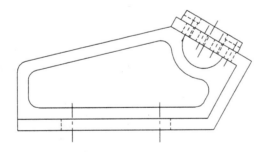

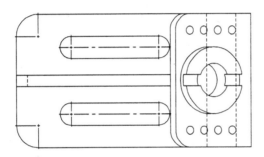

Step #22

Return the grid back to its original orthographic form using the Snap-Rotate option. Use a base point of 0,0 and a rotation angle of 0 degrees. This is especially helpful when adding dimensions to the drawing.

Command: **Snap**
Snap spacing or ON/OFF/Aspect/Rotate/Style <0.50>: **Rotate**
Base point <0,0>: **0,0**
Rotation angle <330>: **0**

Problems for Unit 6

Directions for Problems 6–1 through 6–5
Draw the front view, top view, right-side view and auxiliary view.

Problem 6-1

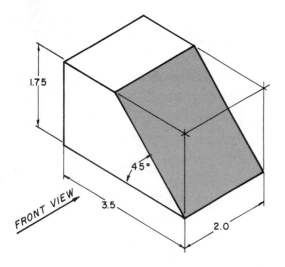

Problem 6-3

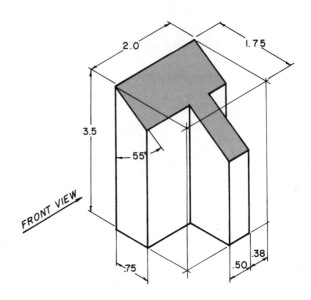

Problem 6-2

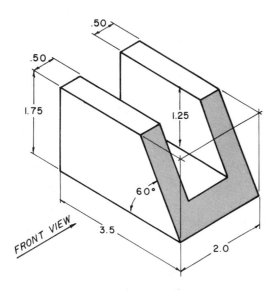

Problem 6-4

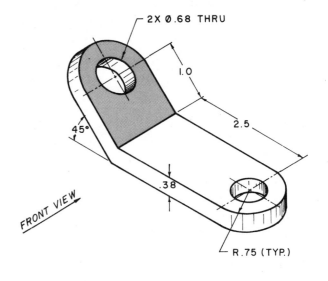

Problem 6-5

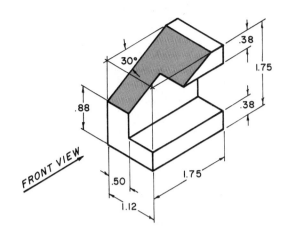

Directions for Problems 6-6 through 6-9
Draw the required views to fully illustrate each object. Be sure to include an auxiliary view.

Problem 6-6

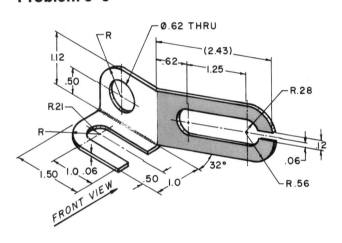

Problem 6-7

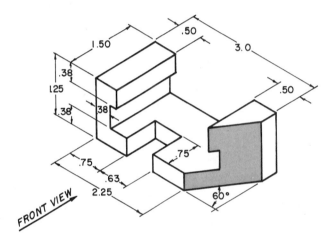

Problem 6-8

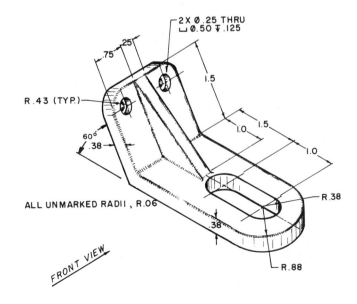

Problem 6-9

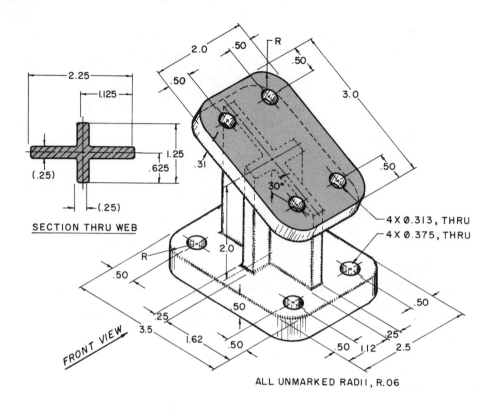

2.25
1.125
1.25
.625
(.25)
(.25)

SECTION THRU WEB

2.0 .50 R .50
.50 3.0
.31 .50
30°
FRONT VIEW
R
.50
.25
3.5 1.62 .50 .50
2.0
.50
.50
.25
.50 1.12 2.5

4 X Ø.313, THRU
4 X Ø.375, THRU

ALL UNMARKED RADII, R.06

Directions for Problems 6-10
Using the background grid as a guide, reproduce this problem on a CAD system.

Problem 6-10

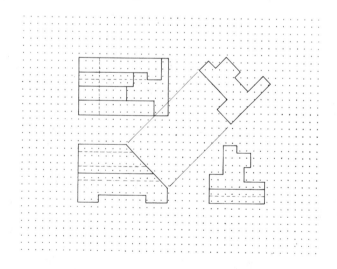

Problem 6-11

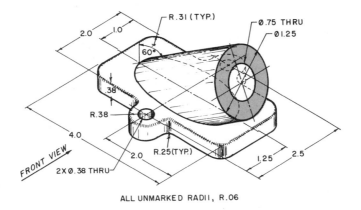

R.31 (TYP.)
Ø.75 THRU
Ø1.25
2.0
1.0
60°
.38
R.38
4.0
2.0
R.25(TYP.)
1.25
2.5
FRONT VIEW
2X Ø.38 THRU

ALL UNMARKED RADII, R.06

Problem 6-12

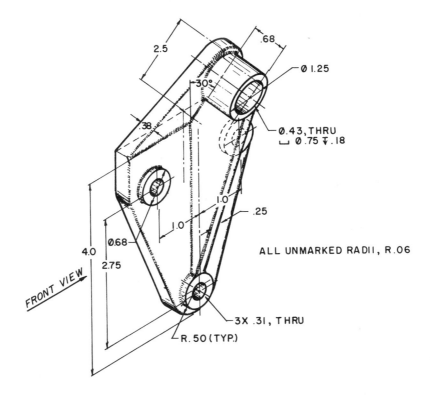

2.5
.68
Ø 1.25
30°
.38
Ø.43, THRU
⌴ Ø .75 ⍖ .18
1.0
.25
Ø.68
1.0
ALL UNMARKED RADII, R.06
4.0
2.75
FRONT VIEW
3X .31, THRU
R.50 (TYP.)

Problem 6-13

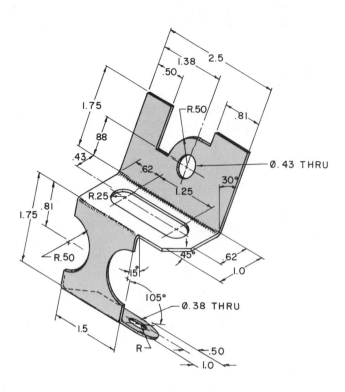

Problem 6-14

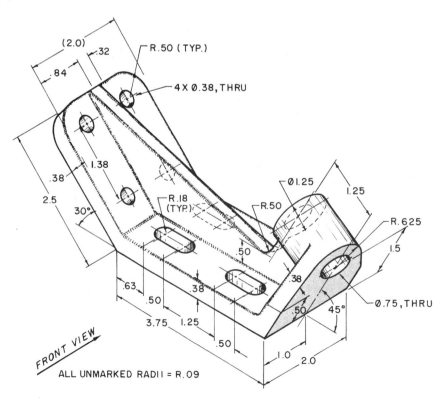

UNIT
7

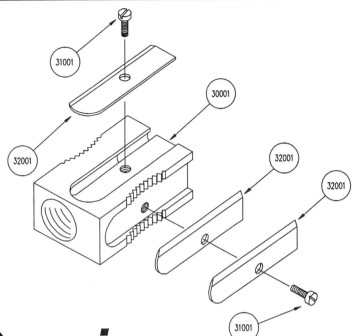

Isometric Drawings

Contents

Multi-view or orthographic projections are necessary to produce parts which go into the construction of all kinds of objects. Skill is involved in laying out the primary views, projecting visible entities into other views, and adding dimensions to describe the size of the object being made. Yet another skill involves reading or interpreting these engineering drawings, which for some individuals, is extremely difficult and complex. If only there existed some type of picture of the object; then the engineering drawing might make sense. Isometric drawings become a means of drawing an object in picture form for better clarification of what the object looks like. These types of drawings resemble a picture of an object that is drawn in two dimensions. As a result, existing AutoCAD commands such as Line and Copy are used for producing isometric drawings. This unit will explain isometric basics including how regular, angular, and circular entities are drawn in isometric. Numerous isometric aids such as snap and isometric axes will be explained to assist in the construction of isometric drawings.

Isometric Basics

Isometric drawings consist of 2D drawings that are tilted at some angle to expose other views and give the viewer the illusion that what he or she is viewing is a 3D drawing. The tilting occurs with two 30-degree angles that are struck from the intersection of a horizontal baseline and a vertical line. The directions formed by the 30-degree angles represent actual dimensions of the object; this may be either the width or depth. The vertical line in most cases represents the height dimension.

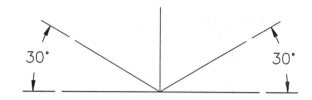

The object illustrated at the right is a very simple example of how an object is aligned to the isometric axis. Once the horizontal baseline and vertical line are drawn, the 30-degree angles are projected from this common point which becomes the reference point of the isometric view. In this example, once the 30-degree lines are drawn, the baseline is no longer needed and is usually discarded through erasing. Depending on how the object is to be viewed, width and depth measurements are made along the 30-degree lines. Height is measured along the vertical line. The example at the right has the width dimension measured off to the left 30-degree line while the depth dimension measures to the right along the right 30-degree line. Once the object is blocked with overall width, depth, and height, details are added, and lines are erased and trimmed, leaving the finished object. Holes no longer appear as full circles but rather as ellipses. Techniques of drawing circles in isometric will be discussed later in this unit.

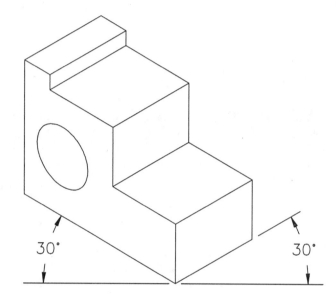

Notice the objects at the right both resemble the previous example except they appear from a different vantage point. The problem with isometric drawings is that if an isometric of an object is drawn from one viewing point and you want an isometric from another viewing point, an entirely different isometric drawing must be generated from scratch. Complex isometric drawings from different views can be very tedious to draw. Another interesting observation concerning the objects at the right is one has hidden lines while the other does not. Usually only the visible surfaces of an object are drawn in isometric leaving out hidden lines. As this is considered good practice, there are always times that hidden lines are needed on very complex isometric drawings.

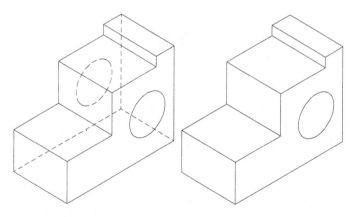

Creating an Isometric Grid

The display illustrated at the right shows the current AutoCAD screen complete with cursor and grid on. In manual drawing and sketching days, an isometric grid was used to lay out all lines before transferring the lines to paper or mylar for pen and ink drawings. An isometric grid may be defined in an AutoCAD drawing using the Snap command. This would be the same grid found on isometric grid paper.

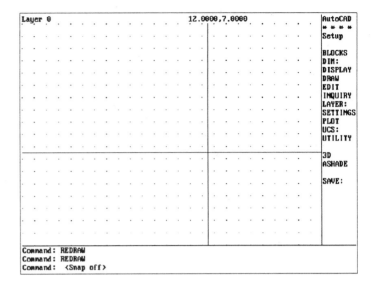

The display screen at the right reflects the use of the Snap command and how this command affects the current grid display:

Command: **Snap**
Snap spacing or ON/OFF/Rotate/Style <0.2500>: **Style**
Standard/Isometric <S>: **Isometric**
Vertical spacing <0.2500>: *(Strike Enter to accept default value)*

Choosing an isometric style of snap changes the grid display from orthographic to isometric, illustrated at the right. The grid distance conforms to a vertical spacing height specified by the user. As the grid changes, notice the display of the typical AutoCAD cursor; it conforms to an isometric axis plane and is used as an aid in constructing isometric drawings.

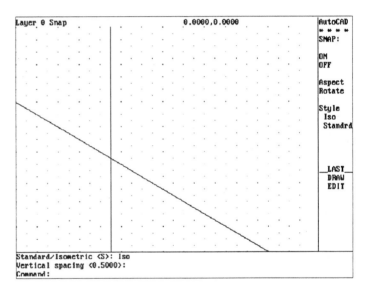

To see how this vertical spacing distance affects the grid changing it to isometric, see the illustration at the right. The grid dot at "A" becomes the reference point where the horizontal baseline is placed followed by the vertical line represented by the dot at "B". At dots "A" and "B", 30-degree lines are drawn; points "C" and "D" are formed where they intersect. This is how an isometric screen display is formed.

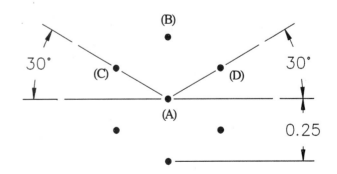

Isoplane Modes

The AutoCAD cursor has always been the vehicle for drawing entities or constructing windows for object selection mode. Once in isometric snap mode, AutoCAD supports three axes to assist in the construction of isometric drawings. This first axis is the Left axis and may control that part of an object falling into the left projection plane. The left axis cursor displays a vertical line intersected by a 30-degree angle line which is drawn to the left. This axis is displayed in the illustration at the right in addition to the drawing below:

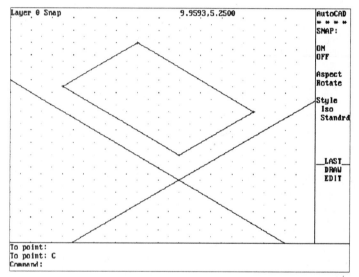

The next isometric axis is the Top mode. Entities falling into the top projection plane may be drawn using this isometric axis. This cursor consists of two 30-degree angle lines intersecting each other forming the center of the cursor. This mode is displayed at the right and below:

The final isometric axis is called the Right mode and is formed by the intersection of a vertical line and a 30-degree angle drawn off to the right. As with the previous two modes, entities that fall along the right projection plane of an isometric drawing may be drawn using this cursor. It is displayed at the right and below. The current Ortho mode affects all three modes. If Ortho is On, and the current isometric axis is Right, lines and other operations requiring direction will be forced to be drawn vertical or at a 30-degree angle to the right as shown below:

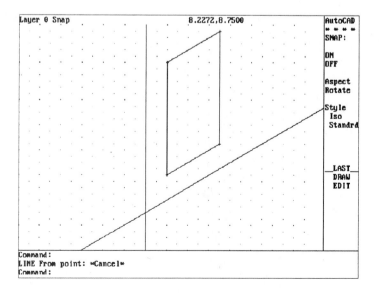

Changing the Grid, Snap, and Cursor Back to their Defaults

Once an isometric drawing is completed, it may be necessary to change the grid, snap, and cursor back to normal. This might result from the need to place text on the drawing and the isometric axis now confuses instead of assists the drawing process. Follow the prompts below for reseting the Snap back to Standard.

Command: **Snap**
Snap spacing or ON/OFF/Rotate/Style <0.2500>: **Style**
Standard/Isometric <I>: **Standard**
Spacing/Aspect <0.2500>: *(Strike Enter to accept default value)*

Notice when changing the snap style to Standard, the AutoCAD cursor changes back to its original display.

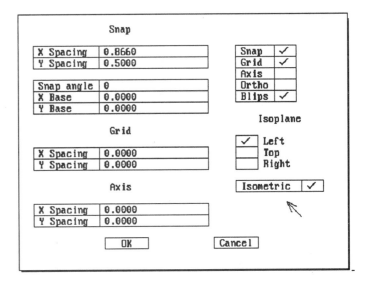

Isoplane Aids

It has been previously mentioned that there are three isometric axis modes to draw on; Left, Top, and Right. It was not mentioned how to make a mode current to draw on. The Settings area of the menu bar exposes the Drawing Aids area where the dynamic dialog box at the right is located. Here, in addition to Snap, Grid, and Axis settings, the three isometric modes may be made current by simply placing a check in the appropriate box. Only one isometric mode may be current at a time. In addition to setting these modes, an isometric area exists to automatically set up an isometric grid by placing a check in the box and put it back to normal by removing the check. This has the same affect as using the Snap-Style-Isometric option. This dialog box may be brought up transparently through the keyboard by entering "'DDRMODES".

A quicker method exists to move from one isometric axis mode to another. By default, after setting up an isometric grid, the Left isometric axis is active. By typing from the keyboard CTRL-E or ^E, the Left axis changes to the Top axis. Typing another ^E changes from the Top axis to the Right axis. Typing a third ^E changes from the Right axis back to the Left axis and the pattern repeats from here. Using this keyboard entry, it is possible to switch or toggle from one mode to another.

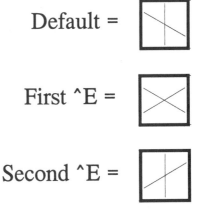

Creating Isometric Circles

Circles appear as ellipses when drawn in any of the three isometric axes. The Ellipse command has a special Isocircle option to assist in drawing isometric circles; the Isocircle option will appear in the Ellipse command only if the current Snap-Style is Isometric. The prompt sequence for this command is:

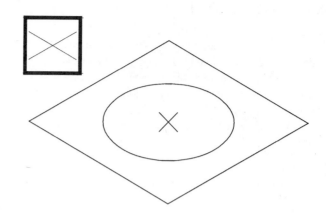

Command: **Ellipse**
<Axis endpoint 1>/Center/Isocircle: **Iso**
Center of circle: *(Select a center point)*
<Circle Radius>/Diameter: *(Enter a value for the radius or type "D" for diameter and enter a value)*

When drawing isometric circles using the Ellipse command, it is important to match the isometric axis with the isometric plane the circle is to be drawn in. Illustrated at the right is a cube displaying all three axes.

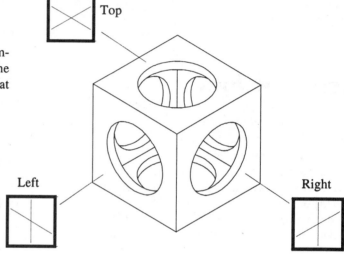

The example at the right is the result of drawing an isometric circle using the wrong isometric axis. The isometric box is drawn in the top isometric plane while the current isometric axis is Left. An isometric circle can be drawn to the correct size; but notice it does not match the box it was designed for. If you notice that halfway through the Ellipse command you are in the wrong isometric axis, type ^E until the correct axis appears.

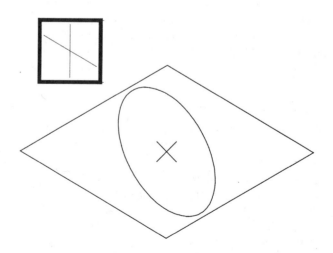

Basic Isometric Construction

Any isometric drawing, no matter how simple or complex, has an overall width, height, and depth dimension. Start laying out the drawing with these three dimensions to create an isometric box illustrated in the example at the right. Some techniques rely on piecing the isometric drawing together by views; unfortunately, it is very easy to get lost in all of the lines using this method. Once a box is created from overall dimensions, somewhere inside the box is the object.

With the box as a guide, begin laying out all visible features in the primary planes. Use the Left, Top, or Right isometric axis modes to assist you in this construction process.

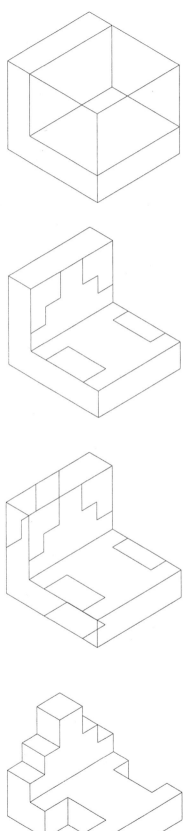

Existing AutoCAD editing commands, especially Copy, may be used to duplicate geometry to show depth of features. Next, use the Trim command to partially delete geometry where entities are not visible. Remember, most isometric objects do not require hidden lines.

Use the Line command to connect intersections of surface corners. The resulting isometric drawing is illustrated at the right.

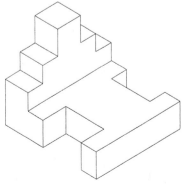

Creating Angles in Isometric - Method #1

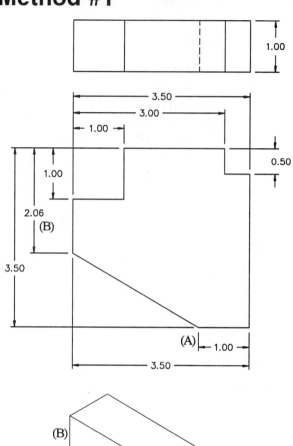

Drawing angles in isometric is a little tricky but not impossible. The two-view drawing at the right has an angle of unknown size; however, one endpoint of the angle measures 2.06 units from the top horizontal line of the front view at "B" and the other endpoint measures 1.00 unit from the vertical line of the front view at "A". This is more than enough information needed to lay out the endpoints of the angle in isometric. The Measure command can be used to easily lay out these distances. The Line command is then used to connect the points to form the angle.

Before using the Measure command, set the Pdmode system variable to a new value of 3. Points will appear as an X instead of a dot. Now use the Measure command to set off the two distances.

Command: **Setvar**
Variable name or ?: **Pdmode**
New value for Pdmode <0>: **3**

Command: **Measure**
Select object to measure: *(Select the inclined line at "A")*
<Segment length>/Block: **1.00**

Command: **Measure**
Select object to measure: *(Select the vertical line at "B")*
<Segment length>/Block: **2.06**

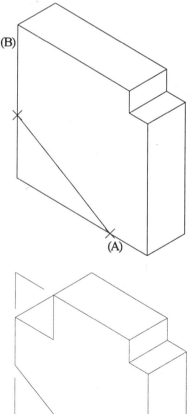

The interesting part of the Measure command is that measuring will occur at the nearest endpoint of the line where the line was selected from. It is therefore important which endpoint of the line is selected. Once the points have been placed, the Line command is used to draw a line from one point to the other using the Osnap-Node option.

Creating Angles in Isometric - Method #2

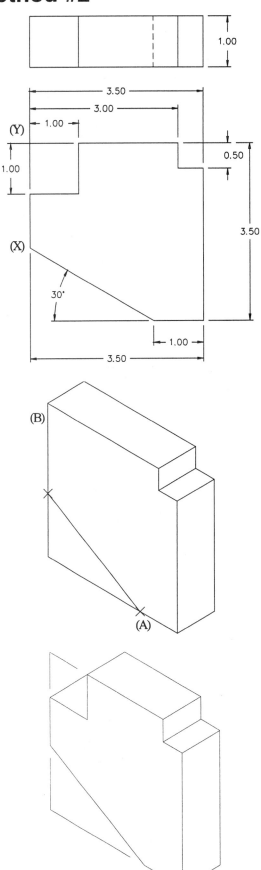

The exact same two-view drawing is illustrated at the right. This time, one distance is specified along with an angle of 30 degrees. Even with the angle given, the position of the isometric axes makes any angle construction by degrees inaccurate. The distance XY is still needed to construct the angle in isometric. The Measure command is used to find distance XY, place a point, and connect the first distance with the second to form the 30-degree angle in isometric. It is always best to set the Pdmode system variable to a new value in order to visibly view the point. A new value of 3 will assign the point as an X.

Command: **Setvar**
Variable name or ?: **Pdmode**
New value for Pdmode <0>: **3**

Command: **Measure**
Select object to measure: *(Select the inclined line at "A")*
<Segment length>/Block: **1.00**

Command: **Measure**
Select object to measure: *(Select the vertical line at "B")*
<Segment length>/Block: **Endp**
of *(Select the endpoint of the line at X in the two-view drawing)*
Second point: **Int**
of *(Select the intersection at Y in the two-view drawing)*

Line "B" is selected as the object to measure. Since this distance is unknown, the Measure command may be used to set off the distance XY by identifying an endpoint and intersection from the front view above. This means the view must be constructed only enough to lay out the angle and project the results to the isometric using the Measure command and the prompts above.

Isometric Construction Using Ellipses

Constructing circles as part of isometric drawing is possible using one of the three isometric axes positions. It is up to the operator to decide which axis to use. Before this, however, an isometric box consisting of overall distances is first constructed. Use the Ellipse command to place the isometric circle at the base. To select the correct axis type CTRL-E until the proper axis appears in the form of the cursor. Place the ellipses. Lay out any other distances.

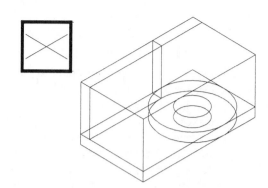

For ellipses at different positions type CTRL-E to select another isometric axis. Remember these axis positions may be selected from the Drawing Aids dialog box from the Settings area of the menu bar.

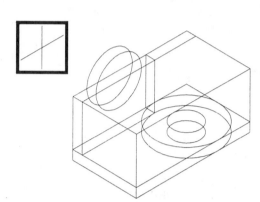

Use the Trim command to trim away any excess entities that are considered unnecessary.

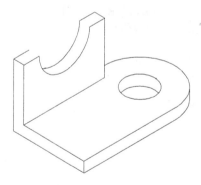

Use the Line command to connect endpoints of edges that form surfaces.

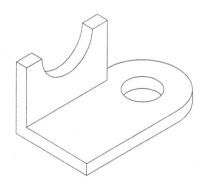

Creating Isometric Sections

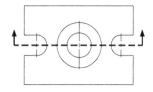

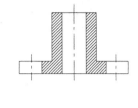

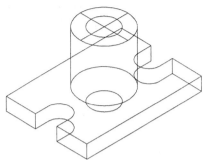

In some cases, it is necessary to cut an isometric drawing to expose internal features. The result is an isometric section similar to a section view formed from an orthographic drawing. The difference with the isometric section, however, is the cutting plane is usually along one of the three isometric axes. The illustration at the right displays an orthographic section in addition to the isometric drawing.

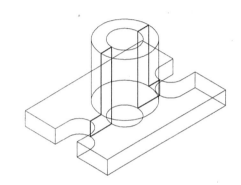

The isometric at the right has additional lines representing surfaces cut by the cutting plane line. The lines to define these surfaces were formed using the Line command in addition to a combination of top and right isometric axes modes. During this process, Ortho mode was toggled On and Off numerous times depending on the axis direction. The Break command was used to break ellipses and lines at their intersection points.

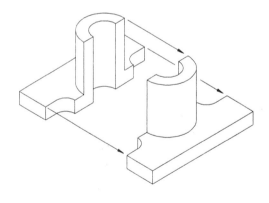

Once ellipses were broken, the front half of the isometric was removed exposing the back half as in the example at the right. The front half is then discarded. This has the same affect as conventional section views where the direction of site dictates which half to keep.

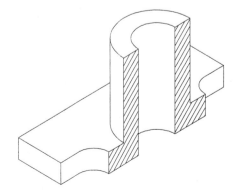

For a full section, the Hatch command is used to crosshatch the surfaces being cut by the cutting plane line. Surfaces designated holes or slots not cut are not crosshatched. This same procedure is followed for converting an object into a half section.

Exploded Isometric Views

Isometric drawings are sometimes grouped together to form an exploded drawing of how a potential or existing product is assembled. This involves aligning parts that fit with line segments, usually in the form of center lines. Bubbles identifying the part number are attached to the drawing. Exploded isometric drawings come in handy for creating bill of material information and for this purpose have an important application to manufacturing. Once the part information is identified in the drawing and title block area, this information is extracted and brought into a third-party business package where important data collection information is able to actually track the status of parts in production in addition to the shipping date for all finished products.

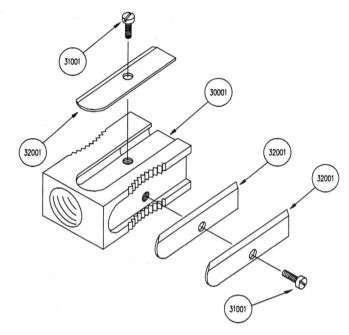

Isometric Assemblies

Assembly drawings show the completed part as if it were to be assembled. Sometimes this drawing has an identifying number placed with a bubble for bill of material needs. Assembly drawings commonly are placed on the same display screen as the working drawing. With the assembly along the side of the working drawing, the user has a pictorial representation of what the final product will look like and can aid in the understanding of the orthographic views.

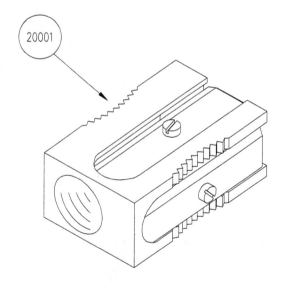

Tutorial Exercise #16
Plate.Dwg

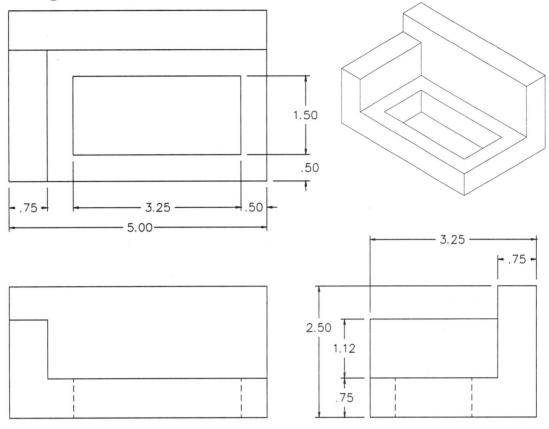

PURPOSE:
The purpose of this tutorial exercise is to use a series of coordinates along with AutoCAD editing commands to construct an isometric drawing of the Plate.

SYSTEM SETTINGS:
Begin a new drawing called "Plate." Use the Units command to change the number of decimal places past the zero from 4 to 2. Keep the remaining default unit values. Using the Limits command, keep 0,0 for the lower left corner and change the upper right corner from 12,9 to 10.50,8.00. Use the Grid command and change the grid spacing from 1.00 to 0.25 units. Do not turn the snap or ortho On.

LAYERS:
Special layers do not have to be created for this tutorial exercise.

SUGGESTED COMMANDS:
Begin this exercise by changing the grid from the standard display to an isometric display using the Snap-Style option. Remember both the grid and snap can be manipulated by the Snap command only if the current grid value is 0. Use Absolute and Polar coordinates to lay out the base of the Plate. Then begin using the Copy command followed by Trim to duplicate entities and clean up or trim unnecessary geometry.

DIMENSIONING:
This object may be dimensioned at a later date using the Dim-Oblique option found only in AutoCAD Release 11. Consult your instructor before continuing.

PLOTTING:
This tutorial exercise may be plotted on "A"-size paper (8.5" x 11"). Use a plotting scale of 1=1 to produce a full size plot.

VERSION OF AUTOCAD:
This tutorial exercise may be completed using either AutoCAD Release 10 or Release 11.

Step #1

Use the Line command to draw the figure at the right.

Command: **Line**
From point: **5.629,0.750**
To point: **@3.25<30**
To point: **@5.00<150**
To point: **@3.25<210**
To point: **C**

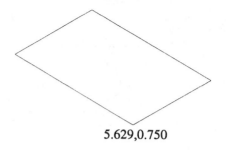

5.629,0.750

Step #2

Copy the four lines drawn in the previous step up at a distance of 2.50 units in the 90-degree direction.

Command: **Copy**
Select objects: *(Select lines "A", "B", "C", and "D")*
Select objects: *(Strike Enter to continue)*
<Base point or displacement>/Multiple: **Endp**
of *(Select the endpoint at "A")*
Second point of displacement: **@2.50<90**

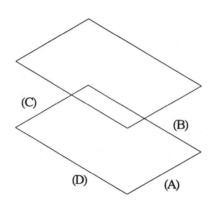

Step #3

Connect the top and bottom isometric boxes with line segments. Draw one segment using the Line command. Use the Copy-Multiple command to duplicate and form the remaining segments. Erase the two dashed lines since they are not visible in an isometric drawing.

Command: **Line**
From point: **Endp**
of *(Select the endpoint of the line at "A")*
To point: **Endp**
of *(Select the endpoint of the line at "B")*
To point: *(Strike Enter to exit this command)*

Copy this line from "A" to "C" and "D".

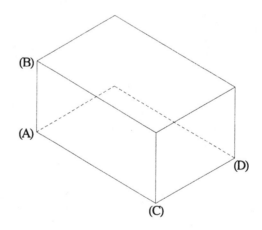

Step #4

Copy the two dashed lines at the right a distance of 0.75 in the 210-degree direction.

Command: **Copy**
Select objects: *(Select the two dashed lines at the right)*
Select objects: *(Strike Enter to continue)*
<Base point or displacement>/Multiple: **Endp**
of *(Select the endpoint at "A")*
Second point of displacement: **@0.75<210**

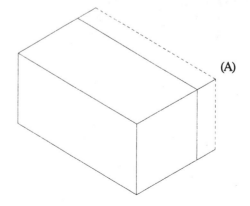

Step #5

Copy the two dashed lines at the right a distance of 0.75 in the 90-degree direction. This forms the base of the plate.

Command: **Copy**
Select objects: *(Select the two dashed lines at the right)*
Select objects: *(Strike Enter to continue)*
<Base point or displacement>/Multiple: **Endp**
of *(Select the endpoint at "A")*
Second point of displacement: **@0.75<90**

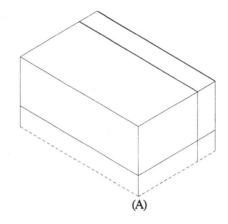

(A)

Step #6

Copy the dashed line at the right a distance of 0.75 in the −30-degree direction.

Command: **Copy**
Select objects: *(Select the dashed line at the right)*
Select objects: *(Strike Enter to continue)*
<Base point or displacement>/Multiple: **Endp**
of *(Select the endpoint at "A")*
Second point of displacement: **@0.75<-30**

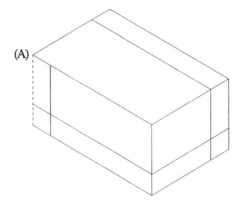

(A)

Step #7

Use the Fillet command to place a corner between the two dashed lines at "A" and "B" and at "C" and "D". The current fillet radius should already be set to a value of 0.

Command: **Fillet**
Polyline/Radius/<Select two objects>: *(Select "A" and "B")*

Command: **Fillet**
Polyline/Radius/<Select two objects>: *(Select "C" and "D")*

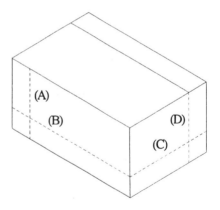

(A)
(B)
(D)
(C)

Step #8

Copy the dashed line at the right to begin forming the top of the base.

Command: **Copy**
Select objects: *(Select the dashed line at the right)*
Select objects: *(Strike Enter to continue)*
<Base point or displacement>/Multiple: **Endp**
of *(Select the endpoint at "A")*
Second point of displacement: **Endp**
of *(Select the endpoint at "B")*

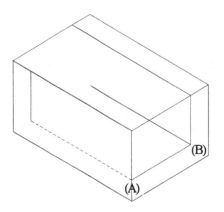

(B)
(A)

Step #9

Copy the dashed line at the right. This forms the base of the plate.

Command: **Copy**
Select objects: *(Select the dashed line at the right)*
Select objects: *(Strike Enter to continue)*
<Base point or displacement>/Multiple: **Endp**
of *(Select the endpoint at "A")*
Second point of displacement: **Endp**
of *(Select the endpoint at "B")*

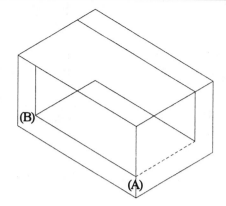

Step #10

Use the Trim command to clean up the excess lines at the right.

Command: **Trim**
Select cutting edge(s)...
Select objects: *(Select the three dashed lines at the right)*
Select objects: *(Strike Enter to continue)*
Select object to trim: *(Select the inclined line at "A")*
Select object to trim: *(Select the inclined line at "B")*
Select object to trim: *(Select the vertical line at "C")*
Select object to trim: *(Strike Enter to exit this command)*

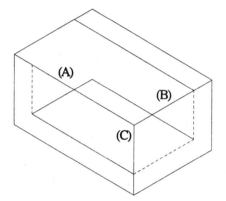

Step #11

Your display should be similar to the illustration at the right.

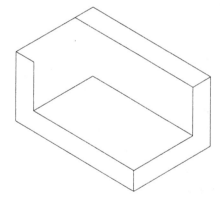

Step #12

Copy the dashed line at the right a distance of 1.12 in the 90-degree direction.

Command: **Copy**
Select objects: *(Select the dashed line at the right)*
Select objects: *(Strike Enter to continue)*
<Base point or displacement>/Multiple: **Endp**
of *(Select the endpoint at "A")*
Second point of displacement: **@1.12<90**

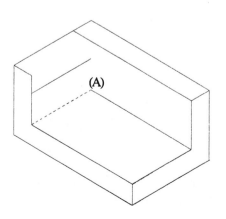

Step #13

Copy the dashed line at "A" to new positions at "B" and "C". Then delete the line at "A" using the Erase command.

Command: **Copy**
Select objects: *(Select the dashed line at the right)*
Select objects: *(Strike Enter to continue)*
<Base point or displacement>/Multiple: **M**
Base point: **Endp**
of *(Select the endpoint at "A")*
Second point of displacement: **Endp**
of *(Select the endpoint at "B")*
Second point of displacement: **Endp**
of *(Select the endpoint at "C")*
Second point of displacement: *(Strike Enter to exit this command)*

Command: **Erase**
Select objects: *(Select the dashed line at "A")*
Select objects: *(Strike Enter to execute this command)*

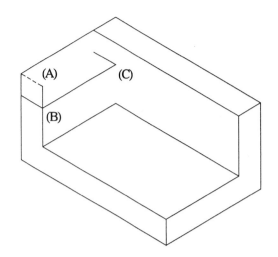

Step #14

Use the Trim command to clean up the excess lines at the right.

Command: **Trim**
Select cutting edge(s)...
Select objects: *(Select the two dashed lines at the right)*
Select objects: *(Strike Enter to continue)*
Select object to trim: *(Select the vertical line at "A")*
Select object to trim: *(Select the vertical line at "B")*
Select object to trim: *(Select the inclined line at "C")*
Select object to trim: *(Strike Enter to exit this command)*

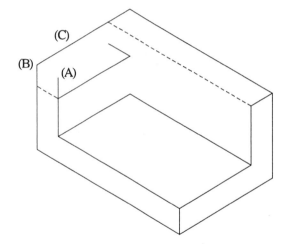

Step #15

Use the Copy command to duplicate the dashed line at the right from the endpoint of "A" to the endpoint at "B".

Command: **Copy**
Select objects: *(Select the dashed line at the right)*
Select objects: *(Strike Enter to continue)*
<Base point or displacement>/Multiple: **Endp**
of *(Select the endpoint at "A")*
Second point of displacement: **Endp**
of *(Select the endpoint at "B")*

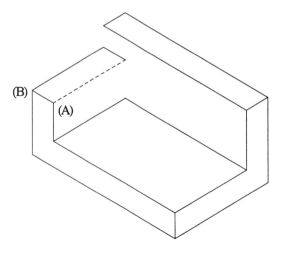

Step #16

Use the Copy command to duplicate the dashed line at the right from the endpoint of "A" to the endpoint at "B".

Command: **Copy**
Select objects: *(Select the dashed line at the right)*
Select objects: *(Strike Enter to continue)*
<Base point or displacement>/Multiple: **Endp**
of *(Select the endpoint at "A")*
Second point of displacement: **Endp**
of *(Select the endpoint at "B")*

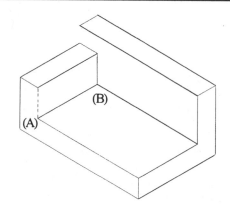

Step #17

Use the Line command to connect the endpoints of the segments at "A" and "B" illustrated at the right.

Command: **Line**
From point: **Endp**
of *(Select the endpoint at "A")*
To point: **Endp**
of *(Select the endpoint at "B")*
To point: *(Strike Enter to exit this command)*

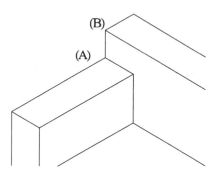

Step #18

Use the Copy command to duplicate the dashed line at the right from the endpoint of "A" to the distance of 0.50 specified by a polar coordinate. This value begins the outline of the rectangular hole through the object.

Command: **Copy**
Select objects: *(Select the dashed line at the right)*
Select objects: *(Strike Enter to continue)*
<Base point or displacement>/Multiple: **Endp**
of *(Select the endpoint at "A")*
Second point of displacement: **@0.50<210**

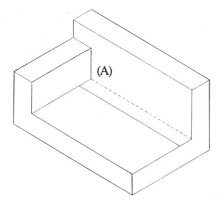

Step #19

Use the Copy command to duplicate the dashed line at the right from the endpoint of "A" to the distance of 0.50 using a polar coordinate.

Command: **Copy**
Select objects: *(Select the dashed line at the right)*
Select objects: *(Strike Enter to continue)*
<Base point or displacement>/Multiple: **Endp**
of *(Select the endpoint at "A")*
Second point of displacement: **@0.50<-30**

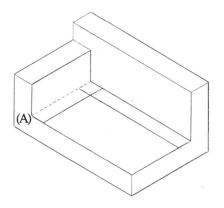

Step #20

Use the Copy command to duplicate the dashed line at the right from the endpoint of "A" to the distance of 0.50 using a polar coordinate.

Command: **Copy**
Select objects: *(Select the dashed line at the right)*
Select objects: *(Strike Enter to continue)*
<Base point or displacement>/Multiple: **Endp**
of *(Select the endpoint at "A")*
Second point of displacement: **@0.50<30**

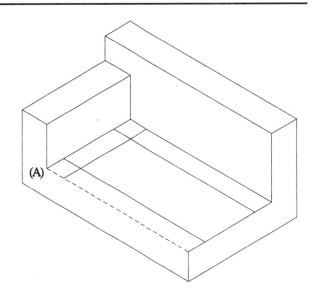

Step #21

Use the Copy command to duplicate the dashed line at the right from the endpoint of "A" to the distance of 0.50 using a polar coordinate.

Command: **Copy**
Select objects: *(Select the dashed line at the right)*
Select objects: *(Strike Enter to continue)*
<Base point or displacement>/Multiple: **Endp**
of *(Select the endpoint at "A")*
Second point of displacement: **@0.50<150**

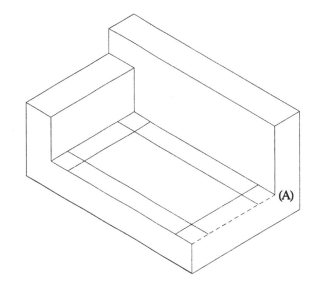

Step #22

Use the Fillet command with a radius of 0 to corner the four dashed lines at the right. Use the Multiple command to remain in the Fillet command. To exit the Fillet command prompts, use the CTRL-C sequence to cancel the command when finished.

Command: **Multiple**
Fillet
Polyline/Radius/<Select two objects>: *(Select lines "A" and "B")*
Polyline/Radius/<Select two objects>: *(Select lines "B" and "C")*
Polyline/Radius/<Select two objects>: *(Select lines "C" and "D")*
Polyline/Radius/<Select two objects>: *(Select lines "D" and "A")*
Polyline/Radius/<Select two objects>: **CTRL-C** *(To cancel)*

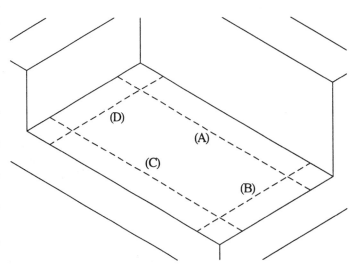

Step #23

Use the Copy command to duplicate the dashed line at the right. This will begin forming the thickness of the base inside of the rectangular hole.

Command: **Copy**
Select objects: *(Select the dashed line at the right)*
Select objects: *(Strike Enter to continue)*
<Base point or displacement>/Multiple: **Endp**
of *(Select the endpoint at "A")*
Second point of displacement: **Endp**
of *(Select the endpoint at "B")*

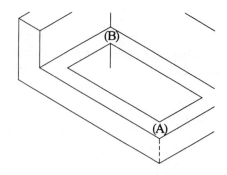

Step #24

Use the Copy command to duplicate the dashed lines at the right. These lines form the inside surfaces to the rectangular hole.

Command: **Copy**
Select objects: *(Select the dashed line at the right)*
Select objects: *(Strike Enter to continue)*
<Base point or displacement>/Multiple: **Endp**
of *(Select the endpoint at "A")*
Second point of displacement: **Endp**
of *(Select the endpoint at "B")*

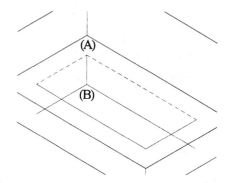

Step #25

Use the Trim command to clean up excess lines illustrated at the right.

Command: **Trim**
Select cutting edges...
Select objects: *(Select the two dashed lines at the right)*
Select objects: *(Strike Enter to continue)*
Select object to trim: *(Select the line at "A")*
Select object to trim: *(Select the line at "B")*
Select object to trim: *(Strike Enter to exit this command)*

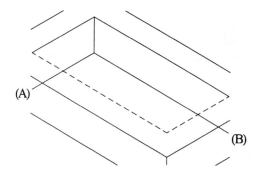

Step #26

The completed isometric is illustrated at the right. AutoCAD Release 11 allows the user to dimension this drawing using the Oblique command. Consult your instructor if this next step is necessary.

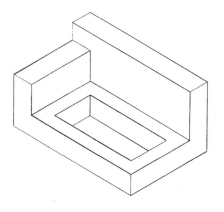

Tutorial Exercise #17
Hanger.Dwg

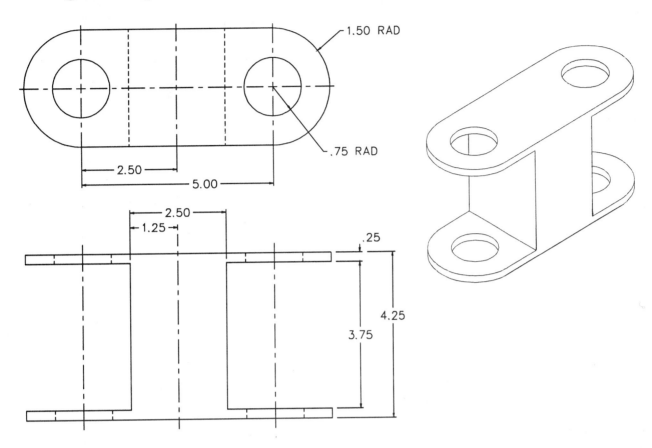

1.50 RAD

.75 RAD

2.50

5.00

2.50

1.25

.25

4.25

3.75

PURPOSE:

The purpose of this tutorial exercise is to use a series of coordinates along with AutoCAD editing commands to construct an isometric drawing of the Hanger.

SYSTEM SETTINGS:

Begin a new drawing called "Hanger." Use the Units command to change the number of decimal places past the zero from 4 to 2. Keep the remaining default unit values. Using the Limits command, keep 0,0 for the lower left corner and change the upper right corner from 12,9 to 15.50,9.50. Use the Grid command and change the grid spacing from 1.00 to 0.25 units. Do not turn the snap or ortho On.

LAYERS:

Special layers do not have to be created for this tutorial exercise.

SUGGESTED COMMANDS:

Begin this exercise by changing the grid from the standard display to an isometric display using the Snap-Style option. Remember both the grid and snap can be manipulated by the Snap command only if the current grid value is 0. Use absolute and polar coordinates to lay out the base of the Plate. Then begin using the Copy command followed by Trim to duplicate entities and clean up or trim unnecessary geometry.

DIMENSIONING:

This object may be dimensioned at a later date using the Dim-Oblique option found only in AutoCAD Release 11. Consult your instructor before continuing.

PLOTTING:

This tutorial exercise may be plotted on "B"-size paper (11" x 17"). Use a plotting scale of 1=1 to produce a full size plot.

VERSION OF AUTOCAD:

This tutorial exercise may be completed using either AutoCAD Release 10 or Release 11.

Step #1

Set the Snap-Style option to Isometric with a vertical spacing of 0.25. Type CTRL-E to switch to the Top Isoplane mode. Use the Line command to draw the rectangular isometric box representing the total depth of the object along with the center-to-center distance of the holes and arcs that will be placed in the next step.

Command: **Snap**
Snap spacing or ON/OFF/Aspect/Rotate/Style <0.25>: **S**
Standard/Isometric <S>: **Iso**
Vertical spacing <1.00>: **0.25**

Command: **^E** *(To switch to the Top Isoplane mode)*

Command: **Line**
From point: **6.28,0.63**
To point: **@5.00<30**
To point: **@ 3.00<150**
To point: **@5.00<210**
To point: **C**

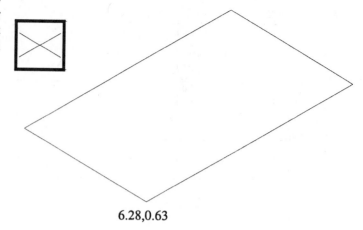

6.28,0.63

Step #2

While in the Top Isoplane mode, use the Ellipse command to draw two isometric ellipses of 0.75 and 1.50 radii each. Identify the midpoint of the inclined line at "A" as the center of the first ellipse. To identify the center of the second ellipse, use the @ option which stands for "last point" and will identify the center of the small circle as the same center as the large circle.

Command: **Ellipse**
<Axis endpoint 1>/Center/Isometric: **Iso**
Center of circle: **Mid**
of *(Select the inclined line at "A")*
<Circle radius>/Diameter: **0.75**

Command: **Ellipse**
<Axis endpoint 1>/Center/Isometric: **Iso**
Center of circle: **@**
<Circle radius>/Diameter: **1.50**

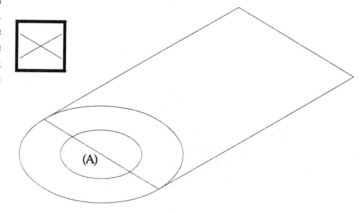

Step #3

Copy both ellipses from the midpoint of the inclined line at "A" to the midpoint of the inclined line at "B".

Command: **Copy**
Select objects: *(Select both ellipses at the right)*
Select objects: *(Strike Enter to continue)*
<Base point or displacement>/Multiple: **Mid**
of *(Select the midpoint of the inclined line at "A")*
Second point of displacement: **Mid**
of *(Select the midpoint of the inclined line at "B")*

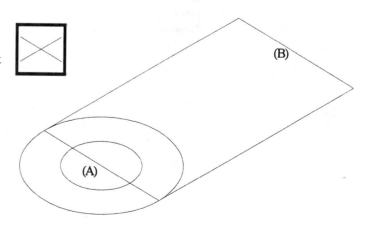

Step #4

Turn the snap Off by striking the F9 function key. Use the Trim command to delete parts of the ellipses.

Command: **Trim**
Select cutting edges...
Select objects: *(Select dashed lines "A" and "B")*
Select objects: *(Strike Enter to continue)*
Select object to trim: *(Select the ellipse at "C")*
Select object to trim: *(Select the ellipse at "D")*
Select object to trim: *(Strike Enter to exit this command)*

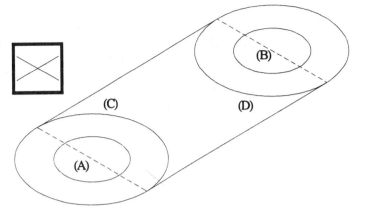

Step #5

Copy all entities at the right up the distance of 0.25 to form the bottom base of the hanger. The Right Isoplane mode can be activated by typing CTRL-E.

Command: **^E**
Right Isoplane

Command: **Copy**
Select objects: *(Select all entities at the right)*
Select objects: *(Strike Enter to continue)*
<Base point or displacement>/Multiple: **Mid**
of *(Select the midpoint of the inclined line at "A")*
Second point of displacement: **@0.25<90**

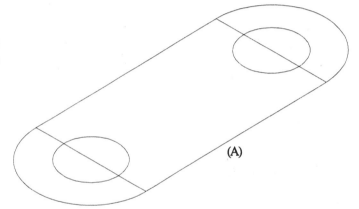

Step #6

Use the Erase command to delete the three dashed lines at the right. These lines are not visible at this point of view and should be erased.

Command: **Erase**
Select objects: *(Select the three dashed lines at the right)*
Select objects: *(Strike Enter to execute this command)*

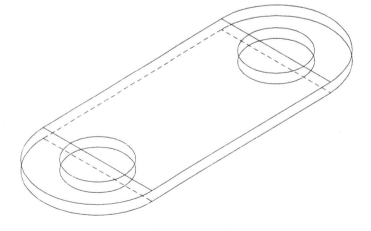

Step #7

Your display should appear similar to the illustration at the right. Begin partially deleting other entities to show only visible features of the isometric drawing. The next few steps that follow refer to the area outlined at the right. Use the Zoom-Window option to magnify this area.

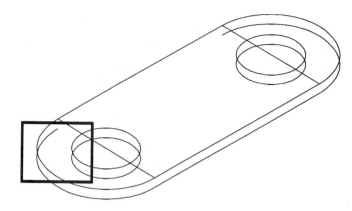

Step #8

Carefully draw a line tangent to both ellipses. Use the Osnap-Quadrant option to assist you in constructing the line.

Command: **Line**
From point: **Qua**
of *(Select the quadrant at "A")*
To point: **Qua**
of *(Select the quadrant at "B")*
To point: *(Strike Enter to exit this command)*

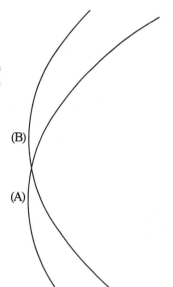

Step #9

Use the Trim command, select the short dashed line as the cutting edge, and select the arc segment at the right to trim.

Command: **Trim**
Select cutting edges...
Select objects: *(Select dashed line at "A")*
Select objects: *(Strike Enter to continue)*
Select object to trim: *(Select the ellipse at "B")*
Select object to trim: *(Strike Enter to exit this command)*

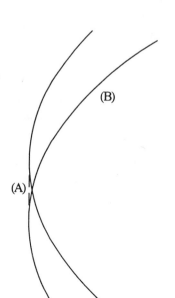

Step #10

The completed operation is illustrated at the right. Use Zoom-Previous to return to the previous display.

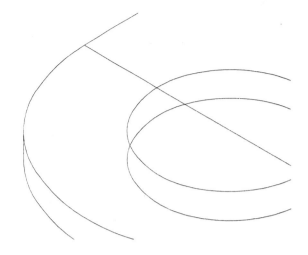

Step #11

Use the Zoom-Window option to magnify the right half of the base. Prepare to construct the tangent edge to the object using the previous steps.

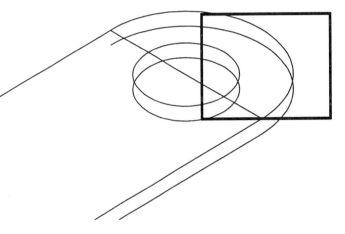

Step #12

Follow the same procedure as in steps 8 and 9 to construct a line from the quadrant point on the top ellipse to the quadrant point on the bottom ellipse. Then use the Trim command to clean up any excess entities. Use the Erase command to delete any elliptical arc segments that may have been left untrimmed.

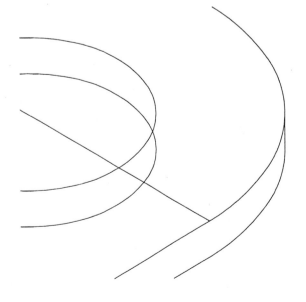

Step #13

Use the Trim command to partially delete the ellipses at the
right to expose the thickness of the base.

Command: **Trim**
Select cutting edges...
Select objects: *(Select dashed ellipses "A" and "B")*
Select objects: *(Strike Enter to continue)*
Select object to trim: *(Select the lower ellipse at "C")*
Select object to trim: *(Select the lower ellipse at "D")*
Select object to trim: *(Strike Enter to exit this command)*

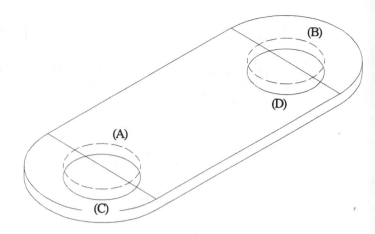

Step #14

Use the Copy command to duplicate the bottom base and
form the upper plate of the hanger. Copy the base a
distance of 4 units straight up.

Command: **Copy**
Select objects: *(Select all dashed entities at the right)*
Select objects: *(Strike Enter to continue)*
<Base point or displacement>/Multiple: **Nea**
of *(Select the nearest point along the bottom arc at "A")*
Second point of displacement: **@4.00<90**

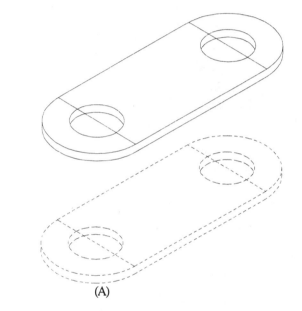

Step #15

Use the Copy command to duplicate the inclined line at
"A" the distance of 2.50 units to form the line represented
by a series of dashes at the right. This line happens to be
located at the center of the object.

Command: **Copy**
Select objects: *(Select the line at "A")*
Select objects: *(Strike Enter to continue)*
<Base point or displacement>/Multiple: **Mid**
of *(Select the midpoint of the inclined line at "A")*
Second point of displacement: **@2.50<30**

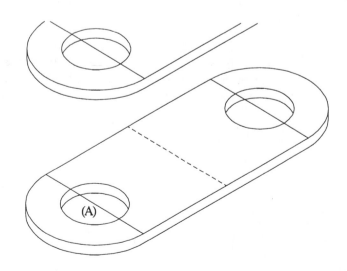

Step #16

Duplicate the line represented by dashes at the right to form the two inclined lines at "B" and "C". These lines will begin the construction of the sides of the hanger. Use the Copy-Multiple option to accomplish this.

Command: **Copy**
Select objects: *(Select the dashed line at the right)*
Select objects: *(Strike Enter to continue)*
<Base point or displacement>/Multiple: **M**
Base point: **Mid**
of *(Select the midpoint of the dashed line at "A")*
Second point of displacement: **@1.25<30**
Second point of displacement: **@1.25<210**
Second point of displacement: *(Strike Enter to exit this command)*

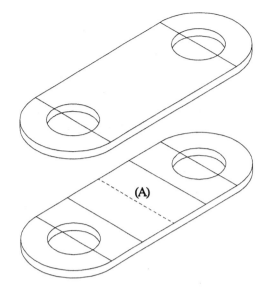

Step #17

Use the Copy command to duplicate the two dashed lines at the right straight up at a distance of 3.75 units. The polar coordinate mode is used to accomplish this.

Command: **Copy**
Select objects: *(Select both dashed lines at the right)*
Select objects: *(Strike Enter to continue)*
<Base point or displacement>/Multiple: **Mid**
of *(Select the midpoint of the inclined line at "A")*
Second point of displacement: **@3.75<90**

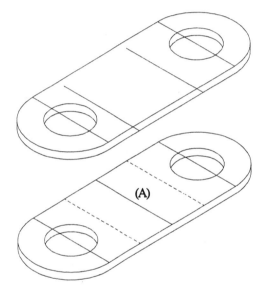

Step #18

Use the Line command along with the Osnap-Endpoint option to draw a line from endpoint "A" to endpoint "B".

Command: **Line**
From point: **Endp**
of *(Select the endpoint of the inclined line at "A")*
To point: **Endp**
of *(Select the endpoint of the inclined line at "B")*
To point: *(Strike Enter to exit this command)*

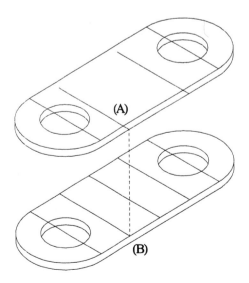

Step #19

Use the Copy command and the Multiple option to copy the dashed line at "A" to "B", "C" and "D".

Command: **Copy**
Select objects: *(Select the dashed line at the right)*
Select objects: *(Strike Enter to continue)*
<Base point or displacement>/Multiple: **M**
Base point: **Endp**
of *(Select the endpoint of the vertical line at "A")*
Second point of displacement: **Endp**
of *(Select the endpoint of the vertical line at "B")*
Second point of displacement: **Endp**
of *(Select the endpoint of the vertical line at "C")*
Second point of displacement: **Endp**
of *(Select the endpoint of the vertical line at "D")*
Second point of displacement: *(Strike Enter to exit the command)*

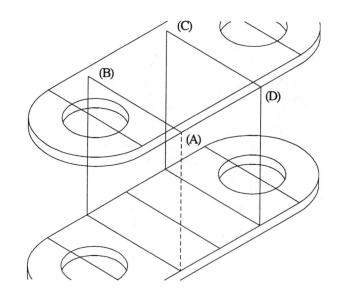

Step #20

Use the Erase command to delete all lines represented as dashed lines at the right.

Command: **Erase**
Select objects: *(Select all dashed entities illustrated at the right)*
Select objects: *(Strike Enter to execute this command)*

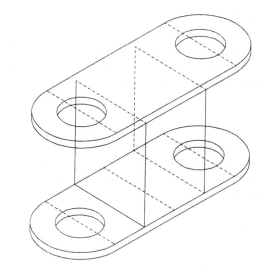

Step #21

Use the Trim command to partially delete the vertical line at the right. The segment to be deleted is hidden and not shown in an isometric drawing.

Command: **Trim**
Select cutting edges...
Select objects: *(Select dashed entities "A" and "B")*
Select objects: *(Strike Enter to continue)*
Select object to trim: *(Select the vertical line at "C")*
Select object to trim: *(Strike Enter to exit this command)*

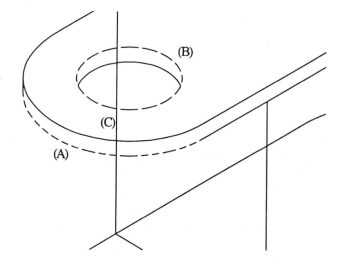

Step #22

Use the Trim command to partially delete the vertical line at the right. The segment to be deleted is hidden and not shown in an isometric drawing.

Command: **Trim**
Select cutting edges...
Select objects: *(Select dashed elliptical arc at "A")*
Select objects: *(Strike Enter to continue)*
Select object to trim: *(Select the vertical line at "B")*
Select object to trim: *(Strike Enter to exit this command)*

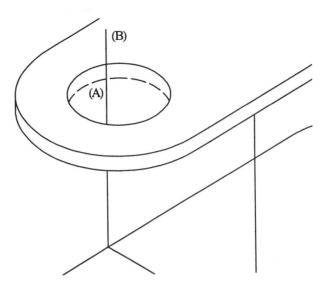

Step #23

Use the Trim command to partially delete the inclined line at the right.

Command: **Trim**
Select cutting edges...
Select objects: *(Select dashed entities "A" and "B")*
Select objects: *(Strike Enter to continue)*
Select object to trim: *(Select the line at "C")*
Select object to trim: *(Select the line at "D")*
Select object to trim: *(Strike Enter to exit this command)*

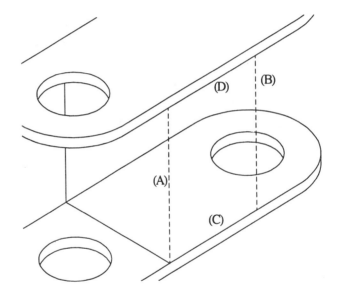

Step #24

Use the Trim command to partially delete the entities at the right. The segments to be deleted are hidden and not shown in an isometric drawing. Use Erase to delete any leftover elliptical arc segments.

Command: **Trim**
Select cutting edges...
Select objects: *(Select dashed entities "A" and "B")*
Select objects: *(Strike Enter to continue)*
Select object to trim: *(Select the line at "C")*
Select object to trim: *(Select the arc at "D")*
Select object to trim: *(Select the ellipse at "E")*
Select object to trim: *(Select the arc at "F")*
Select object to trim: *(Strike Enter to exit this command)*

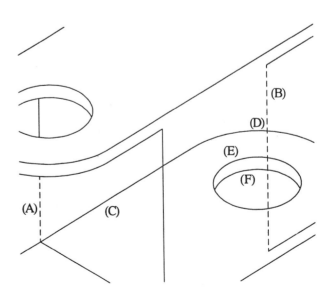

Step #25

The completed isometric is illustrated at the right. AutoCAD Release 11 allows the user to dimension this drawing using the Oblique command. Consult your instructor if this next step is necessary.

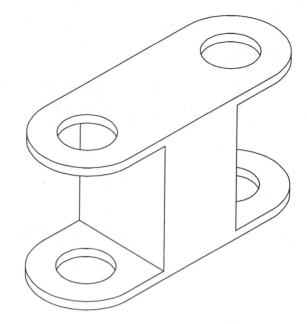

Step #26

As an extra step, convert the completed hanger into an object with a rectangular hole through it. Begin by copying lines "A", "B", "C", and "D" at a distance of 0.25 units to form the inside rectangle using polar coordinates. Since the lines will overlap at the corners, use the Fillet command set to a radius of 0 to create corners of the rectangle. Use the Line command to draw the inclined line "E" at any distance with a 150-degree angle. Use either Trim or Extend to complete the new version of the hanger.

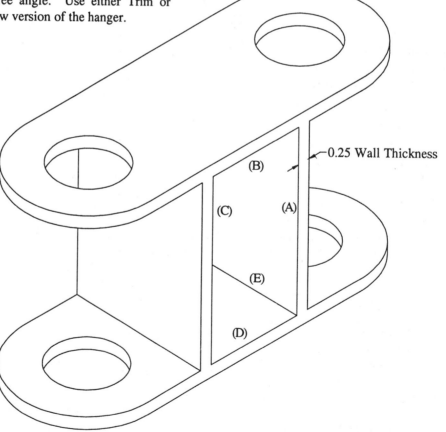

0.25 Wall Thickness

(A) (B) (C) (D) (E)

Problems for Unit 7

Directions for Problem 7-1
Construct an isometric drawing of the object.

Problem 7-1

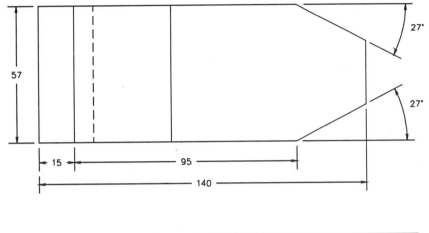

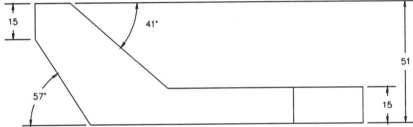

Directions for Problem 7-2
Construct an isometric drawing of the object. Begin the corner of the isometric at "A".

Problem 7-2

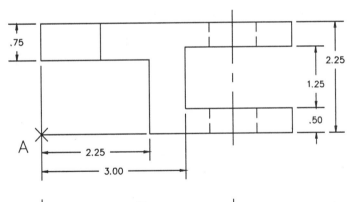

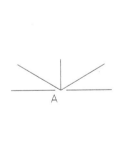

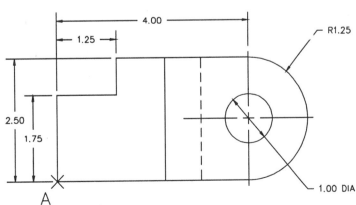

Directions for Problems 7–3 through 7–13
Construct an isometric drawing of the object. (These problems have only straight lines.)

Problem 7–3

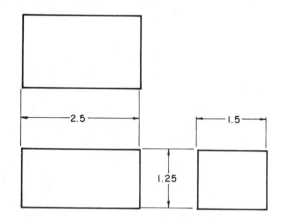

Problem 7–6

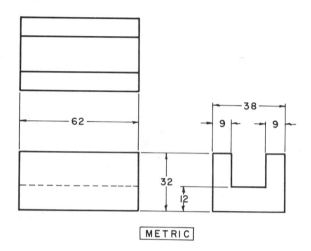

METRIC

Problem 7–4

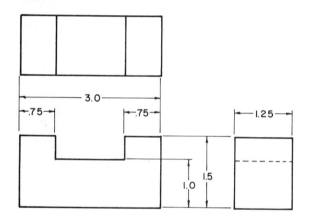

Problem 7–7

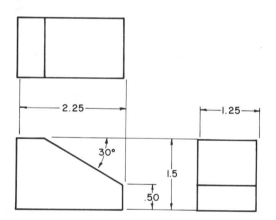

Problem 7–5

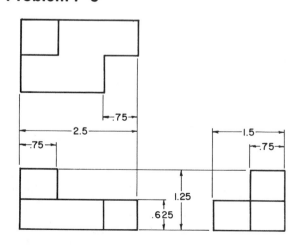

Problem 7–8

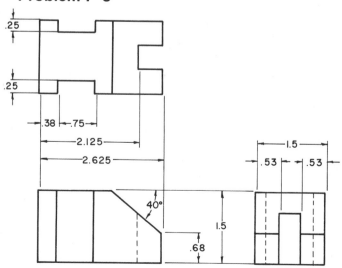

Problem 7-9

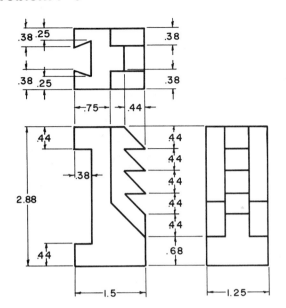

Problem 7-10

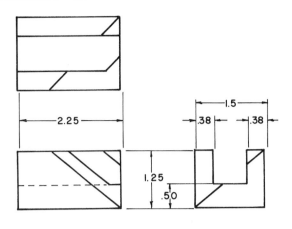

Problem 7-11

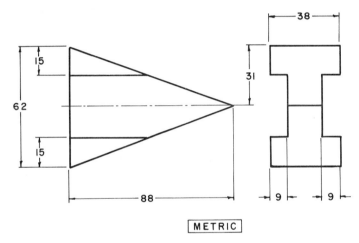

METRIC

Problem 7-12

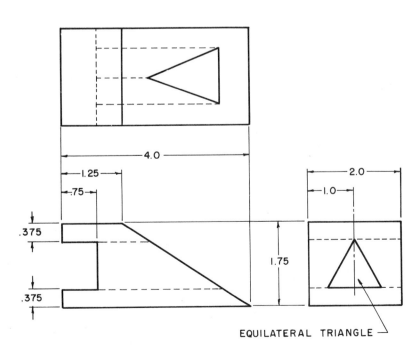

EQUILATERAL TRIANGLE

Problem 7-13

Directions for Problems 7-14 through 7-25
Construct an isometric drawing of the object. (These problems have straight lines, arcs, and circles.)

Problem 7-14

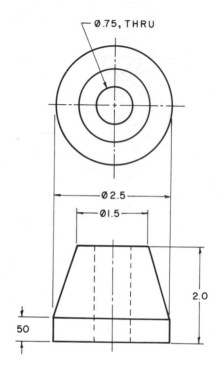

Problem 7-15

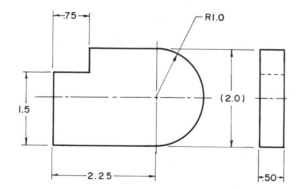

Problem 7-16

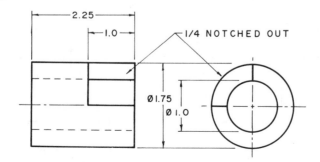

Problem 7-17

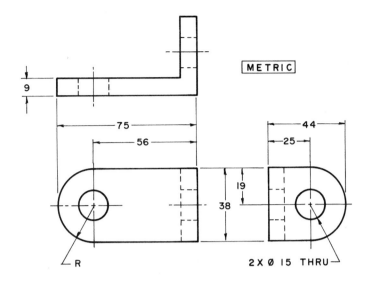

Problem 7-18

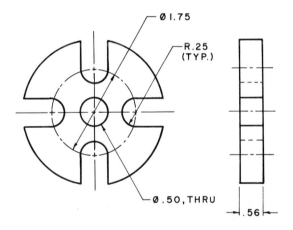

Problem 7-19

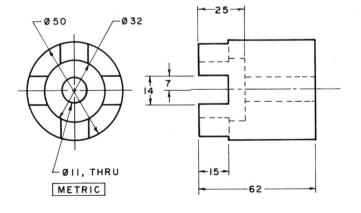

Problem 7-20

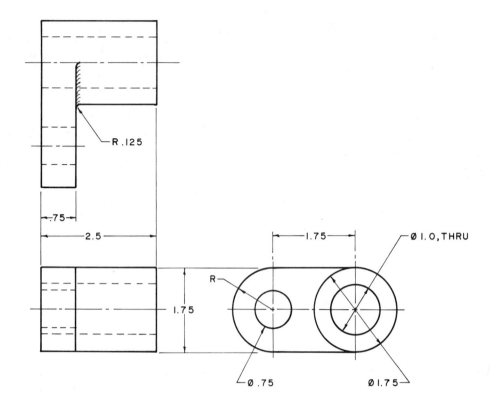

R .125

.75

2.5

1.75

1.75

R

Ø .75

Ø 1.0, THRU

Ø 1.75

Problem 7-21

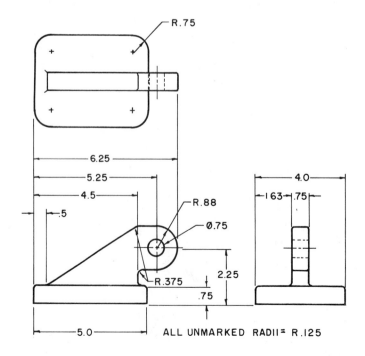

R.75

6.25

5.25

4.5

.5

R.88

Ø.75

R.375

2.25

.75

4.0

1.63 .75

5.0

ALL UNMARKED RADII = R .125

Problem 7-22

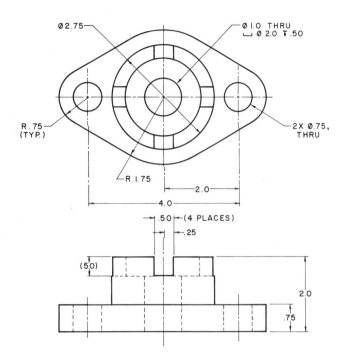

Ø 2.75

Ø 1.0 THRU
⌴ Ø 2.0 ⊤ .50

R .75
(TYP.)

2X Ø .75,
THRU

R 1.75

2.0

4.0

.50 ⊢ (4 PLACES)

.25

(50)

2.0

.75

Problem 7-23

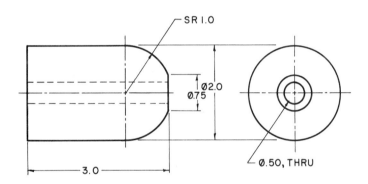

SR 1.0

Ø 2.0

Ø .75

Ø .50, THRU

3.0

Problem 7-24

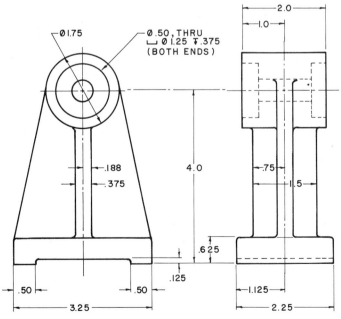

Ø 1.75

Ø .50, THRU
⌴ Ø 1.25 ⊤ .375
(BOTH ENDS)

2.0

1.0

.188

.375

4.0

.75

1.5

.625

.125

.50

.50

3.25

1.125

2.25

ALL UNMARKED RADII = R .09

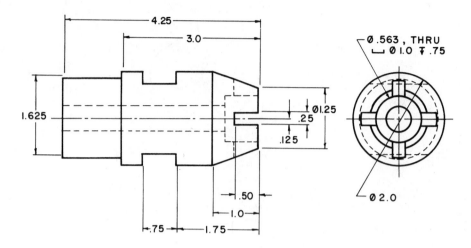

UNIT
8

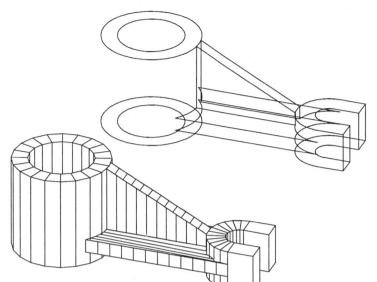

Three-Dimensional Modeling

Contents

It is said that humans see, hear, and exist in a 3D world. Why not draw in three dimensions; this means visualizing the object in the designer's mind before placing entities on the computer screen. Part of learning the art of visualization is in the study and construction of models in 3D in order to obtain as accurate as possible an image of an object undergoing design. In this unit, you will be exposed to creating a model first in the form of a wireframe representation, then the wireframe will be surfaced using a few of the many surfacing tools AutoCAD has to offer. Once the model is surfaced, it can be viewed from any angle, similar to the wireframe. The surfaced model, however, can have hidden lines removed to aid in the visualization of the final design.

Orthographic Projection

It is no secret that the heart of any engineering drawing is the ability to break up the design into three main views of front, top, and right side views representing what is called orthographic or multi-view projection. The engineer or designer is then required to interpret the views and their dimensions to paint a mental picture of what a pictorial version of the object would look like if already made or constructed.

As most engineering individuals have the skill to convert the multi-view drawing into a pictorial drawing in their mind, a vast majority of people would be confused by the numerous hidden and center lines of a drawing and their meaning to the overall design. The individuals need some type of picture to help them interpret the multi-view drawing and get a feel for what the part looks like, including the functionality of the part. This may be the major advantage of constructing an object in 3D.

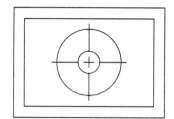

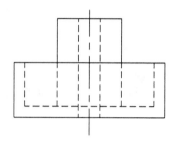

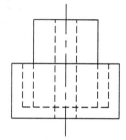

Isometric Drawings

The isometric drawing is the easiest 3D representation to produce. It is based on tilting two axes at 30-degree angles along with a vertical line to represent the height of the object. As easy as the isometric is to produce, it is also one of the most inaccurate methods of producing a 3D image.

The illustration at the right shows a view aligned to see the top of the object, front view, and right side view. If we wished to see a different view of the object in isometric, a new drawing would have to be produced from scratch. This is the major disadvantage of using isometric drawings. Still, because of ease in drawing, isometrics remain popular in many school and technical drafting rooms.

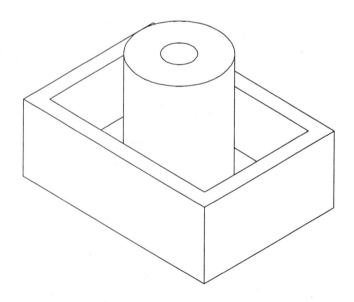

3D Extruded Models

AutoCAD versions in the past afforded the user the capability of drawing entities in the Z direction and then viewing the model at any angle. This method was called 3D Visualization. Entities are assigned an elevation for starting the surface and a thickness which extrudes the entity and produces opaque sides. The thickness can be entered in either a positive direction, (Up), or negative direction, (Down).

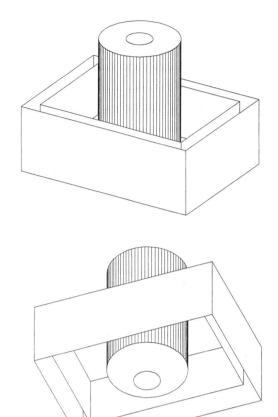

To view the model, the Vpoint command is used to identify a position on a 2D globe which serves as a means of identifying a 3D position. Advantages of this type of construction are the use of the Vpoint command and how the model is able to be viewed from any angle.

A disadvantage of this method was the inability to place top and bottom surfaces on the model. As a result, the model looks hollow when viewed from above or underneath.

3D Wireframe Models

The evolution of AutoCAD Release 10 made it possible to design in a true 3D data base. Using a user-definable coordinate system, previously defined as the UCS, a new coordinate system can be defined along any plane to place entities along. Depth can be controlled by a number of aids, especially XYZ point filters. In this method, values are temporarily saved for later use inside a command. The values can be retrieved and a coordinate value entered to complete the command.

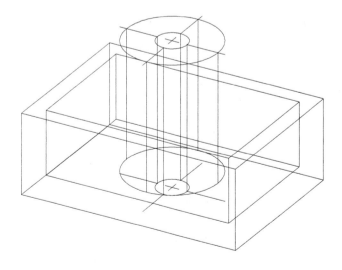

Wireframes are fundamental types of models. However, as lines intersect from the front and back of the object, it is sometimes difficult to interpret the true design of the wireframe.

3D Surfaced Models

Surfacing picks up where the wireframe model leaves off. Because of the complexity of the wireframe and the number of intersecting lines, surfaces are applied to the wireframe. The surfaces are in the form of opaque entities, called 3Dfaces. As always, 3Dfaces are placed using Osnap options to assist in point selection. A hidden line removal is performed to view the model without the interference of other entities. Again, the Vpoint command is used to view the model in different positions to make sure all sides of the model have been surfaced. The format of the 3Dface command is outlined using the prompts below and the illustration at the right.

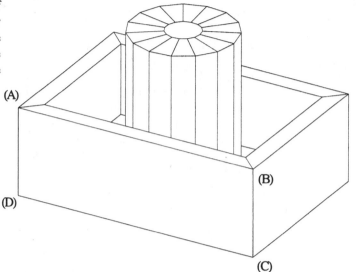

Command: **Osnap**
Object snap modes: **Endp**
Command: **3Dface**
First point: *(Select the point at "A")*
Second point: *(Select the point at "B")*
Third point: *(Select the point at "C")*
Fourth point: *(Select the point at "D")*
Third point: *(Strike Enter to exit this command)*

Ruled Surfaces

3Dfaces work where four endpoints of an entity can be selected. What if a surface needs to be produced between two different entities such as the two arcs illustrated at the right? In this case, a Ruled surface would be used. In the example at the right, selecting one arc at "A" as the first defining curve and the other arc at "B" as the second defining curve will produce a surface mesh between the two entities. The density of the mesh is controlled by the system variable Surftab1. Assigning a large value will produce a smoother surface but will take a longer time to regenerate for such commands as Hide. Assigning a small value will regenerate the screen faster but will produce a curve appearing as a series of polygons. Ruled surfaces may be placed in between any combination of lines, arcs, circles, and/or points. The format of the Rulesurf command is outlined using the prompts below and the illustration at the right.

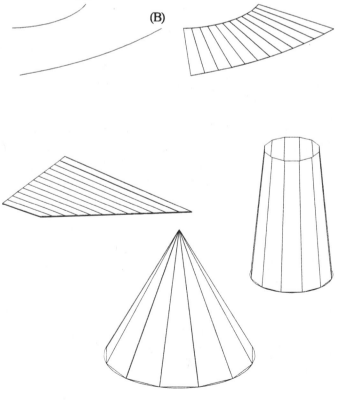

Command: **Rulesurf**
Select first defining curve: *(Select the arc segment at "A")*
Select second defining curve: *(Select the arc segment at "B")*

Tabulated Surfaces

Tabulated surfaces are defined as surfaces generated from a defining curve and direction vector. In the examples at the right, a defining curve is selected at "A" followed by the direction vector at "B". It is important to note that selecting the direction vector also determines the start of the tabulated surface. The direction vector at "B" signifies the beginning of the surface and directs the defining curve at "A" to proceed upward and to the right. A similar problem is illustrated at the right with the defining curve labeled as "C" and the direction vector "D". This time, selecting the direction vector at "D" begins the defining curve and proceeds below and to the right. The results of each curve are illustrated to the right of the individual components. The system variable Surftab1 controls the density of the surfaces. A low number displays the curve as a series of large surfaces. A large number displays smaller surfaces that will show a better outline of a curve and also require more time for regenerations. The format of the Tabsurf command is outlined using the prompts below and the illustrations at the right.

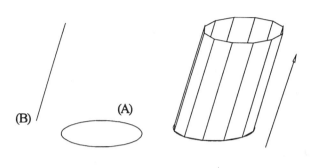

Command: **Tabsurf**
Select path curve: *(Select the circle at "A")*
Select direction vector: *(Select the line segment at "B")*

Command: **Tabsurf**
Select path curve: *(Select the curve at "C")*
Select direction vector: *(Select the line segment at "D")*

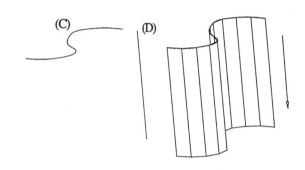

Edge Surfaces

The Edge surface or Coons surface patch is a surface constructed from four adjoining entity edges. One requirement of this surface is that the edges of the four entities must touch at their exact endpoints. Entities may be any combination of lines, arcs, or polylines. Since the surface mesh is generated in two directions, two system variables control the density of this mesh, namely, Surftab1 and Surftab2. As with all geometry generated surfaces, entering large values for the two system variables will construct a very smooth surface that will require extra time for system regeneration. Entering small values for the two system variables will construct a very rough curve; but the system performance will be enhanced and the surface will appear quickly. The format of the Edgesurf command is outlined using the prompts below and the illustration at the right.

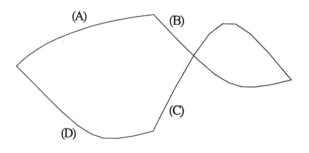

Command: **Edgesurf**
Select edge 1: *(Select the entity at "A")*
Select edge 2: *(Select the entity at "B")*
Select edge 3: *(Select the entity at "C")*
Select edge 4: *(Select the entity at "D")*

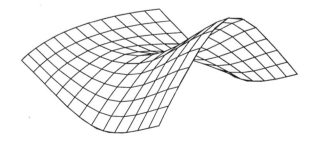

Revolved Surfaces

A surface of revolution can be created by rotating a path curve around a selected axis by using the Revsurf command. The path curve can take the form of a line, arc, circle, 2D polyline, or 3D polyline. The axis of revolution can be either a line or polyline. As with the Edge surface patch, two system variables control the density of the revolved surface, namely, Surftab1 and Surftab2. The format of the Revsurf command is outlined using the prompts below and the illustration at the right.

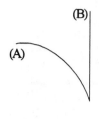
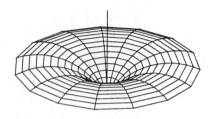

Command: **Revsurf**
Select path curve: *(Select the arc at "A")*
Select axis of revolution: *(Select the line segment at "B")*
Start angle <0>: *(Strike Enter to accept this default value)*
Included angle (+=ccw, -=cw) <Full circle>: *(Strike Enter)*

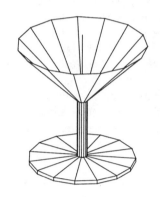

Command: **Revsurf**
Select path curve: *(Select the polyline segment at "C")*
Select axis of revolution: *(Select the line segment at "D")*
Start angle <0>: *(Strike Enter to accept this default value)*
Included angle (+=ccw, -=cw) <Full circle>: *(Strike Enter)*

Solid Models

The creation of a solid model remains the most exact way to represent a model in 3D. It is also the most versatile representation of an object. Solid models may be viewed as identical to wireframe and surfaced construction models. Wireframe models and surfaced models may be analyzed by taking distance readings and identifying key points of the model. Key orthographic views such as front, top, and right side views may be extracted from wireframe models. Surfaced models may be imported into shading packages such as AutoShade for increased visualization. Solid models do all of the above operations and more. This is because the solid model, as it is called, is a solid representation of the actual object. From cylinders to slabs, wedges to boxes, all entities that go in the creation of a solid model have volume. This allows a model to be constructed of what are referred to as primitives. These primitives are then merged into one using addition and subtraction operations. What remains is the most versatile of 3D drawings. This method of creating models will be discussed in greater detail in Unit 10.

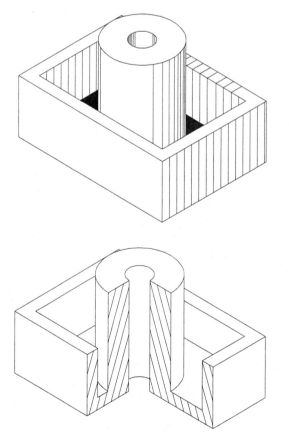

Viewing Models Using the Vpoint Command

The object at the right represents the plan view of a 3D model already created. Unfortunately, it becomes very difficult to understand what the model looks like in its present condition. Use the Vpoint command to view a wireframe, surfaced, or solid model in three dimensions. The following is a typical prompt sequence for the Vpoint command:

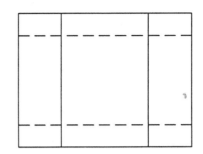

Command: **Vpoint**
Rotate/<View point> <0.0000,0.0000,0.0000>:

The Vpoint command stands for "View point." A point is identified in 3D space. This point becomes a location where the model is viewed from. The point may be entered in at the keyboard as an X,Y,Z coordinate or picked with the aid of a two-dimensional globe. The object illustrated at the right is a typical result of using the Vpoint command.

When executing the Vpoint command, three methods are available to the user for viewing a model in 3D.

The first method is defining a view point using an X,Y,Z coordinate. In the command sequence below, a new viewing point is established 1 unit in the positive X direction, 1 unit in the negative Y direction, and 1 unit in the positive Z direction.

Command: **Vpoint**
Rotate/<Viewpont><0.0000,0.0000,0.0000>: **1,–1,1**

A second method of defining a viewing point is to define two angular axes by rotation. The first angle defines the view point in the X-Y axis. However this is only two-dimensional. The second angle defines the view point from the X-Y axis. This tilts the viewing point up for a positive angle or down for a negative angle.

Command: **Vpoint**
Rotate/<Viewpont><0.0000,0.0000,0.0000>: **Rotate**
Enter angle in X-Y plane from X axis < >: **45**
Enter angle from X-Y plane < >: **30**

If the command sequence is followed by the Enter key, a graphic image consisting of globe and tripod appear illustrated at the right. Although the globe appears two-dimensional, it provides the user the ability to pick a view point depending on how the globe is read. The intersection of the horizontal and vertical lines form the North Pole of the globe. The inner circle forms the equator and the outer circle the South Pole. The examples that follow illustrate numerous viewing points.

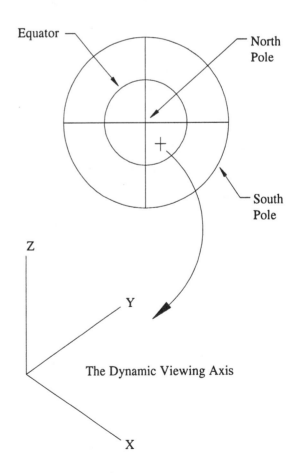

The Dynamic Viewing Axis

Viewing along the Equator

To obtain the results at the right, use the Vpoint command and mark a point at "A" to view the front view; mark a point at "B" to view the top view; and mark a point at "C" to view the right side view. Coordinates could also have been entered to achieve the same results:

Command: **Vpoint**
Rotate/<View point><0.0000,0.0000,0.0000>: **0,-1,0** (For "A")

Command: **Vpoint**
Rotate/<View point><0.0000,0.0000,0.0000>: **0,0,1** (For "B")

Command: **Vpoint**
Rotate/<View point><0.0000,0.0000,0.0000>: **1,0,0** (For "C")

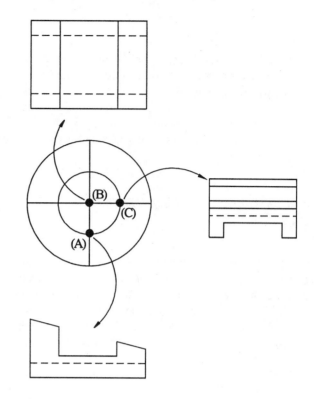

Viewing near the North Pole

Picking the four points illustrated at the right results in the different viewing points for the object. Since all points are inside the inner circle, the results are aerial views, or views from above. Remember that the Equator is symbolized by the inner circle. Depending on which quadrant you select, you will look up at the object from the right corner, left corner, or either of the rear corners.

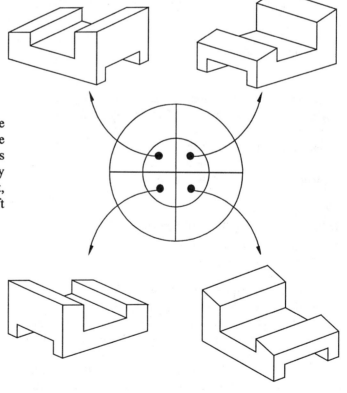

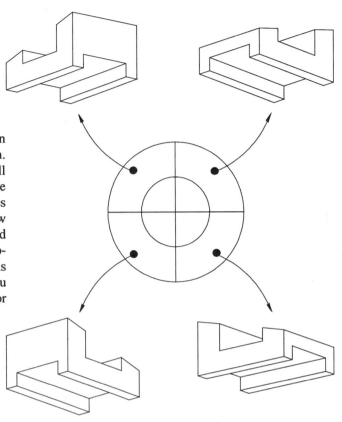

Viewing near the South Pole

Picking the four points illustrated at the right results in underground views, or viewing the object from underneath. This is true since all points selected lie between the small and large circles. Again, remember that the small circle symbolizes the Equator, while the large circle symbolizes the South Pole. A more graphical method of selecting view points would be to select the Display command strip and pull down the menu holding the view point command options illustrated below. Once a particular viewing point is selected, use the screen menu to determine whether you desire an aerial view (+10 to +80), normal view (0), or underground view (−10 to −80).

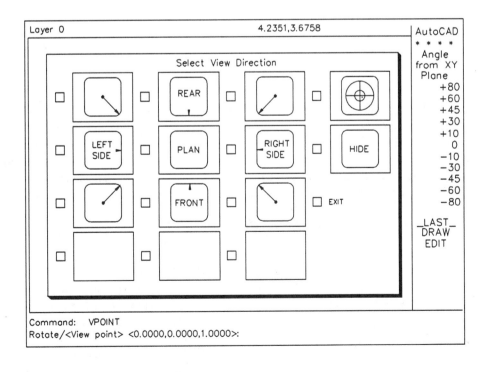

Tutorial Exercise #18
Wedge.Dwg

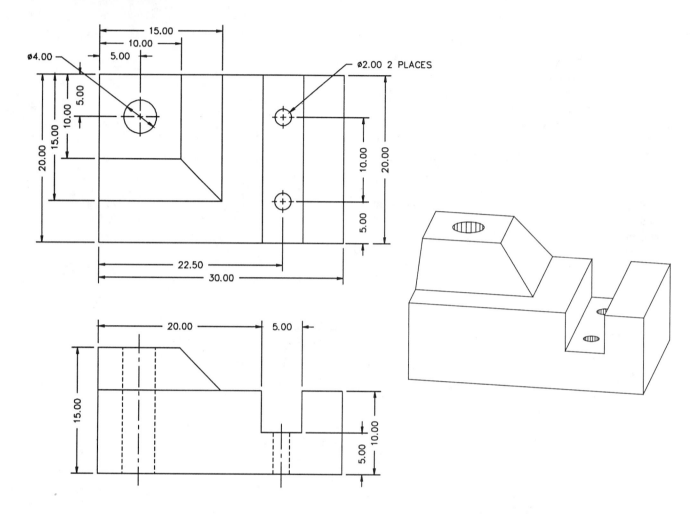

PURPOSE:

This tutorial is designed to construct a true 3D wireframe model of the Wedge illustrated above.

SYSTEM SETTINGS:

Begin a new drawing called "Wedge." Use the Units command to change the number of decimal places past the zero from 4 to 0. Keep the remaining default unit values. Using the Limits command, keep 0,0 for the lower left corner and change the upper right corner from 12,9 to 60,45.

LAYERS:

Create the following layers with the format:

Name-Color-Linetype
Wireframe - Yellow - Continuous

SUGGESTED COMMANDS:

The Line command is used exclusively for this drawing of the Wedge. A combination of absolute and polar coordinates are used for precision placement of entities. The holes of the wedge will be represented by circles. The wedge may be surfaced using the 3Dface command. This command will be covered in a later tutorial exercise.

DIMENSIONING:

This object does not require any dimensioning.

PLOTTING:

This tutorial exercise may be plotted on "B"-size paper (11" x 17"). Be sure to set the units of measure from inches to millimeters. Plot the object to at a scale value of 1=1.

VERSION OF AUTOCAD:

This tutorial exercise may be completed using either AutoCAD Release 10 or Release 11.

Step #1

Begin the drawing by drawing the base of the Wedge. Use the line command and absolute coordinates to perform this operation. Follow the prompts below:

Command: **Line**
From point: **0,0,0**
To point: **30,0,0**
To point: **30,20,0**
To point: **0,20,0**
To point: **C**

Instead of typing in a value for the last coordinate, enter the letter **C** which stands for "Close". This will close the box and exit the line command.

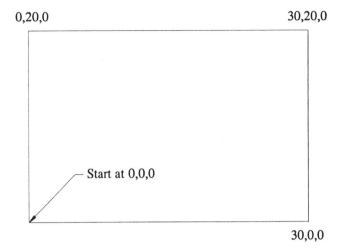

Step #2

Use the Vpoint command and the Rotate option to view the box in 3D. As you construct the next lines, they will show up in 3D. Follow the prompts below:

Command: **Vpoint**
Rotate/<View point><0,0,1>: **R**
Enter angle in X-Y plane from X axis<270>: **-45**
Enter angle from X-Y plane <90>: **30**

Command: **Zoom**
All/Center/Dynamic/Extents/Left/Previous/Window/<Scale(X)>: **0.6X**

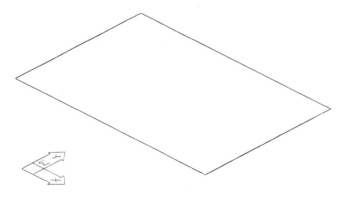

Step #3

Begin creating the upper surface of the Wedge. Again, use the Line command with an absolute coordinate value to begin the line followed by polar coordinates. Follow the prompts below:

Command: **Line**
From point: **0,10,15**
To point: **@10<0**
To point: **@10<90**
To point: **@10<180**
To point: **C**

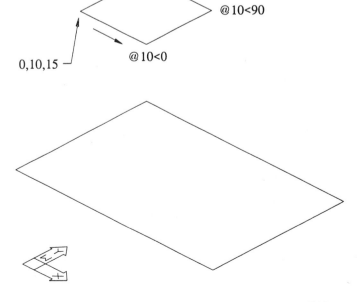

Step #4

Begin work in the next surface of the Wedge. Use a combination of absolute and polar coordinates with the Line command to construct the geometry at the right. Follow the prompts below:

Command: **Line**
From point: **0,0,10**
To point: **@20<0**
To point: **@20<90**
To point: **@5<180**
To point: **@15<270**
To point: **@15<180**
To point: **C**

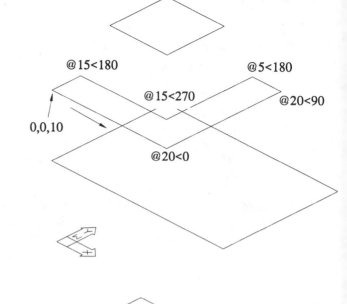

Step #5

Construct the next surface of the wedge using the following prompts of the Line command:

Command: **Line**
From point: **20,0,5**
To point: **@5<0**
To point: **@20<90**
To point: **@5<180**
To point: **C**

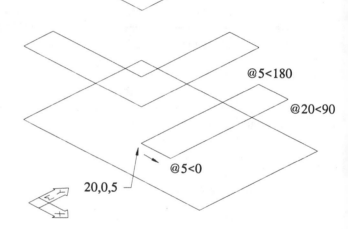

Step #6

Construct the last level of the Wedge with the Line command and a combination of absolute and relative coordinates. Follow the prompts below:

Command: **Line**
From point: **25,0,10**
To point: **@5<0**
To point: **@20<90**
To point: **@5<180**
To point: **C**

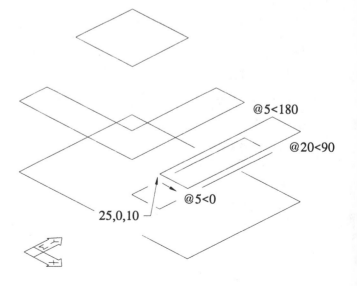

Step #7

The wireframe should appear similar to the illustration at the right. The next series of steps consist of connecting all levels with lines and adding circles to complete the wireframe.

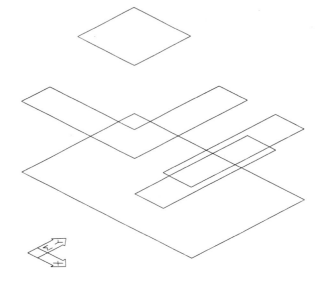

Step #8

Use the Line command to connect all levels to complete the outline of the Wedge. Be sure to use the Osnap-Endpoint option at all times.

Command: **Osnap**
Object snap modes: **Endp**

Command: **Line**
From point: *(Select the endpoint of the line at "A")*
To point: *(Select the endpoint of the line at "B")*
To point: *(Strike Enter to exit this command)*

Repeat the above procedure to connect the remaining edges of the wireframe model. When completed, set the Osnap command to None.

Command: **Osnap**
Object snap modes: **None**

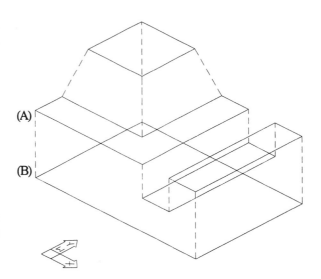

Step #9

Begin representing the holes in the Wedge by constructing circles at the beginning of the hole and end of the hole. Use the Circle command and enter the absolute coordinate 5,15,15 for the center of the circle. (This will place the center of the circle 5 units in the X direction, 15 units in the Y direction, and 15 units in the Z direction.) Enter **D** for diameter of the circle followed by the number 4.00.

Command: **Circle**
3P/2P/TTR/<Center point>: **5,15,15**
Diameter or <Radius>: **D**
Diameter: **4**

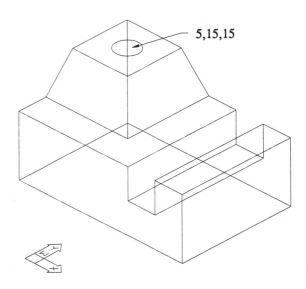

5,15,15

Step #10

Use the Copy command to copy the circle from the top of the wireframe to the bottom of the wireframe. Again, use Osnap-Endpoint and existing geometry to accomplish this.

Command: **Copy**
Select objects: *(Select the circle on the top surface)*
Select objects: *(Strike Enter to continue)*
<Base point or displacement>/Multiple: **Endp**
of *(Select the endpoint of the corner at "A")*
Second point of displacement: **Int**
of *(Select the intersection at "B")*

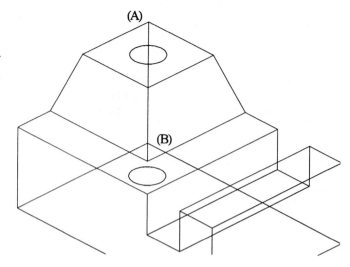

Step #11

Use the Circle command again to place the circles of diameter 2.00 in the slot as illustrated at the right. Use the absolute coordinates of 22.5,5,5 and 22.5,15,5 to identify the centers of the circles.

Command: **Circle**
3P/2P/TTR/<Center point>: **22.5,5,5**
Diameter or <Radius>: **D**
Diameter: **2**

Command: **Circle**
3P/2P/TTR/<Center point>: **22.5,15,5**
Diameter or <Radius>: **D**
Diameter: **2**

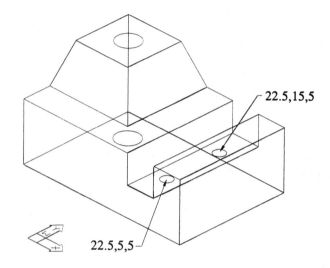

Step #12

Use the Copy command again to copy the two holes from the top of the slot to the bottom of the base of the Wedge. A hint to performing this on existing geometry is to use the Osnap-Endpoint and Osnap-Perpendicular options.

Command: **Copy**
Select objects: *(Select the 2 small circles at the right)*
Select objects: *(Strike Enter to continue)*
<Base point or displacement>/Multiple: **Endp**
of *(Select the endpoint of the corner at "A")*
Second point of displacement: **Per**
of *(Select the perpendicular point to "A" at "B")*

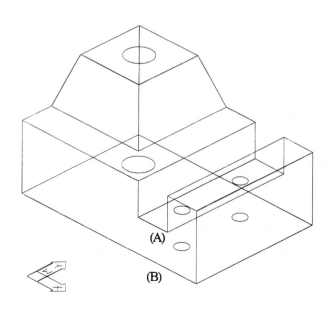

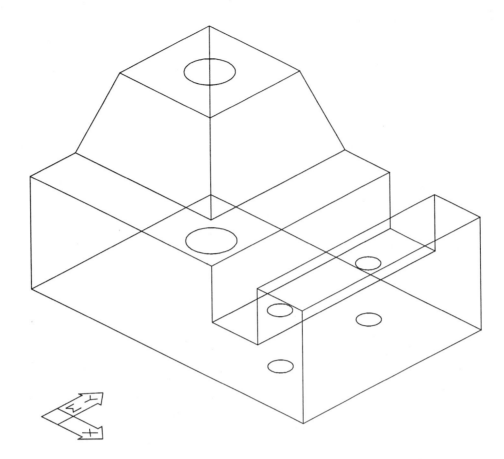

An alternate step would be to use the 3Dface command to surface the model for import into shading packages such as AutoShade. To display the circles as holes that show depth, the "Slot" AutoLisp routine is used. Use of the 3Dface command will be detailed in Tutorial Exercise #20.

Tutorial Exercise #19
Lever.Dwg

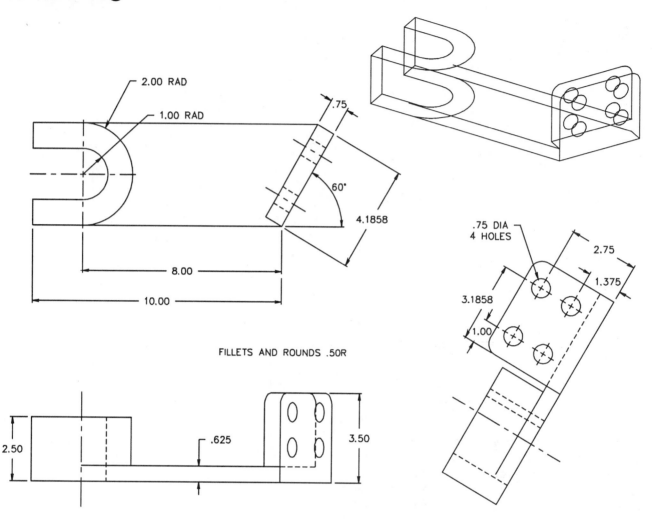

2.00 RAD

1.00 RAD

.75

60°

4.1858

8.00

10.00

FILLETS AND ROUNDS .50R

.75 DIA
4 HOLES

2.75

1.375

3.1858

1.00

2.50

.625

3.50

PURPOSE:
This tutorial is designed to use the UCS command to construct a 3D model of the Lever.

SYSTEM SETTINGS:
Begin a new drawing called "Lever." Use the Units command to change the number of decimal places past the zero from 4 to 2. Keep all default values for the Units command. Using the Limits command, keep 0,0 for the lower left corner and change the upper right corner from 12,9 to 15.50,9.50. Use the Grid command and change the grid spacing from 1.00 to 0.50 units. Do not turn the snap or ortho On.

LAYERS:
Create the following layers with the format:
Name-Color-Linetype
Object - White - Continuous

SUGGESTED COMMANDS:
Begin layout of this problem by constructing the plan view of the Lever. Use the UCS command to manipulate, create, and save numerous UCS' to complete details of the Lever.

DIMENSIONING:
Dimensions will not be added to this problem.

PLOTTING:
This tutorial exercise may be plotted on "B"-size paper (11" x 17"). Use a plotting scale of 1=1 to produce a scaled plot.

VERSION OF AUTOCAD:
This tutorial exercise may be completed using either AutoCAD Release 10 or Release 11.

Step #1

Begin this drawing by constructing two circles of radius values 1.00 and 2.00 using the Circle command and 4.00,5.00 as the center of both circles.

Command: **Circle**
3P/2P/TTR/<Center point>: **4.00,5.00**
Diameter/<Radius>: **1.00**

Command: **Circle**
3P/2P/TTR/<Center point>: **4.00,5.00**
Diameter/<Radius>: **2.00**

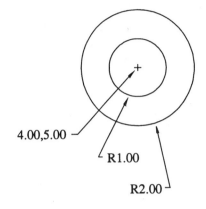

Step #2

Draw a vertical line 2 units to the left of the center of the circles and 4 units long. This can easily be accomplished by using the .XYZ filters.

Command: **Line**
From point: **.Y**
of **Qua**
of *(Select the quadrant of the circle at "Y")*
(need XZ): **Qua**
of *(Select the quadrant of the circle at "X")*
To point: **@4<270**
To point: *(Strike Enter to exit this command)*

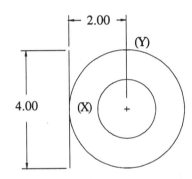

Step #3

Draw two horizontal lines from quadrant points on the two circles a distance of 2 units. These lines should intersect with the vertical line drawn in the previous step.

Command: **Line**
From point: **Qua**
of *(Select the quadrant of the large circle at "A")*
To point: **@2<180**
To point: *(Strike Enter to exit this command)*

Repeat the above procedure and draw three more lines from points "B", "C", and "D" using the same polar coordinate value for the length, namely, @2<180. An alternate method would be to use the Copy-Multiple command to duplicate the three remaining lines.

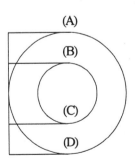

Step #4

Use the Trim command, select the two dashed lines at the right as cutting edges, and trim the left side of the large circle.

Command: **Trim**
Select cutting edges...
Select objects: *(Select the two dashed lines at the right)*
Select objects: *(Strike Enter to continue)*
Select object to trim: *(Select the large circle at "A")*
Select object to trim: *(Strike Enter to exit this command)*

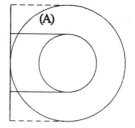

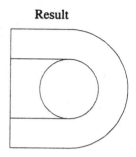

Result

Step #5

Use the Trim command again and select the two dashed lines at the right as cutting edges. Select the left side of the small circle and the middle of the vertical line as the objects to trim.

Command: **Trim**
Select cutting edges...
Select objects: *(Select the two dashed lines at the right)*
Select objects: *(Strike Enter to continue)*
Select object to trim: *(Select the vertical line at "A")*
Select object to trim: *(Select the small circle at "B")*
Select object to trim: *(Strike Enter to exit this command)*

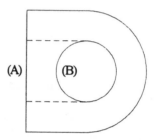

Result

Step #6

Complete a partial plan view by using the Line command and the illustration at the right to draw the four lines. Use Osnap-Endpoint whenever possible; also use polar coordinates.

Command: **Line**
From point: **Endp**
of *(Select the endpoint of the line or arc labeled "Start")*
To point: **@8.00<0**
To point: **@4.1858<60**
To point: **@.75<150**
To point: **Endp**
of *(Select the endpoint of the line or arc labeled "End")*
To point: *(Strike Enter to exit this command)*

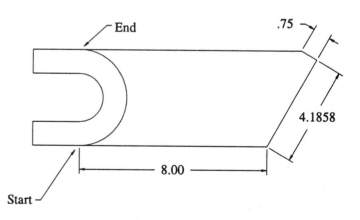

Step #7

Your display should be similar to the illustration at the right.

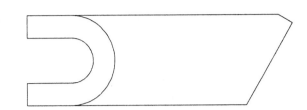

Step #8

The next feature to be drawn of the lever will be the bracket that consists of four holes. Before this can be drawn, a new UCS must be made that will allow entities to be drawn on the new user-specified plane. The next series of steps outlines manipulating the UCS icon to form the new UCS. Use the Osnap-Endpoint option whenever possible. Update the UCS icon using the Ucsicon command.

Command: **UCS**
Origin/ZAxis/3point/Entity/View/X/Y/Z/Prev/Restore/Save/
    Del/?/<World>: **O**
Origin point <0,0,0>: **Endp**
of *(Select the endpoint of the horizontal line at "A")*

Command: **Ucsicon**
ON/OFF/All/Noorigin/ORigin<ON>: **Or**

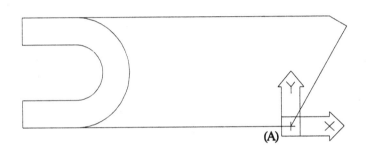

Step #9

Use the Vpoint command and Rotate option to generate a view that is rotated 300 degrees in the X-Y plane and 30 degrees from the X-Y plane. Use the View command and save the display under the name "Iso". Use the UCS command and rotate the UCS icon 60 degrees about the Z axis.

Command: **Vpoint**
Rotate/<View point><0.0000,0.0000,0.0000,1.0000>: **Ro-
    tate**
Enter angle in X-Y plane from X axis <>: **300**
Enter angle from X-Y plane < >: **30**

Command: **View**
?/Delete/Restore/Save/Window: **S**
View name to save: **Iso**

Command: **UCS**
Origin/ZAxis/3point/Entity/View/X/Y/Z/Prev/Restore/Save/
    Del/?/<World>: **Z**
Rotation angle about Z axis <0.0>: **60**

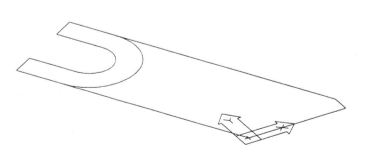

Step #10

Use the UCS command to rotate the UCS icon 90 degrees about the X axis. Save this UCS under the name "Bracket".

Command: **UCS**
Origin/ZAxis/3point/Entity/View/X/Y/Z/Prev/Restore/Save/
    Del/?/<World>: **X**
Rotation angle about X axis <0.0>: **90**

Command: **UCS**
Origin/ZAxis/3point/Entity/View/X/Y/Z/Prev/Restore/Save/
    Del/?/<World>: **S**
UCS name to save: **Bracket**

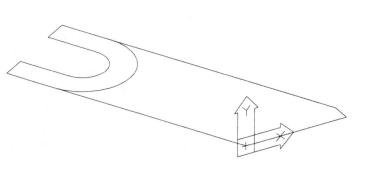

Step #11

Begin drawing the bracket part of the lever in the new UCS that you already defined. To assist you in this operation, use the Plan command to view the model in plan view to the current UCS. Then use the Line command and polar coordinates to draw the outline of the bracket.

Command: **Plan**
<Current UCS>/Ucs/World: *(Strike Enter)*

Command: **Line**
From point: **0,0**
To point: **@3.50<90**
To point: **@4.1858<0**
To point: **@3.50<270**
To point: *(Strike Enter to exit this command)*

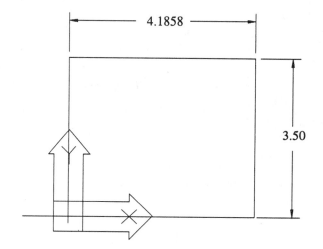

Step #12

Add the four circles of .75 diameter to the lever's bracket. To locate the centers, use the Offset command to offset one horizontal and one vertical line using the distances at the right. Now use the circle command in combination with the Osnap-Intersection option to draw the circles from the intersection of the four offset lines. Erase the lines used to locate the centers of the circles.

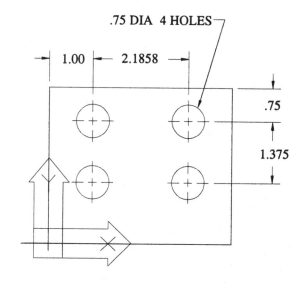

Step #13

Use the Fillet command and change the current fillet radius value to 0.50. Then use the Fillet command to place arcs between lines "A" and "B" and between lines "B" and "C".

Command: **Fillet**
Polyline/Radius/<Select two objects>: **R**
Enter fillet radius <0.0000>: **0.50**

Command: **Fillet**
Polyline/Radius/<Select two objects>: *(Select lines "A" and "B")*

Command: **Fillet**
Polyline/Radius/<Select two objects>: *(Select lines "B" and "C")*

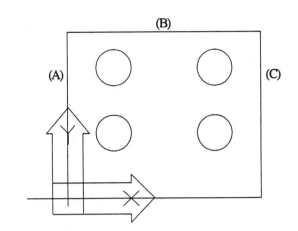

Step #14

Your display should be similar to the one illustrated at the right.

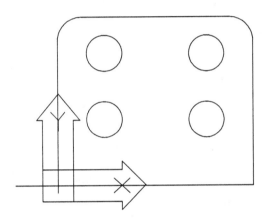

Step #15

Use the View command to restore the previously saved view called "Iso". Your display should be similar to the illustration at the right.

Command: **View**
?/Delete/Restore/Save/Window: **R**
View name to restore: **Iso**

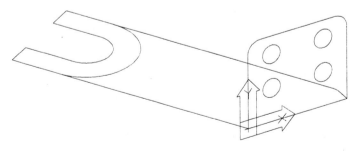

Step #16

Before performing any drawing, reset the current coordinate system to the World coordinate system. Then use the illustration at the right to guide you in copying the small line at "A" to the new position at "B".

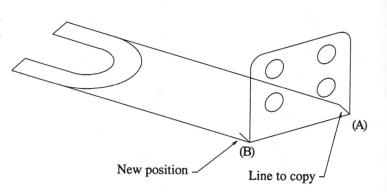

Command: **UCS**
Origin/ZAxis/3point/Entity/View/X/Y/Z/Prev/Restore/Save/
    Del/?/<World>: **W**

Command: **Copy**
Select objects: *(Select the line at the right to copy)*
Select objects: *(Strike Enter to continue)*
<Base point or displacement>/Multiple: **Endp**
of *(Select the endpoint of the line at "A")*
Second point of displacement: **Endp**
of *(Select the endpoint of the line at "B" labeled "New
    position")*

Step #17

Move the dashed line the distance .625 in the Z direction using the Osnap-Endpoint option and XYZ filters.

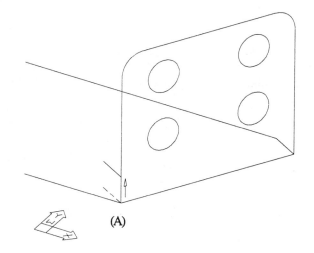

Command: **Move**
Select objects: *(Select the dashed line at the right)*
Select objects: *(Strike Enter to continue)*
Base point or displacement: **Endp**
of *(Select the endpoint of the dashed line at "A")*
Second point of displacement: **.XY**
of **Endp**
of *(Select the endpoint of the dashed line at "A")*
(Need Z): **.625**

Step #18

Copy the two dashed lines at the right a distance of .625 using the Copy command, Osnap-Endpoint option, and XYZ filters.

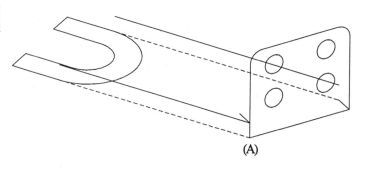

Command: **Copy**
Select objects: *(Select the two dashed lines at the right)*
Select objects: *(Strike Enter to continue)*
<Base point or displacement>/Multiple: **Endp**
of *(Select the endpoint of the dashed line at "A")*
Second point of displacement: **.XY**
of **Endp**
of *(Select the endpoint of the dashed line at "A")*
(Need Z): **.625**

Step #19

Use the UCS command and restore the UCS called "Bracket".

Command: **UCS**
Origin/ZAxis/3point/Entity/View/X/Y/Z/Prev/Restore/Save/
    Del/?/<World>: **R**
Ucs to restore: **Bracket**

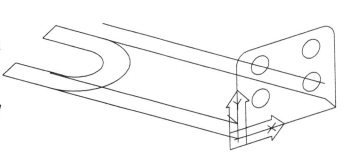

Step #20

Use the UCS command with the Origin option to move the UCS icon to the endpoint of line "A". If the icon is not displayed on the line, use Zoom and a value of .8x.

Command: **UCS**
Origin/ZAxis/3point/Entity/View/X/Y/Z/Prev/Restore/Save/
    Del/?/<World>: **O**
Origin point <0,0,0>: **Endp**
of   *(Select the endpoint of line "A")*

Command: **Zoom**
All/Center/Dynamic/Extents/Left/Previous/Window/
    <Scale(x)>: **0.8x**

Step #21

Draw four lines representing the back side of the bracket using the Line command and the current position of the UCS. Draw the lines from "A" , to "B", to "C", to "D", and close the rectangle.

Command: **Line**
From point: **0,0** *(at "A")*
To point: **@4.1858<180** *(at "B")*
To point: **@2.875<90** *(at "C")*
To point: **@4.1858<0** *(at "D")*
To point: **C** *(at "A")*

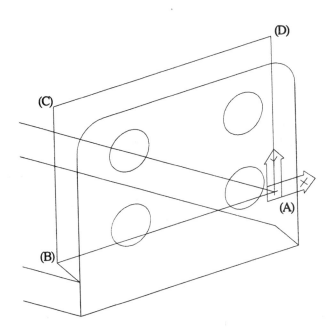

Step #22

Your display should appear similar to the illustration at the right.

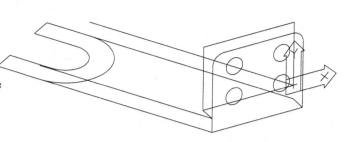

Step #23

Copy the four circles from the front face of the bracket to the rear face using the Copy command and the Osnap-Endpoint option.

Command: **Copy**
Select objects: *(Select the four dashed circles)*
Select objects: *(Strike Enter to continue)*
<Base point or displacement>/Multiple: **Endp**
of *(Select the endpoint of the line at "A")*
Second point of displacement: **Endp**
of *(Select the endpoint of the line at "B")*

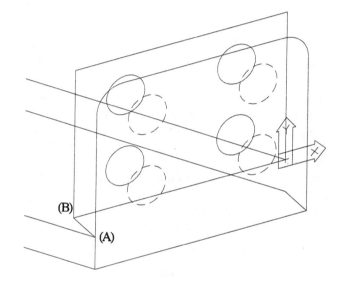

Step #24

Use the Fillet command to place arcs between lines "A" and "B" and between lines "B" and "C" (the fillet radius should already be set at 0.50).

Command: **Fillet**
Polyline/Radius/<Select two objects>: *(Select lines "A" and "B")*

Command: **Fillet**
Polyline/Radius/<Select two objects>: *(Select lines "B" and "C")*

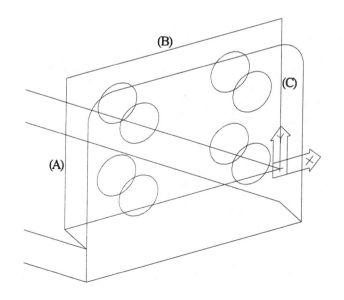

Step #25

Your display should appear similar to the illustration at the right.

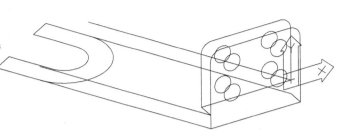

Step #26

Change back to the World coordinate system using the UCS command. Copy the dashed entities at the right the distance of 2.50 in the Z direction using the Osnap-Endpoint option and XYZ filters.

Command: **UCS**
Origin/ZAxis/3point/Entity/View/X/Y/Z/Prev/Restore/Save/
    Del/?/<World>: **W**

Command: **Copy**
Select objects: *(Select all dashed entities at the right)*
Select objects: *(Strike Enter to continue)*
<Base point or displacement>/Multiple: **Endp**
of *(Select the endpoint of the line at "A")*
Second point of displacement: **.XY**
of **Endp**
of *(Select the endpoint of the line at "A")*
(Need Z): **2.50**

(A)

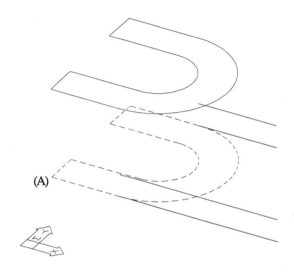

Step #27

Your display should appear similar to the illustration at the right.

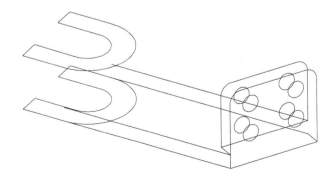

Step #28

Draw a line from "A" to "B" using the Line command and the Osnap-Endpoint option. Copy this line to points "C", "D", and "E" using the Copy-Multiple option.

Command: **Line**
From point: **Endp**
of *(Select the endpoint of the line at "A")*
To point: **Endp**
of *(Select the endpoint of the line at "B")*
To point: *(Strike Enter to exit this command)*

Command: **Copy**
Select objects: **L**
Select objects: *(Strike Enter to continue)*
<Base point or displacement>/Multiple: **M**
First point: **Endp**
of *(Select the endpoint of the line at "B")*
Second point of displacement: **Endp**
of *(Select the endpoint of the line at "C")*
Second point of displacement: **Endp**
of *(Select the endpoint of the line at "D")*
Second point of displacement: **Endp**
of *(Select the endpoint of the line at "E")*
Second point of displacement: *(Strike Enter to exit this command)*

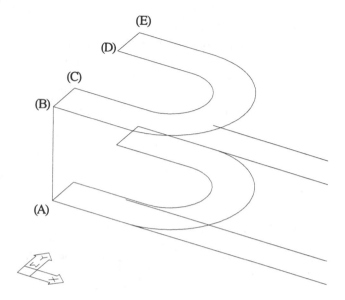

Step #29

Move the dashed circle at the right a distance of .625 in the Z direction. Use the Osnap-Endpoint option and XYZ filters.

Command: **Move**
Select objects: *(Select the dashed circle at the right)*
Select objects: *(Strike Enter to continue)*
Base point or displacement: **Endp**
of *(Select the endpoint of the line at "A")*
Second point of displacement: **.XY**
of **Endp**
of *(Select the endpoint of the line at "A")*
(Need Z): **.625**

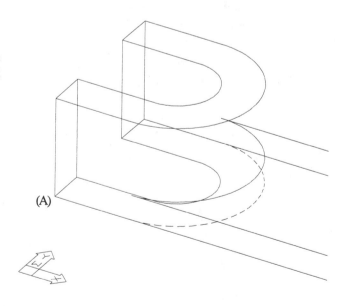

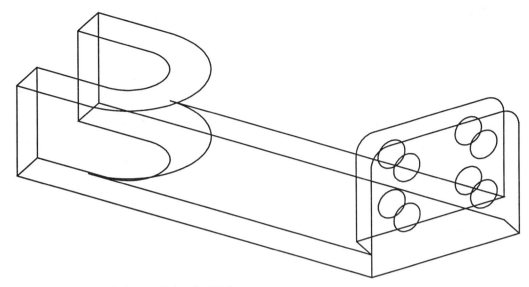

The completed object is illustrated above. Using the 3Dface command and geometry generated surface commands, the Lever may be surfaced as an optional step. These commands will be explained in more detail in the tutorial exercises to follow.

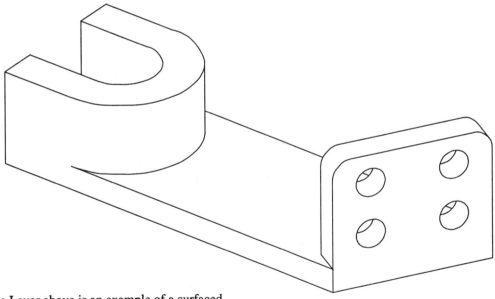

The display of the Lever above is an example of a surfaced model that has had the hidden lines removed to view the object as it would appear in real life. However, even though the object appears as a solid image, it is actually hollow on the inside. Converting the image to a solid model will be discussed in Unit 10.

Tutorial Exercise #20
Gear.Dwg

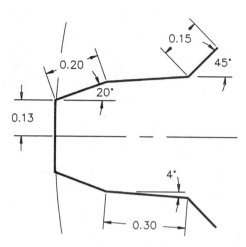

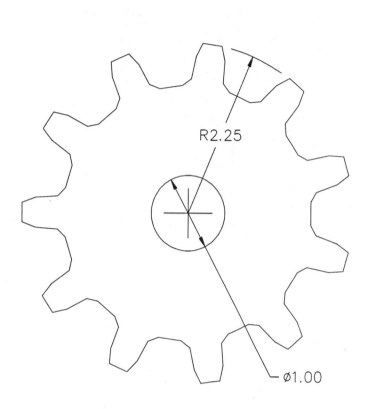

Detail of Gear Tooth

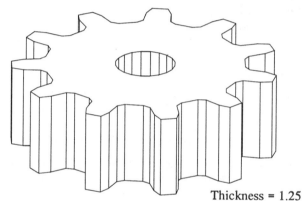

Thickness = 1.25

PURPOSE:
This tutorial is designed to use the UCS and 3DFace commands to produce a model of the Gear.

SYSTEM SETTINGS:
Begin a new drawing called "Gear." Use the Units command to change the number of decimal places past the zero from 4 to 2. Keep the remaining default unit values. Using the Limits command, keep 0,0 for the lower left corner and change the upper right corner from 12,9 to 10.50,8.00. Use the Grid command and change the grid spacing from 1.00 to 0.25 units. Do not turn the snap or ortho On. Use the Setvar command and change the Pdmode variable from 0 to 2. Use the Setvar command again to change the Pdsize variable from 0 to 0.25.

LAYERS:
Create the following layers with the format:
Name-Color-Linetype
Wireframe - Yellow - Continuous
Surface - Magenta - Continuous

SUGGESTED COMMANDS:
Use the Line and Array commands to layout the plan view of the Gear. Use .XYZ filters to copy the gear profile in the Z direction to produce the thickness of the Gear. Begin surfacing one gear tooth using the 3dface command. Use the Array to copy and duplicate the 3dfaces to the produce the other gear teeth. The Slot.Lsp routine is used to produce the hole through the Gear. Use the Rulesurf command if Slot.Lsp is not available.

DIMENSIONING:
Dimensions will not be added to this drawing.

PLOTTING:
This tutorial exercise may be plotted on "B" size paper, (11" x 17"). Use a plotting scale of 1=1 to produce a full size plot.

VERSION OF AUTOCAD:
This tutorial exercise may be completed using either AutoCAD Release 10 or Release 11.

Step #1

Begin construction of the gear by drawing two circles using coordinate 4.50,4.00 as the center of both circle. Use the Point command to place a point at the center of the small circle using the Osnap-Center option.

Command: **Circle**
3P/2P/TTR/<Center point>: **4.50,4.00**
Diameter/<Radius>: **2.25** *(For the large circle)*

Command: **Circle**
3P/2P/TTR/<Center point>: **4.50,4.00**
Diameter/<Radius>: **0.50** *(For the small circle)*

Command: **Point**
Point: **Cen**
of *(Select the circle at point "A")*

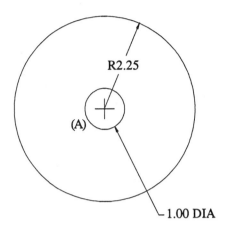

Step #2

Begin drawing the profile of one gear tooth using the Line command. Begin the gear tooth at the 270 degree quadrant of the circle (left side of the circle). Use the Osnap-Quadrant mode to begin the line. Then use the following coordinates to form the gear tooth illustrated at the right. To better view the construction of the gear tooth, use the Zoom-Window command and magnify the left side of the circle.

Command: **Line**
From point: **Qua**
of *(Select the quadrant of the circle at point "A")*
To point: **@0.13<90** *(Point "B")*
To point: **@0.20<20** *(Point "C")*
To point: **@0.30<4** *(Point "D")*
To point: **@0.15<45** *(Point "E")*
To point: *(Strike Enter to exit this command)*

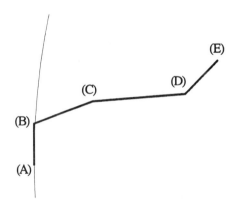

Step #3

Use the Mirror command to copy and flip the highlighted entities at the right to form the other half of the gear. Use a polar coordinate value of @1.00<0 as the second point of the mirror line.

Command: **Mirror**
Select objects: *(Select all dashed entities at the right)*
Select objects: *(Strike Enter to continue)*
First point of mirror line: **Qua**
of *(Select the quadrant of the circle at point "A")*
Second point: **@1.00<0**
Delete old objects?<N>: *(Enter to accept the default and exit the command)*

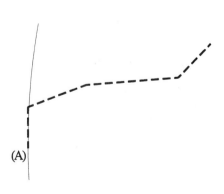

Step #4

Use the Array command to copy all entities representing the gear tooth in a circular pattern. Copy the tooth 11 times. When completed, use the Erase command to delete the large circle.

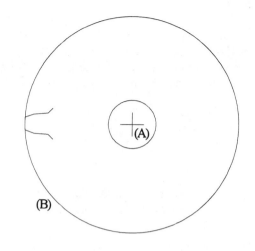

Command: **Array**
Select objects: *(Select all gear tooth entities at the right)*
Select objects: *(Strike Enter to continue)*
Rectangular or Polar array (R/P): **P**
Center point of the array: **Nod**
of *(Select the point illustrated at "A")*
Number of items: **11**
Angle to fill (+=CCW, -+CW) <360>: *(Strike Enter to accept)*
Rotate items as they are copied?<Y>: *(Strike Enter to accept)*

Command: **Erase**
Select objects: *(Select the large circle at "B")*
Select objects: *(Strike Enter to execute this command)*

Command: **Redraw**

Step #5

Use the Zoom-Window command to magnify the area of the gear illustrated at the right. Draw a line connecting one gear tooth with the other using the Line command. After the line is drawn, perform a Zoom-Previous to return to the previous screen display.

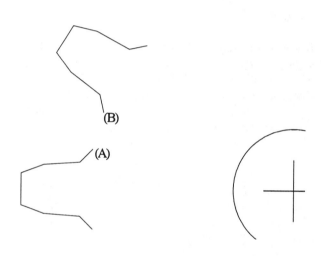

Command: **Line**
From point: **Endp**
of *(Select the endpoint of the line at "A")*
To point: **Endp**
of *(Select the endpoint of the line at "B")*
To point: *(Strike Enter to exit this command)*

Step #6

Use the Array command to copy the last line drawn 11 times in a circular pattern. Use the View command and save the view at the right under the name of "Plan".

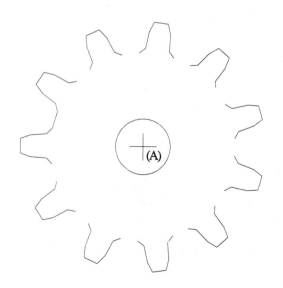

Command: **Array**
Select objects: **L**
Select objects: *(Strike Enter to continue)*
Rectangular or Polar array (R/P): **P**
Center point of the array: **Nod**
of *(Select the point illustrated at "A")*
Number of items: **11**
Angle to fill (+=CCW, -+CW) <360>: *(Strike Enter to accept)*
Rotate items as they are copied?<Y>: *(Strike Enter to accept)*

Command: **View**
?/Delete/Restore/Save/Window: **S**
View name to save: **Plan**

Step #7

Use the Vpoint command and type in a coordinate value of 1.00, -0.50, 0.50 to view the gear in 3D. Then use the Zoom-Window command to view the gear in an enlarged size. Save this view under the name of "Iso".

Command: **Vpoint**
Rotate/<View point><0.00,0.00,0.00>: **1, -0.50, 0.50**

Command: **View**
?/Delete/Restore/Save/Window: **S**
View name to save: **Iso**

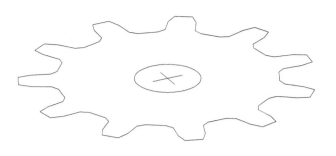

Step #8

Use .XYZ filters to copy the point of the gear a distance of 1.25 in the Z direction. This point will be used in a later step.

Command: **Copy**
Select Object: *(Select the point at "A")*
Select Objects: *(Strike Enter to continue)*
<Base point or displacement>/Multiple: **Cen**
of *(Select the center of the circle at "B")*
Second point of displacement: **.XY**
of **Cen**
of *(Select the center of the circle at "B")*
need (Z): **1.25**

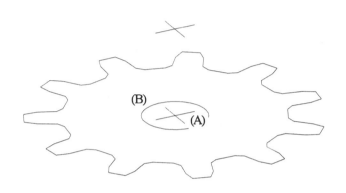

Step #9

Use the Slot Lisp routine to draw a hole in 3D with a diameter of 1.00 and depth of 1.25. Use the Setvar command to set the Splframe variable to a value of (1). This will display or "turn on" the square box outlining the hole at the base and top as illustrated in Step #10. This box will be used later for construction purposes.

Command: **Setvar**
Variable name or ? <Pdsize>: **Splframe**
New value for Splframe<0>: **1**

Command: **(load "slot")**
C:SLOT

Command: **Slot**
Hole or Slot? H/S <S>: **H**
Center point: **Node**
of *(Select the point at "A")*
Radius: **0.50**
Depth: **1.25**
Please wait...done

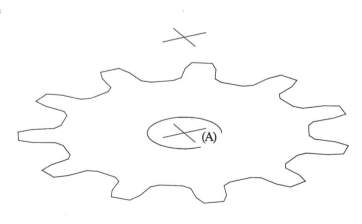

Step #10

The completed hole using the Slot Lisp routine is illustrated at the right. It is very important not to be locked into an Osnap mode such as Endpoint or Intersec. The Slot.Lsp routine will display unfavorable results if locked into running Osnap mode. The Slot.Lsp routine is available on the Bonus diskette supplied with AutoCAD Release 10. The routine is not available in future AutoCAD releases; however Slot.Lsp has properly funtioned when copied and loaded into a Release 11 AutoCAD drawing.

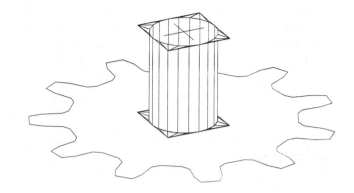

Step #11

Use the Chprop command, select the profile of the gear, and change the thickness of the gear profile from 0 to 1.25. The 1.25 distance represents the total height or thickness of the gear.

Command: **Chprop**
Select objects: *(Select the profile of the gear)*
Select objects: *(Strike Enter to continue)*
Change what property (Color/LAyer/LType/Thickness)? **T**
NewThickness <0.00>: **1.25**
Change what property(Color/LAyer/LType/Thickness)?
    (Strike Enter to exit this command)

Command: **Hide**
Regenerating drawing.
Removing hidden lines: **xxx**

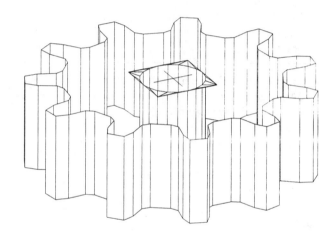

Step #12

Use the Zoom-Window command to magnify one tooth of the gear as illustrated at the right. The top of the gear tooth will be surfaced using the 3Dface command.

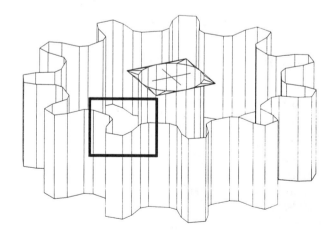

Step #13

Use the 3Dface command to create 3 surfaces on the top of one gear tooth. Be sure to have the Osnap command set to the "Endpoint" mode to assist you in drawing the 3D faces.

Command: **Osnap**
Object snap modes: **Endp**

Command: **3Dface**
First point: *(Select the endpoint at "A")*
Second point: *(Select the endpoint at "B")*
Third point: *(Select the endpoint at "C")*
Fourth point: *(Select the endpoint at "D")*
Third point: *(Select the endpoint at "E")*
Fourth point: *(Select the endpoint at "F")*
Third point: *(Select the endpoint at "G")*
Fourth point: *(Select the endpoint at "H")*
Third point: *(Strike Enter to exit this command)*

Command: **Osnap**
Object snap modes: **Non**

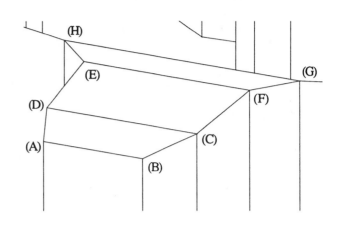

Step #14

Use the Array command to copy the three 3D faces 11 times in a circular pattern. Use the absolute coordinate 4.50,4.00 as the center of the array.

Command: **Array**
Select objects: *(Select the 3Dfaces "A", "B", and "C")*
Select objects: *(Strike Enter to continue)*
Rectangular or Polar array (R/P): **P**
Center point of array: **4.50,4.00**
Number of items: **11**
Angle to fill (+=CCW, -+CW)<360>:*(Strike Enter to accept)*
Rotate objects as they are copied?<Y>*(Strike Enter to accept)*

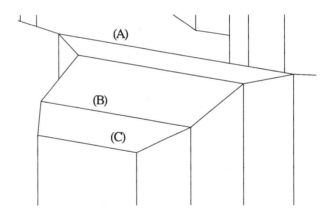

Step #15

Use the View command and restore the save view "Iso". Your drawing of the gear should be similar to the illustration at the right.

Command: **View**
?/Delete/Restore/Save/Window: **R**
View name to restore: **Iso**

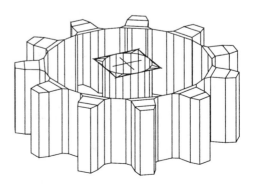

Step #16

Complete the surfacing operation on the gear by using the 3Dface command create the remaining surfaces on the top of the gear. Restore the view "Plan" to accomplish this. To be sure the 3Dfaces are placed on the top surface of the gear, use the Elev (Elevation) command and assign a new elevation of 1.25. Lock yourself in the Osnap-Endpoint mode to further assist in the construction of the faces.

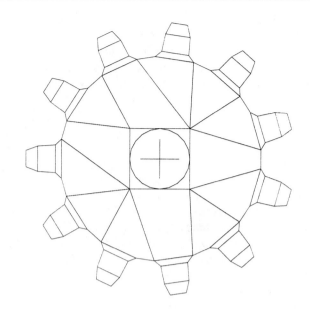

Command: **View**
?/Delete/Restore/Save/Window: **R**
View name to save: **Plan**

Command: **Elev**
New current elevation <0.00>: **1.25**
New current thickness <0.00>: *(Strike Enter to accept the default value and exit the command)*

Step #17

Restore the view called "Iso" to look at the 3D model created. Check that all surfaces have been placed at their appropriate elevations. Reset the Elev command back to 0.00. Remove hidden lines using the Hide command.

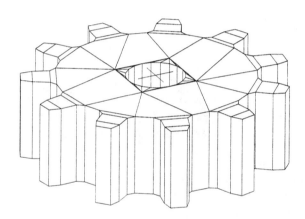

Command: **View**
?/Delete/Restore/Save/Window: **R**
View name to save: **Iso**

Command: **Elev**
New current elevation <1.25>: **0.00**
New current thickness <0.00>: *(Strike Enter to accept the default value and exit the command)*

Step #18

An alternate step would be to load the AutoLisp routine called "Edge". This routine will make visible 3Dface lines invisible when the system variable "Splframe" is set to "0" or off. Use this routine by calling up the "Plan" view, loading the "Edge" routine by entering **(load "Edge")** at the command prompt. Now simply use the Edge command, follow the prompts, and make all visible 3Dface lines invisible.

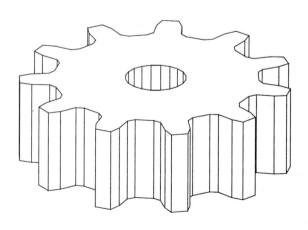

Command: **Edge**
Display/<Select edge>: *(Select a 3Dface edge; it will turn invisible with the system variable Splframe set to "0")*

This procedure has been described in great detail on how to construct the gear in 3D. You will find that 3D constructions become easier with experience.

Tutorial Exercise #21
Column.Dwg (Wireframe Model)

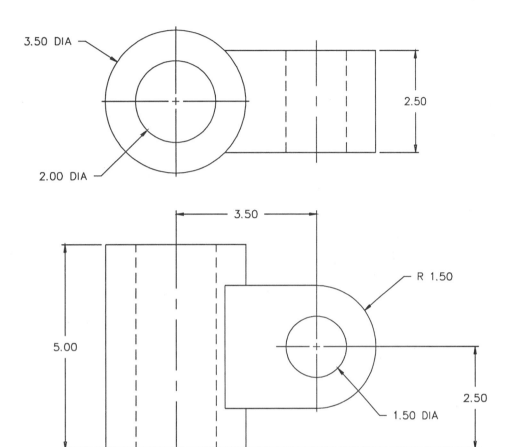

PURPOSE:
This tutorial is designed to use the UCS command to produce a 3D wireframe model of the Column. This model will be surfaced in the next tutorial segment.

SYSTEM SETTINGS:
Begin a new drawing called "Column." Use the Units command to change the number of decimal places past the zero from 4 to 2. Keep the remaining default unit values. Using the Limits command, keep 0,0 for the lower left corner and change the upper right corner from 12,9 to 10.50,8.00. Use the Grid command and change the grid spacing from 1.00 to 0.25 units. Do not turn the snap or ortho On.

LAYERS:
Create the following layers with the format:

Name-Color-Linetype
Wireframe - Yellow - Continuous
Surface - Magenta - Continuous

SUGGESTED COMMANDS:
Begin drawing the Column by laying out the plan view using the circle command. Use XYZ filters to copy the circles in the Z direction to form the column. Create new coordinate systems to lay out the flange that attaches to the column. Use the Trim command to edit the Column before surfacing.

DIMENSIONING:
Dimensions do not have to be added to this problem.

PLOTTING:
This tutorial exercise may be plotted on "B"-size paper (11" x 17"). Use a plotting scale of 1=1 to produce a scaled plot.

VERSION OF AUTOCAD:
This tutorial exercise may be completed using either AutoCAD Release 10 or Release 11.

Step #1

Begin this tutorial by drawing two circles in plan view using the Circle command. These Circles represent the main cylinder and hole going through it.

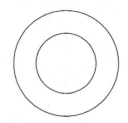

Command: **Circle**
3P/2P/TTR/<Center point>: **4,4**
Diameter/<Radius>: **D**
Diameter: **3.5**

Command: **Circle**
3P/2P/TTR/<Center point>: **4,4**
Diameter/<Radius>: **D**
Diameter: **2.00**

Step #2

Use the Vpoint command to view the circles in 3D. Most constructions will take place from this view point. Next, use the Zoom command to reduce the current display by a factor of 0.7. Finally, save this display using the View command and name the view "Iso". Update the current UCS icon with the Ucsicon-Origin command.

Command: **Vpoint**
Rotate/<View point><0.0000,0.0000,1.0000>: **0.5,-0.75,0.5**

Command: **Zoom**
All/Center/Dynamic/Extents/Left/Previous/Window/
    <Scale(X)> **0.3X**

Command: **View**
?/Delete/Restore/Save/Window: **S**
View name to save: **Iso**

Command: **Ucsicon**
ON/OFF/All/Noorigin/ORigin<ON>: **Or**

Step #3

Use the Copy command and XYZ point filters to copy the bottom circles 5 units in the Z direction.

Command: **Copy**
Select objects: *(Select the two circles)*
Select objects: *(Strike Enter to continue)*
<Base point or displacement>/Multiple: **Cen**
of: *(Select the large circle)*
Second point of displacement: **.XY**
of: **Cen** *(Select the large circle)*
(Need Z): **5**

Step #4

The next series of entities drawn are merely for construction purposes only. They will be deleted at later steps. First, draw a line from the quadrant of the bottom large circle at point "A" to the quadrant of the top large circle at point "B". Next, draw a line from the center of the large circle to a point 3.5 units in the 0 direction using polar coordinates.

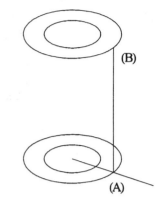

Command: **Line**
From point: **Qua**
of *(Select bottom large circle at point "A")*
To point: **Qua**
of *(Select top large circle at point "B")*
To point: *(Strike Enter to exit this command)*

Command: **Line**
From point: **Cen**
of *(Select bottom large circle)*
To point: **@3.5<0**
To point: *(Strike Enter to exit this command)*

Step #5

Move the construction line 3.5 units in length to a new height 2.5 units from its origin. XYZ filters will be used to accomplish this.

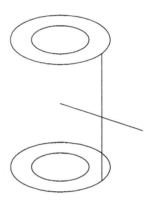

Command: **Move**
Select objects: *(Select the 3.5 unit line)*
Select objects: *(Strike Enter to continue)*
Base point or displacement: **Cen**
of *(Select the bottom large circle)*
Second point of displacement: **.XY**
of **Cen** *(Select the bottom large circle again)*
(Need Z): **2.5**

Step #6

Create a new UCS using the UCS command and the 3point option. (Be sure to use Osnap-Intersec and Osnap-Endpoint to snap onto intersections and endpoints of entities.) Use the prompts below and illustration at the right to guide you in this procedure. Use the UCS command again to save the position under the name "Front".

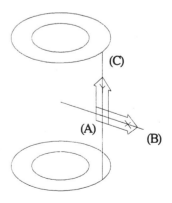

Command: **UCS**
Origin/ZAxis/3point/Entity/View/X/Y/Z/Prev/Restore/Save/
     Del/?/<World>: **3**
Origin point <0,0,0>: *(Select the intersection of Point "A")*
Point on positive portion of the X axis<>: *(Select the endpoint Point "B")*
Point on positive-Y portion of the UCS X-Y plane < >:
     (Select the endpoint Point "C")

Command: **UCS**
Origin/ZAxis/3point/Entity/View/X/Y/Z/Prev/Restore/Save/
     Del/?/<World>: **S**
Name to save: **Front**

Step #7

Draw two circles of radius 1.50 and diameter 1.50. Use point "A" as the center of both circles. (Use Osnap-Endpoint to snap to the endpoint of the line at point "A".)

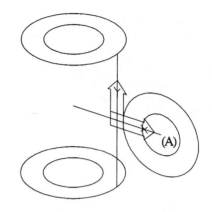

Command: **Circle**
3P/2P/TTR/<Center point>: **Endp**
of *(Select the endpoint of the line at point "A")*
Diameter/<Radius>: **1.50**

Command: **Circle**
3P/2P/TTR/<Center point>: **Endp**
of *(Select the endpoint of the line at point "A")*
Diameter/<Radius>: **D**
Diameter: **1.50**

Step #8

Copy line "A" 1.50 units above and below its current position using the Copy command.

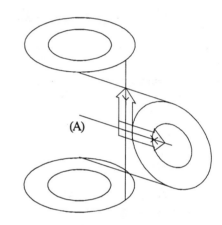

Command: **Copy**
Select objects: *(Select line "A")*
Select objects: *(Strike Enter to continue)*
<Base point or displacement>/Multiple: **Endp**
of *(Select a line near point "A")*
Second point of displacement: **@1.50<90**

Command: **Copy**
Select objects: *(Select line "A")*
Select objects: *(Strike Enter to continue)*
<Base point or displacement>/Multiple: **Endp**
of *(Select a line near point "A")*
Second point of displacement: **@1.50<270**

Step #9

Use the Plan command to obtain a plan view based on the current UCS. The purpose of this step is to use the Trim command to trim the large circle in between the two horizontal lines.

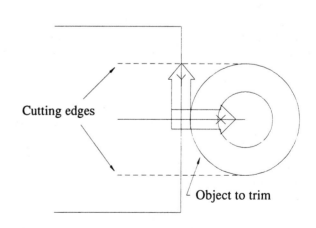

Cutting edges

Object to trim

Command: **Plan**
<Current UCS>/Ucs/World: *(Strike Enter to continue)*

Your display should be similar to the illustration at the right. Now trim the large circle.

Command: **Trim**
Select cutting edge(s)...
Select objects: *(Select the two horizontal dashed lines)*
Select objects: *(Strike Enter to continue)*
Select object to trim: *(Select the large circle at the right)*
Select object to trim: *(Select Enter to exit this command)*

Step #10

Your display should be similar to the illustration at the right. Continue on to Step #11.

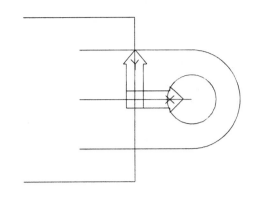

Step #11

Restore the view called "Iso". Then select the dashed entities at the right and move the entities -1.25 units in the Z direction. Use .XYZ filters and the Osnap-Endpoint option to accomplish this.

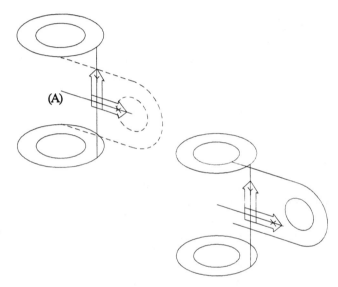

Command: **View**
?/Delete/Restore/Save/Window: **R**
View name to restore: **Iso**

Command: **Move**
Select objects: *(Select the four dashed entities)*
Select objects: *(Strike Enter to continue)*
Base point or displacement: *(Select the endpoint of the line at "A")*
Second point of displacement: **.XY**
of **Endp**
of *(Select the endpoint of line "A" at point "A")*
(Need Z): **-1.25**

Step #12

Select the dashed entities at the right and copy the entities 1.25 units in the Z direction. Use XYZ filters and the Osnap-Endpoint option to accomplish this.

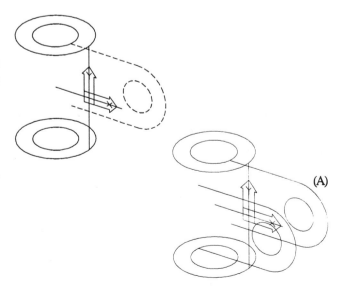

Command: **Copy**
Select objects: *(Select the four dashed entities)*
Select objects: *(Strike Enter to continue)*
<Base point or displacement>/Multiple: *(Select the endpoint of the line at "A")*
Second point of displacement: **.XY**
of **Endp**
of *(Select the endpoint of the line at "A")*
(Need Z): **1.25**

Step #13

Change from the UCS to the World coordinate system by issuing the UCS command and striking the Enter key at the prompt (the World coordinate system is the default).

Command: **UCS**
Origin/ZAxis/3point/Entity/View/X/Y/Z/Prev/Restore/Save/
 Del/?/<World>: **W**

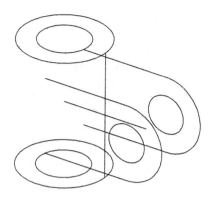

Step #14

Use the Plan command to obtain a plan view based on the current World coordinate system. The purpose of this step is to use the Trim command to trim the two horizontal lines using the large circle as the cutting edge.

Command: **Plan**
<Current UCS>/Ucs/World: *(Strike Enter to continue)*

Your display should be similar to the illustration at the right. Now trim the two horizontal lines.

Command: **Trim**
Select cutting edge(s)...
Select objects: *(Select the large circle)*
Select objects: *(Strike Enter to continue)*
Select object to trim: *(Select the two horizontal lines)*
Select object to trim: **'Redraw** *(Transparent redraw)*
Resuming TRIM command.
Select object to trim: *(Select the two horizontal lines again)*
Select object to trim: *(Strike Enter to exit this command)*

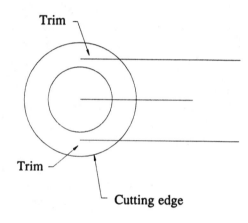

Step #15

Your display should appear similar to the illustration at the right.

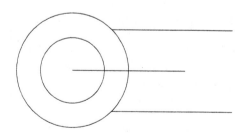

Step #16

Use the View command to restore the view named "Iso". Next, use the Erase command to delete the two construction lines illustrated at the right (lines "A" and "B").

Command: **View**
?/Delete/Restore/Save/Window: **R**
View name to restore: **Iso**

Command: **Erase**
Select objects: *(Select line "A")*
Select objects: *(Select line "B")*
Select objects: *(Strike Enter to execute this command)*

Command: **Redraw**

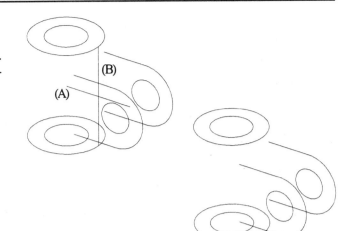

Step #17

Use the Copy command to place two additional circles 1 and 4 units above circle "A". Use the Osnap-Center option in combination with XYZ filters.

Command: **Copy**
Select objects: *(Select circle "A")*
Select objects: *(Strike Enter to continue)*
<Base point or displacement>/Multiple: **Cen**
of *(Select circle "A" using the Osnap-Center option)*
Second point of displacement: **.XY**
of **Cen**
of *(Select Circle "A" again using Osnap-Center)*
(Need Z): **1**

Repeat the above procedure exactly as before; however, type in a value of **4** instead of **1** for the new Z value.

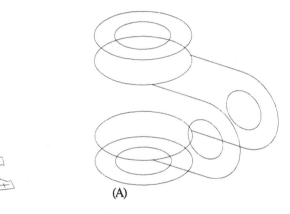

Step #18

The two large circles will be trimmed to form arcs that match the curvature of the cylinder. Use the Trim command, select the four dashed lines as cutting edges, and select one circle at "A" and the other circle at "B" to trim the circles and form arcs.

Command: **Trim**
Select cutting edge(s)...
Select objects: *(Select the four dashed lines)*
Select objects: *(Strike Enter to continue)*
Select object to trim: *(Select the large circle at "A")*
Select object to trim: *(Select the large circle at "B")*
Select object to trim: *(Strike Enter to exit this command)*

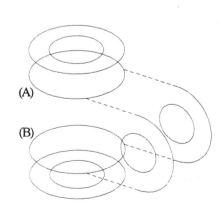

Step #19

Your screen should be similar to the illustration at the right.

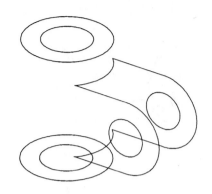

Step #20

This step is meant for construction purposes only and will be used at a later step where surfacing is required. One line is drawn from point "A" to point "B". The line is then copied from point "B" to point "C" using Osnap-Endpoint.

Command: **Line**
From point: **Endp**
of *(Select the endpoint of the line or arc at "A")*
To point: **Endp**
of *(Select the endpoint of the line or arc at "B")*
To point: *(Strike Enter to exit this command)*

Command: **Copy**
Select objects: **L**
Select objects: *(Strike Enter to continue)*
<Base point or displacement>/Multiple: **Endp**
of *(Select the endpoint of the line at "B")*
Second point of displacement: **Endp**
of *(Select the endpoint of the line or arc at "C")*

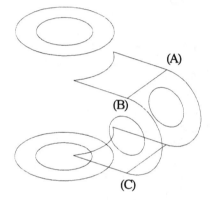

Step #21

Draw another line using the Line command from the end-point of the line or arc at point "A" to the endpoint of the line or arc at point "B".

Command: **Line**
From point: **Endp**
of *(Select the endpoint of the line or arc at "A")*
To point: **Endp**
of *(Select the endpoint of the line or arc at "B")*
To point: *(Strike Enter to exit this command)*

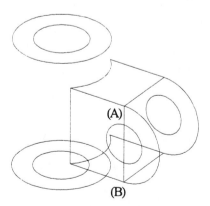

Step #22

Use the UCS command and restore the view named "Front".
Next, use the Trim command, select the circle and vertical
line illustrated at the right as cutting edges, and trim half of
the circle and the middle of the line.

Command: **UCS**
Origin/ZAxis/3point/Entity/View/X/Y/Z/Prev/Restore/Save/
Del/?/<World>: **Restore**
?/Name of UCS to restore: **Front**

Command: **Trim**
Select cutting edge(s)...
Select objects: *(Select line "A" and circle "B")*
Select objects: *(Strike Enter to continue)*
Select object to trim: *(Select line "A" inside the circle)*
Select object to trim: *(Select the right half of the circle)*
Select object to trim: *(Strike Enter to exit this command)*

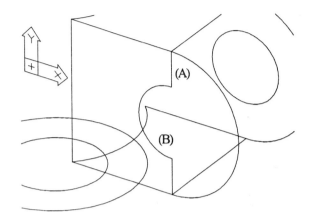

Step #23

Use the Mirror command to copy and flip the dashed arc at
the right. Then use the Pedit command to convert all
dashed entities at the right to a polyline. This polyline will
be used later for adding surfaces to the model.

Command: **Mirror**
Select objects: *(Select the dashed arc at the right)*
Select objects: *(Strike Enter to continue)*
First point of mirror line: *(Select the endpoint at "A")*
Second point: *(Select the endpoint at "B")*
Delete old objects? <N> *(Strike Enter to exit this command)*

Command: **Pedit**
Select polyline: *(Select the dashed line near "A")*
Entity selected is not a polyline.
Do you want to turn it into one?<Y>: **Y**
Close/Join/Width/Edit vertex/Fit curve/Spline curve/
    Decurve/Undo/eXit <X>: **J**
Select objects: *(Select the dashed arc)*
Select objects: *(Select the dashed line near "B")*
Select objects: *(Strike Enter to exit this segment of Pedit)*
2 segments added to polyline
Close/Join/Width/Edit vertex/Fit curve/Spline curve/
    Decurve/Undo/eXit <X>: **X**

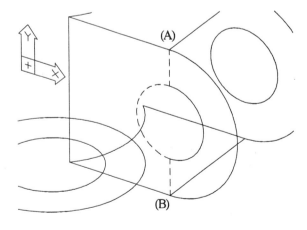

Step #24

Change back to the World coordinate system using the UCS command. Draw a line using the line command from the endpoint of the arc at point "A" to the endpoint of the arc at point "B".

Command: **UCS**
Origin/ZAxis/3point/Entity/View/X/Y/Z/Prev/Restore/Save/
 Del/?/<World>: **W**

Command: **Line**
From point: **Endp**
of *(Select the endpoint of the arc at "A")*
To point: **Endp**
of *(Select the endpoint of the arc at "B")*
To point: *(Strike Enter to exit this command)*

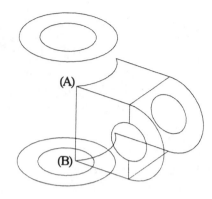

Step #25

Use the copy command to duplicate the last line drawn. Copy this entity from "A" to "B", as illustrated at the right.

Command: **Copy**
Select objects: **L**
Select objects: *(Strike Enter to continue)*
<Base point or displacement>/Multiple: **Endp**
of *(Select the endpoint of the line at "A")*
Second point of displacement: **Endp**
of *(Select the endpoint of the line at "B")*

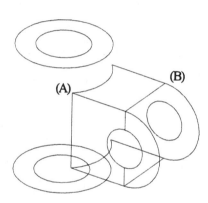

The Wireframe model becomes the basis if the need to surface the model is required. Existing geometry is used for creating the following surfaces; Ruled, Tabulated, Revolved, and Edge surfaces. Follow the next series of steps for surfacing the Column.

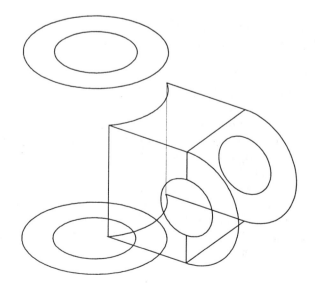

Tutorial Exercise #22
Column.Dwg (Surfaced Model)

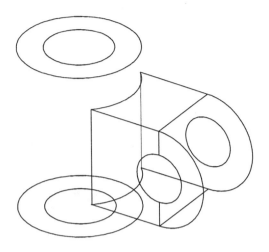

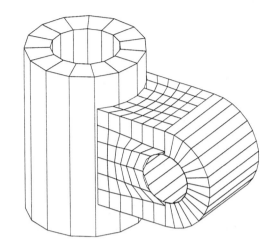

Step #1

This exercise is a continuation of the previous wireframe problem. All necessary entities have been added to the wireframe drawing to prepare the model to be surfaced using the geometry generated surfacing commands outlined in the next series of steps. A few other tasks need to be performed before surfacing. First, make a new layer using the Layer command; call the layer Mesh with a magenta color. Next, change two system variables, namely, Surftab1 and Surftab2, from a value of 6 to a value of 15. These two variables control the density of the surfaces; a value of 15 will be adequate for this model.

Command: **Layer**
?/Make/Set/New/ON/OFF/Color/Ltype/Freeze/Thaw: **M**
New current layer <0>: **Mesh**
?/Make/Set/New/ON/OFF/Color/Ltype/Freeze/Thaw: **C**
Color: **Magenta**
Layer name(s) for color 6 <MESH>: *(Strike Enter to accept the default)*
?/Make/Set/New/ON/OFF/Color/Ltype/Freeze/Thaw: *(Strike Enter to exit this command)*

Observe the top of your screen to see that the new current layer is MESH.

Command: Setvar
Variable name or ?: **Surftab1**
New value for Surftab1 <6>: **15**

Command: Setvar
Variable name or ?: **Surftab2**
New value for Surftab2 <6>: **15**

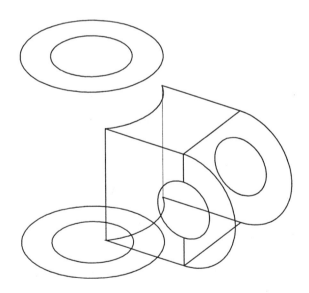

Step #2

Begin surfacing the wireframe model of the column by using the Rulesurf command. This command will place surfaces in between the top and bottom circles to form the cylinder. Pick the outside edge of the large bottom circle at "A" for the first defining curve; pick the outside edge of the large top circle at "B" for the second defining curve. Repeat the procedure for the small circles.

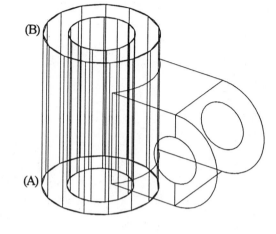

Command: **Rulesurf**
Select first defining curve: *(Select the circle at "A")*
Select second defining curve: *(Select the circle at "B")*

Notice the surface is formed between the two large circles. Now repeat the above procedure for the smaller circles.

Step #3

As the surfacing segment of a wireframe gets more complex, selecting entities may become difficult. Use the Move command to reposition the mesh away from the wireframe. This is accomplished by entering a displacement to move the mesh patterns. Follow the prompts below to see how this is accomplished.

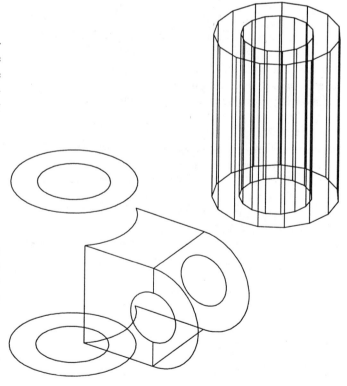

Command: **Move**
Select objects: *(Select both surface mesh patterns)*
Select objects: *(Strike Enter to continue)*
Base point or displacement: **0,10,0**
Second point of displacement: *(Strike Enter)*

Notice both mesh patterns move a distance away from the wireframe based on the move displacement coordinate value of 0,10,0. This same value will be used for moving other surface mesh patterns. Perform a Zoom-All to view the wireframe and mesh pattern.

Command: **Zoom**
All/Center/Dynamic/Extents/Left/Previous/Window/
        <Scale(X)>: **All**
Regenerating Drawing.

Step #4

Use the Rulesurf command to place ruled surfaces on wireframe at the right. Select points "A" and "B" as defining curves to surface the top of the cylinder. Perform the same steps for the bottom cylinder. Select points "C" and "D" as defining curves to begin surfacing the front projection. Copy the front projection to the rear.

Command: **Rulesurf**
Select first defining curve: *(Select the circle at "A")*
Select second defining curve: *(Select the circle at "B")*

Follow the same steps for the bottom circles of the wireframe.

Command: **Rulesurf**
Select first defining curve: *(Select the arc at "C")*
Select second defining curve: *(Select the arc at "D")*

Command: **Copy**
Select objects: **L**
Select objects: *(Strike Enter to continue)*
<Base point or displacement>/Multiple: **Endp**
of *(Select the endpoint of the line at "E")*
Second point of displacement: **Endp**
of *(Select the endpoint of the line at "F")*

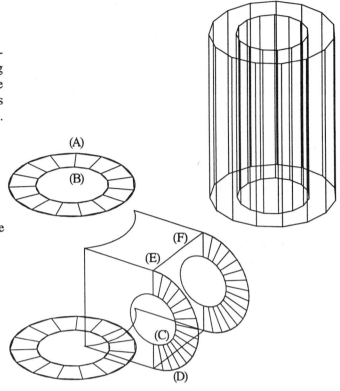

Step #5

Move the four surface mesh patterns at a displacement of 0,10,0 using the Move command.

Command: **Move**
Select objects: *(Select the four surface mesh patterns in the wireframe)*
Select objects: *(Strike Enter to continue)*
Base point or displacement: **0,10,0**
Second point of displacement: *(Strike Enter)*

Command: **Redraw**

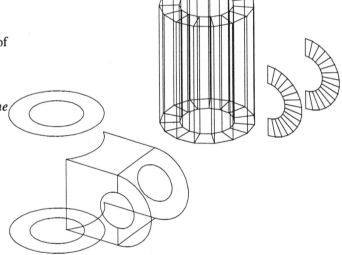

Step #6

Continue using the Rulesurf command by surfacing the curved outer surface of the projection. Select points "A" and "B" as defining curves.

Command: **Rulesurf**
Select first defining curve: *(Select the arc at "A")*
Select second defining curve: *(Select the arc at "B")*

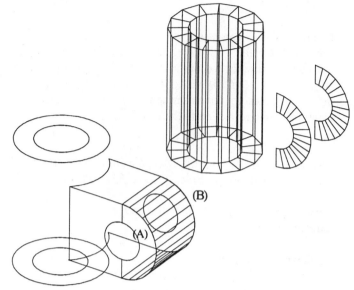

Step #7

Move the last surface mesh pattern at a displacement of 0,10,0 using the Move command.

Command: **Move**
Select objects: **L**
Select objects: *(Strike Enter to continue)*
Base point or displacement: **0,10,0**
Second point of displacement: *(Strike Enter)*

Command: **Redraw**

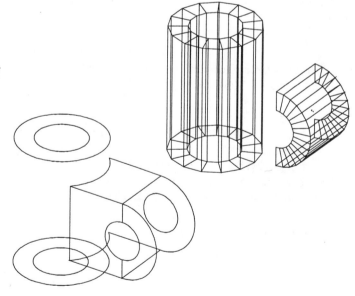

Step #8

A cylinder in the projection needs to be created using the Tabsurf command. This command requires a path curve and direction vector to form the desired mesh pattern. Select the full circle as the path curve; select the line at the right as the direction vector. A mesh pattern will form consisting of information in the path curve; length of the mesh is based on the direction vector.

Command: **Tabsurf**
Select path curve: *(Select the full circle at the right)*
Select direction vector: *(Select the line at the right)*

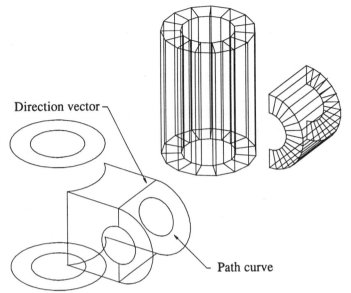

Direction vector

Path curve

Step #9

Your display should be similar to the illustration at the right.

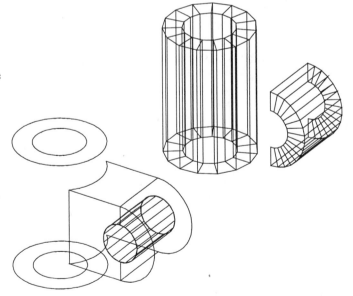

Step #10

Move the last surface mesh pattern at a displacement of 0,10,0 using the Move command.

Command: **Move**
Select objects: **L**
Select objects: *(Strike Enter to continue)*
Base point or displacement: **0,10,0**
Second point of displacement: *(Strike Enter)*

Command: **Redraw**

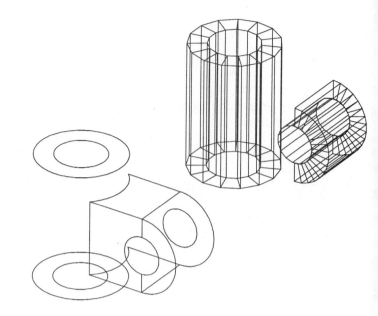

Step #11

Reset the system variables Surftab1 to 6 and Surftab2 to 10. This will reduce the amount of surface meshes created and help with computer regeneration time. Next, use the Edgesurf command, select the four entities labeled at the right, and create a surface similar to the second illustration at the right. Use the illustration in step 12 as a guide in performing this operation.

Command: **Setvar**
Variable name or ?: **Surftab1**
New value for Surftab1 <15>: **6**

Command: **Setvar**
Variable name or ?: **Surftab2**
New value for Surftab2 <15>: **10**

Command: **Edgesurf**
Select edge 1: *(Select the entity at "A")*
Select edge 2: *(Select the entity at "B")*
Select edge 3: *(Select the entity at "C")*
Select edge 4: *(Select the entity at "D")*

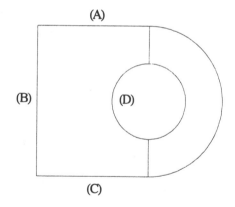

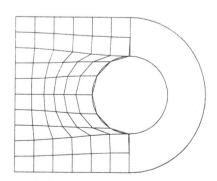

Step #12

Copy the surface just created using the Edgesurf command to create the rear surface.

Command: **Copy**
Select objects: **L**
Select objects: *(Strike Enter to continue)*
<Base point or displacement>/Multiple: **Endp**
of *(Select the endpoint at "A")*
Second point of displacement: **Endp**
of *(Select the endpoint at "B")*

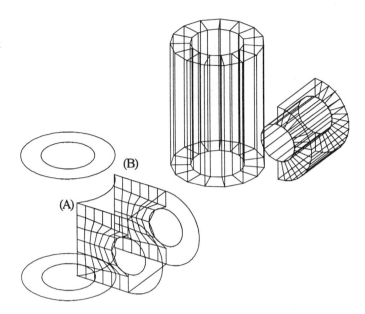

Step #13

Move the two surface mesh patterns at a displacement of 0,10,0 using the Move command.

Command: **Move**
Select objects: *(Select the last two surface mesh patterns)*
Select objects: *(Strike Enter to continue)*
Base point or displacement: **0,10,0**
Second point of displacement: *(Strike Enter)*

Command: **Redraw**

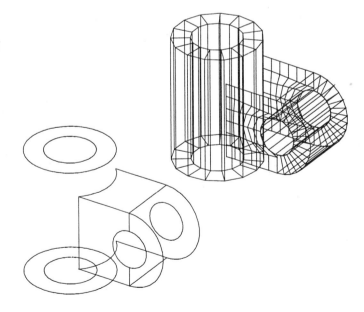

Step #14

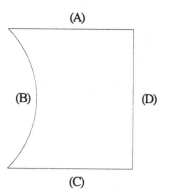

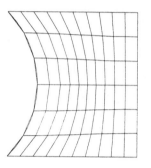

Use the Edgesurf command, select the four entities labeled at the right, and create a surface similar to the second illustration at the right. Use the illustration in step 15 as a guide in performing this operation.

Command: **Edgesurf**
Select edge 1: *(Select the entity at "A")*
Select edge 2: *(Select the entity at "B")*
Select edge 3: *(Select the entity at "C")*
Select edge 4: *(Select the entity at "D")*

Step #15

Copy the surface just created using the Edgesurf command to create the bottom surface.

Command: **Copy**
Select objects: **L**
Select objects: *(Strike Enter to continue)*
<Base point or displacement>/Multiple: **Endp**
of *(Select the endpoint at "A")*
Second point of displacement: **Endp**
of *(Select the endpoint at "B")*

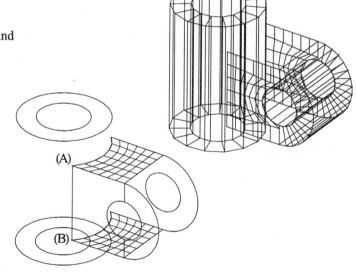

Step #16

Move the two surface mesh patterns at a displacement of 0,10,0 using the Move command.

Command: **Move**
Select objects: *(Select the last two surface mesh patterns)*
Select objects: *(Strike Enter to continue)*
Base point or displacement: **0,10,0**
Second point of displacement: *(Strike Enter)*

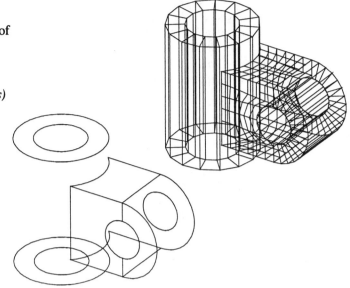

Step #17

The figure at the right represents the Column that has been completely surfaced using the previous steps. Use this model to add lights and cameras for importing into AutoShade. Do not, however, discard the wirframe representation of the Column. This model may have key views such as front, top, and right side views extracted from it. This extraction process will not occur on a surfaced model.

Command: **Hide**
Regenerating drawing.
Removing hidden lines: **XXX**

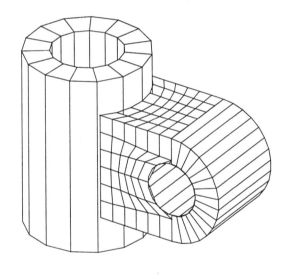

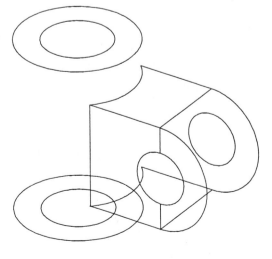

Problems for Unit 8

Directions for Problems 8-1 through 8-2

1. *Use a grid spacing of 0.50 to determine all dimensions.*
2. *Create two layers for each object. Call the layers "Wireframe" and "Surface."*
3. *Create a 3D wireframe model of each object on the layer "Wireframe."*
4. *Surface the wireframe model and place all surfaces on the layer "Surface."*

Problem 8-1

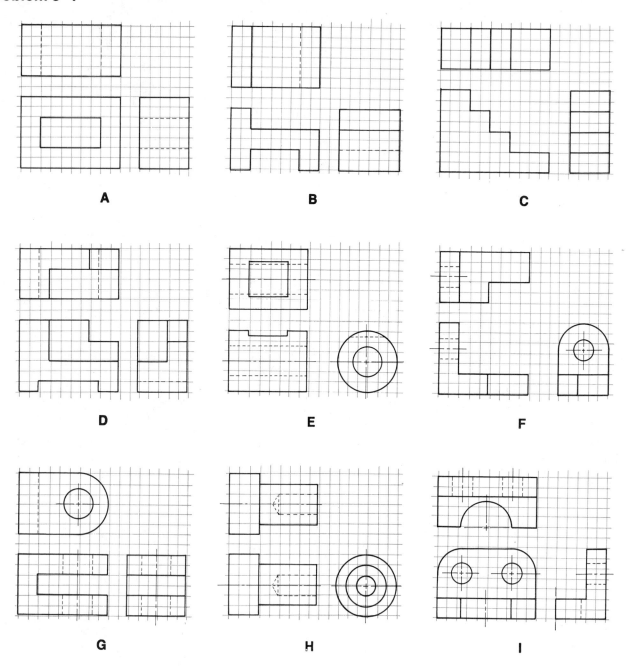

Problem 8-2

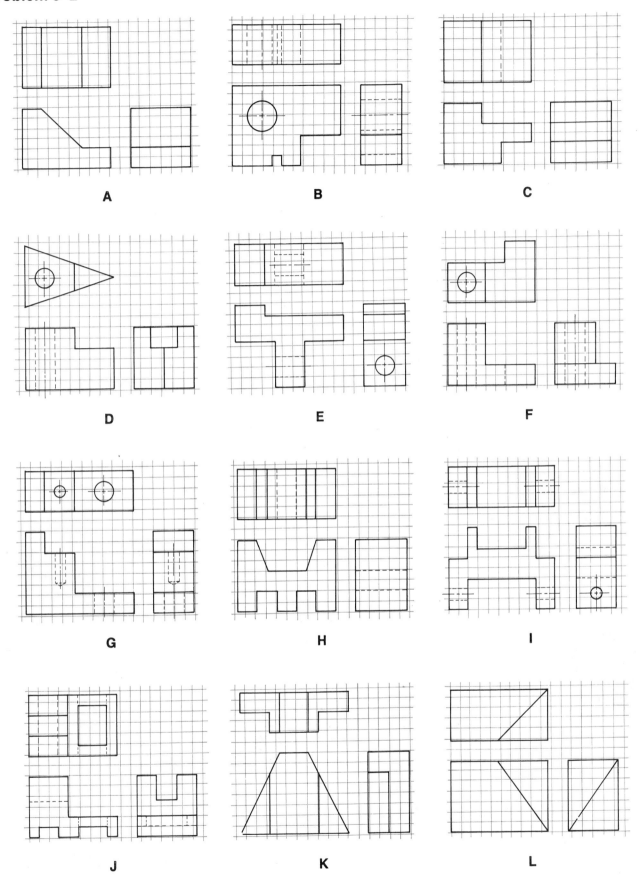

A

B

C

D

E

F

G

H

I

J

K

L

Create a 3D wireframe model of each object. Surface each wireframe using a series of 3Dfaces, ruled, tabulated, revolved, and edge surfaces.

Problem 8-3

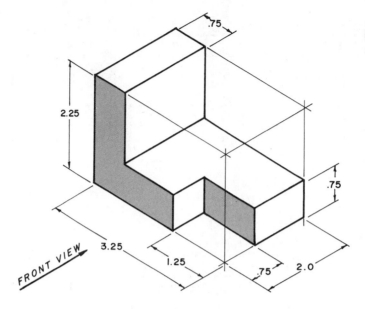

Problem 8-4

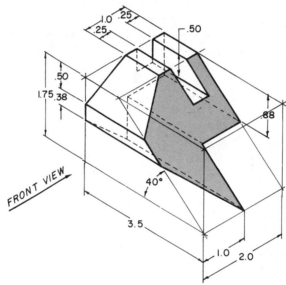

Problem 8-5

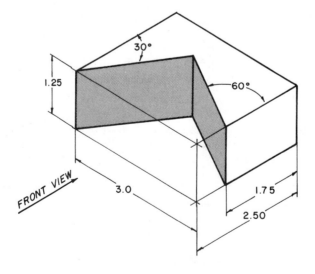

Problem 8-6

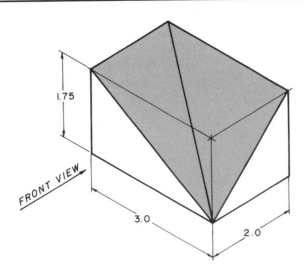

FRONT VIEW

1.75

3.0

2.0

Problem 8-7

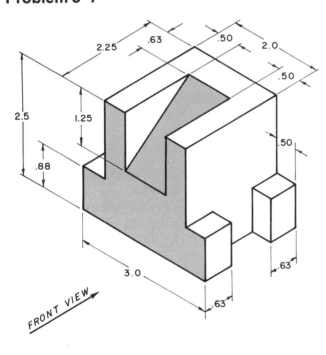

2.25   .63   .50   2.0

.50

2.5   1.25

.50

.88

3.0

.63

FRONT VIEW   .63

Problem 8-8

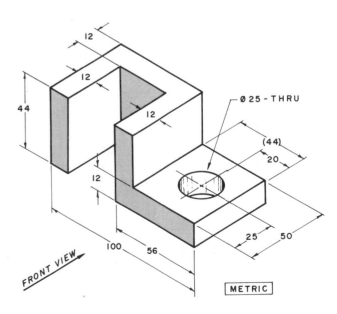

12

12

44

12

Ø 25 - THRU

(44)

20

12

100   56

25   50

FRONT VIEW

METRIC

Problem 8-9

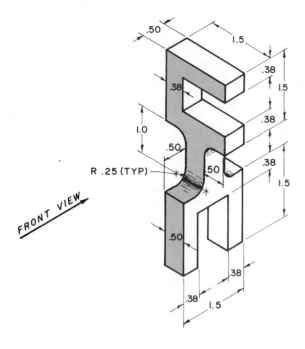

.50
1.5
.38
1.5
.38
.38
1.0
.50
.50
R .25 (TYP)
.38
1.5
FRONT VIEW
.50
.38
.38
1.5

Problem 8-10

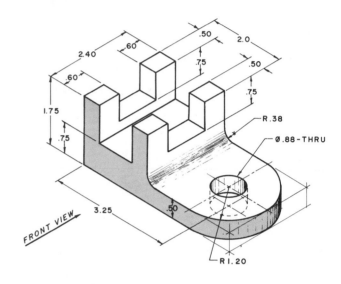

2.40
.60
.50
2.0
.60
.75
.50
1.75
.75
.75
.75
R .38
Ø .88 – THRU
FRONT VIEW
3.25
.50
R 1.20

Problem 8-11

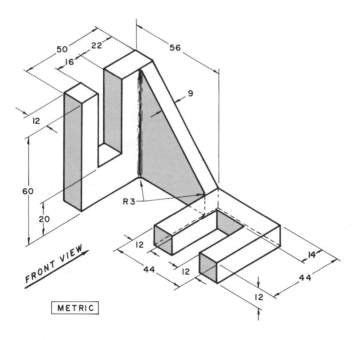

50
22
56
16
9
12
60
20
R 3
FRONT VIEW
12
14
44
12
44
12

METRIC

Problem 8-12

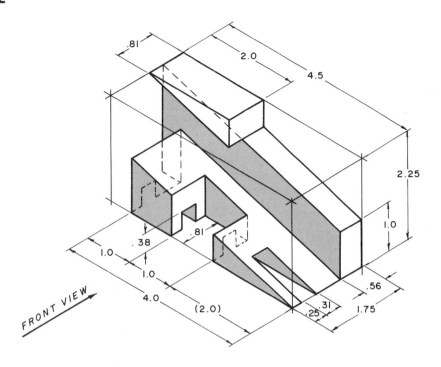

FRONT VIEW

Problem 8-13

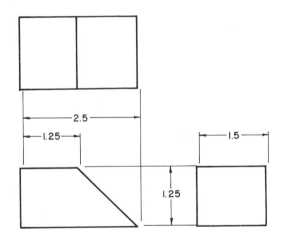

Problem 8-14

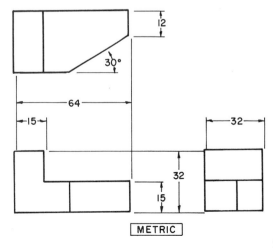

METRIC

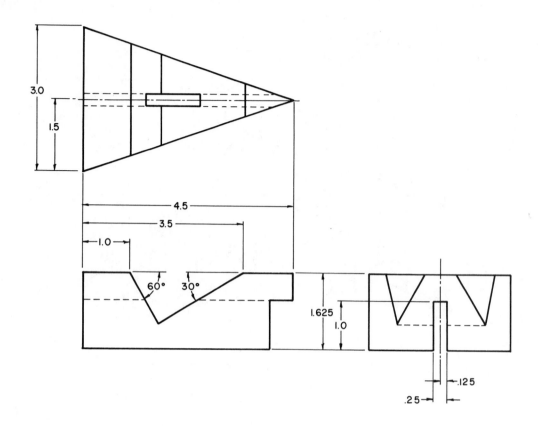

UNIT
9

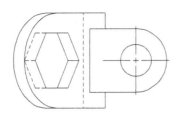

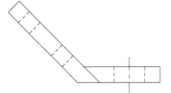
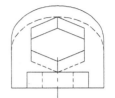

Projection Techniques Using Wire Models

Contents

Previous units focused on 2D projection techniques toward the creation of multi-view drawings. Other units concentrated on the formation of 3D models in the wireframe and shaded varieties. This unit will allow a 3D wireframe model to be used as the basis for the creation of a multi-view drawing using an existing AutoCAD AutoLISP file called Project.LSP. Through this routine, selected entities are projected to a UCS parallel to one of the primary viewing locations, namely front, top, and right side views. This unit will explain the advantages of performing a wireframe model and how to load the Project.LSP file before discussing how the front, top, and right side views are projected from the model. An explanation on projecting auxiliary views is also included. A tutorial at the end of the unit will review laying out five User Coordinate Systems before focusing on projecting five views of a selected object. Problems at the end of the unit are on diskette designed in wireframe mode for immediate projection, insertion onto a flat 2D drawing screen, and dimensioning.

Advantages of Using Wireframe Models

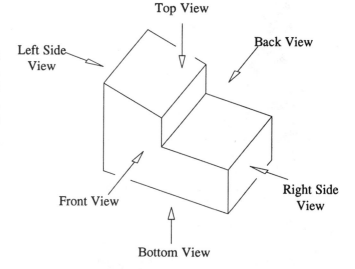

Top View

Left Side View

Back View

Front View

Right Side View

Bottom View

The art of performing multi-view drawing has already been discussed in Unit 3 regarding the illustration at the right. This reviews the six basic ways of viewing an object, not counting special views such as sections and auxiliaries. Depending on the complexity of the object, two or three views may accurately describe most objects.

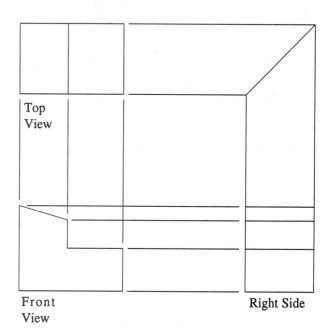

Top View

Front View

Right Side

Once the views are determined, they are laid out on a 2D surface with the front view placed in the lower left corner followed by the top view above the front and the right side view to the right of the front. In Unit 3, orthographic projection techniques were devised to use the relationship between views as an advantage for drawing one view and projecting up and over to partially form adjacent views. This procedure, however, may seem time consuming with all of the constructions required for projection. Then comes the clean up required to display just the completed views. The Trim and Erase commands were used to perform these operations.

We have just discussed in Unit 8 the art of drawing wireframe constructions with multiple User Coordinate Systems which allows us to draw practically any shape and surface as a 3D mode. Unfortunately, this procedure takes practice. However, if a wireframe is made, your efforts may be rewarded by not having to produce the orthographic solution. Instead, main views can be extracted or projected from the wireframe and saved as a Block. The blocks of all individual views are then inserted onto a flat drawing screen in the standard order of views. To perform these projections, AutoCAD provides special programs called LISP files which compliment the main body of commands that come with the software. The routine that allows entities to be projected from a wireframe model is called Project.LSP and should be located in your ACAD subdirectory if running Release 10 or in the AutoCAD Sample subdirectory if using Release 11.

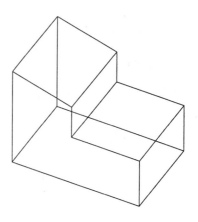

Using the Project.LSP Routine

The illustration at the right shows the mechanics of the Project.LSP routine. Entities are selected for projection from a wireframe model to match a primary view (the example displays the entities that would describe the right side view). Before any projections take place, the current UCS must be in the correct position to accept the entities that make up the view. All selected entities are projected to an imaginary plane set up by the current UCS. This plane may be anywhere, even on the object itself. With the entities projected to the UCS, the user has the option of making these entities into a Block or Wblock. This allows the user to collect the desired views of a wireframe model and insert the blocks on a flat display screen. The prompts for the Project.LSP routine vary somewhat depending on whether a Block or Wblock is being defined. In either case, this routine needs to be loaded into the current drawing where it becomes a valid AutoCAD command. To load the routine automatically into any drawing, make this code part of an ACAD.LSP file.

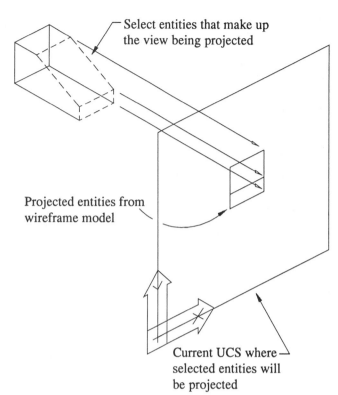

Select entities that make up the view being projected

Projected entities from wireframe model

Current UCS where selected entities will be projected

Command: **(load "project")**
C: Project

For merging entities into a Block:

Command: **Project**
Extrusion and mesh information will not be projected.
Select objects: *(Select the entities to project to the current UCS)*
Select objects: *(Strike Enter to begin the projection process)*
Project more entities? <N>: *(Strike Enter to continue or Yes to continue the selection process)*
Make projected entitie(s) into block? <N>: **Yes**
Block name: *(Enter the name of the Block)*
Insertion point <UCS 0,0,0>: *(Strike Enter to accept this default or type in coordinates for a new insertion point)*

For merging entities into a Wblock:

Command: **Project**
Extrusion and mesh information will not be projected.
Select objects: *(Select the entities to project to the current UCS)*
Select objects: *(Strike Enter to begin the projection process)*
Project more entities? <N>: *(Strike Enter to continue or Yes to continue the selection process)*
Make projected entitie(s) into block? <N>: *(Strike Enter)*
Write projected entities to disk as a DWG file <N>: **Yes**
File name: *(Enter the name of a standard DOS file)*
Insertion point <UCS 0,0,0>: *(Strike Enter to accept this default or type in coordinates for a new insertion point)*

Projecting a Front View

For projecting a front view, have a previously defined UCS similar to the example at the right. Load Project.LSP and follow the prompts below. For the insertion point of the block, it is best to identify a point somewhere on the object, usually the lower left corner of the front view.

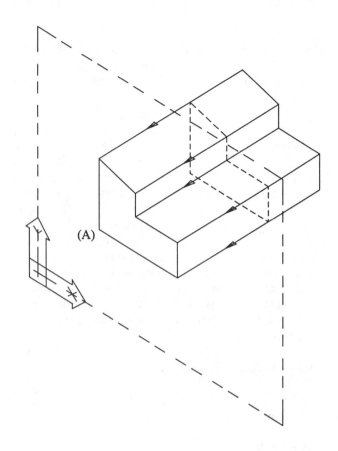

Command: **Project**
Extrusion and mesh information will not be projected.
Select objects: *(Select all dashed entities illustrated at the right)*
Select objects: *(Strike Enter to begin the projection process)*
0 entities not projected
Project more entities? <N>: *(Strike Enter to continue)*
Make projected entitie(s) into block? <N>: **Yes**
Block name: **Front**
Insertion point <UCS 0,0,0>: **Endp**
of *(Select the endpoint of the line at "A" for the insertion point)*

Projecting a Right Side View

Repeat the same above procedure for projecting the right side view. If the LISP routine has already been loaded and the drawing not exited, the routine does not have to be reloaded.

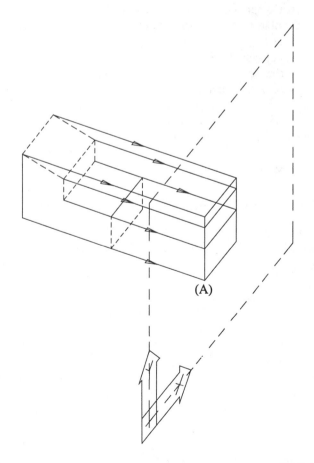

Command: **Project**
Extrusion and mesh information will not be projected.
Select objects: *(Select all dashed entities illustrated at the right)*
Select objects: *(Strike Enter to begin the projection process)*
0 entities not projected
Project more entities? <N>: *(Strike Enter to continue)*
Make projected entitie(s) into block? <N>: **Yes**
Block name: **R_side**
Insertion point <UCS 0,0,0>: **Endp**
of *(Select the endpoint of the line at "A" as the insertion point)*

Projecting a Top View

Follow the same procedure and the prompts below for projecting the Top view.

Command: **Project**
Extrusion and mesh information will not be projected.
Select objects: *(Select all dashed entities illustrated at the right)*
Select objects: *(Strike Enter to begin the projection process)*
0 entities not projected
Project more entities? <N>: *(Strike Enter to continue)*
Make projected entitie(s) into block? <N>: **Yes**
Block name: **Top**
Insertion point <UCS 0,0,0>: **Endp**
of *(Select the endpoint of the line at "A" for the insertion point)*

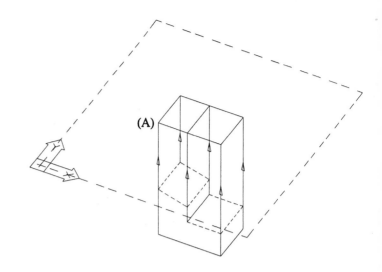

Projecting a Pictorial View

Follow the same procedure for projecting an isometric view. The trick here is to use the UCS-View option for the new UCS.

Command: **Project**
Extrusion and mesh information will not be projected.
Select objects: *(Select all dashed entities illustrated at the right)*
Select objects: *(Strike Enter to begin the projection process)*
0 entities not projected
Project more entities? <N>: *(Strike Enter to continue)*
Make projected entitie(s) into block? <N>: **Yes**
Block name: **Iso**
Insertion point <UCS 0,0,0>: **Endp**
of *(Select the endpoint of the line at "A" for the insertion point)*

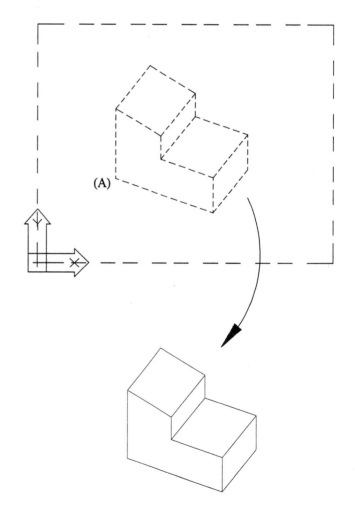

Projecting an Auxiliary View

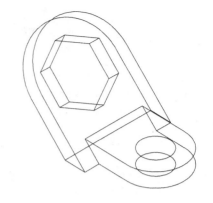

The Project.LSP routine is especially helpful when an auxiliary view of an object is needed. If a wireframe drawing similar to the object at the right exists, the UCS used to create the inclined surface complete with any features may be used to project the entities that make up the auxiliary view.

Projection of an auxiliary view begins with the creation of a UCS parallel to the inclined surface. As in multi-view projection, an object is visualized to be placed in a glass box complete with inclined plane. A new UCS is created consisting of this plane.

Command: **UCS**
Origin/ZAxis/3point/Entity/View/X/Y/Z/Prev/Restore/Save/
    Del/?/<World>: **3point**
Origin point <0,0,0>: **Endp**
of *(Select the endpoint at "A")*
Point on positive portion of the X axis <0.000>: **Endp**
of *(Select the endpoint at "B")*
Point on positive-Y portion of the UCS X-Y plane <0.000>:
    Endp
of *(Select the endpoint at "C")*

Command: **UCS**
Origin/ZAxis/3point/Entity/View/X/Y/Z/Prev/Restore/Save/
    Del/?/<World>: **S**
?/Name of UCS: **Aux**

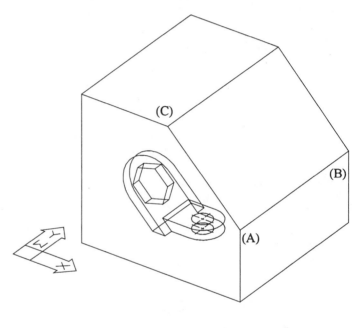

The UCS that went into the construction of the wireframe model is restored using the UCS-Restore option. This UCS may reside directly on the object or may be positioned away from the object as in the illustration at the right. Regardless of the distance away from the object, as long as the UCS is parallel to the inclined surface, the results will be identical if the UCS were to be placed on the object itself.

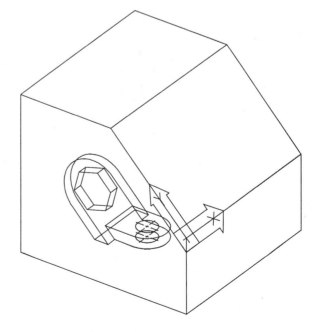

Projecting an Auxiliary View Continued

With the UCS defined, load the Project.LSP routine (see
the prompt sequence below). When loading, it is not
necessary to add the extension to the file. With Project
now a valid AutoCAD command, follow the prompts to
select all dashed entities illustrated at the right, make the
entities into a block, and insert the block into a drawing.

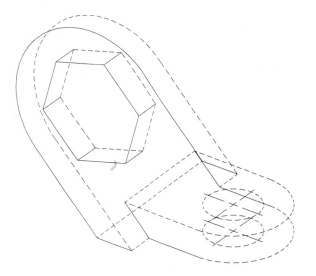

Command: **(load "project")**
C: Project

Command: **Project**
Extrusion and mesh information will not be projected.
Select objects: *(Select all dashed entities illustrated at the
right)*
Select objects: *(Strike Enter to begin the projection process)*
0 entities not projected
Project more entities? <N>: *(Strike Enter to continue)*
Make projected entitie(s) into block? <N>: **Yes**
Block name: **Aux**
Insertion point <UCS 0,0,0>: *(Strike Enter to accept this
default)*

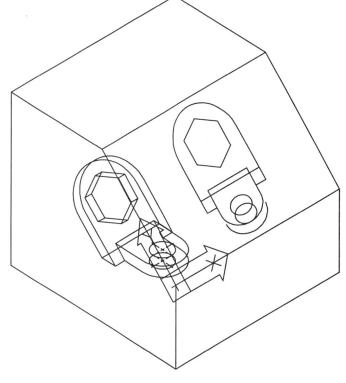

The illustration at the right shows the affects of projecting
to a UCS away from the object. If the UCS were on the
object, the projected entities would be directly on the
wireframe model. This does not really matter since the
projected entities will disappear when creating as a block.
Do not use the Oops command after the entities depart the
screen. The screen display will, however, require a Redraw.

Projecting Hidden Features

The illustrations at the right display a common problem for projecting visible entities in one view and invisible entities in another. The wireframe model shows a square base with a hole going through it. Using the Project.LSP routine, two blocks were created, namely, front and top views. (Remember, only two views of this object were needed to accurately describe the object.) The results of the projection and insertion are shown at "A". The top view is complete with square shape with circle in the center. The front view is incomplete since it does not show any hidden lines for the depth of the hole. Actually, there was nothing in the form of height entities to project pertaining to the hole. The model simply has two circles. The size of the hole in the top view at "B" is used to project vertical lines into the front view. These lines are then trimmed to size and changed to the hidden line type. Another important consideration to take note of is that the curves in the wireframe are actual circles. Once the circles are projected, the Project.LSP routine converted these entities into a polyline. As a result, once views containing visible circles are inserted, the size in bytes of the drawing may increase dramatically. One method is to delete the polyline and draw a regular circle.

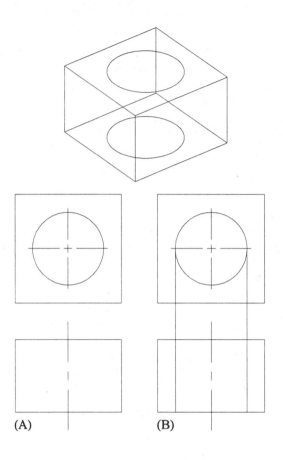

(A)   (B)

The illustrations at the right represent the identical wireframe model as above. The difference in this model are the four vertical lines from top circle to bottom circle. The purpose of these lines is to project them using Project.LSP to display the depth of the hole in the front view. The vertical lines representing the holes in the model were drawn using the Line command along with the Osnap-Quadrant option. Also, these entities were drawn with the World coordinate system being the current coordinate system. Once the lines appear in the front view, they need to be changed to the hidden linetype. (Since all projected views are inserted as blocks, any changes require the block to be first exploded before being edited.) In case you were wondering why not add the hidden lines to the model; in this way, the lines would be projected and appear as hidden lines in the block definition. As this appears to be a better solution, the Project.LSP routing converts lines of different linetypes back to continuous lines.

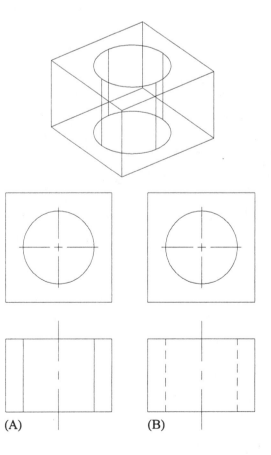

(A)   (B)

Editing Projected Views

Both examples at the right represent projected views. One appears incomplete, the other complete. As views are projected, in some cases editing of the view needs to be done. The continuous line at "A" needs to be hidden. The bottom ellipse at "B" needs to be partially visible, partially hidden. The sides of the object at "C" and "D" need to be connected with the main object. This is usually the case for some views when using Project.LSP. Before any editing can take place, the view is exploded and the changes made. Yet another method is to insert a second block of the identical shape, explode it, make changes, and redefine the block under the same name. This would keep the block definition intact. In either case, the Chprop command becomes invaluable for converting entities to different colors or linetypes.

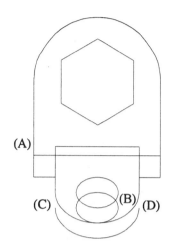 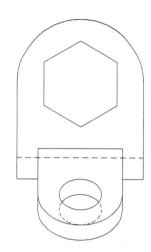

Illustrated below is an engineering drawing complete with projected front, top, right side, and isometric views. All views were extracted from a model using Project.LSP and edited to conform to standard engineering linetypes.

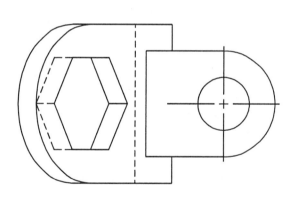 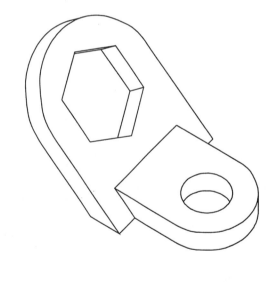

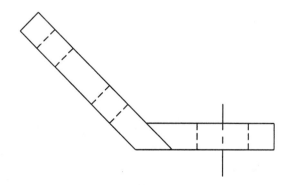 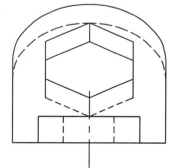

Tutorial Exercise #23
Incline.Dwg

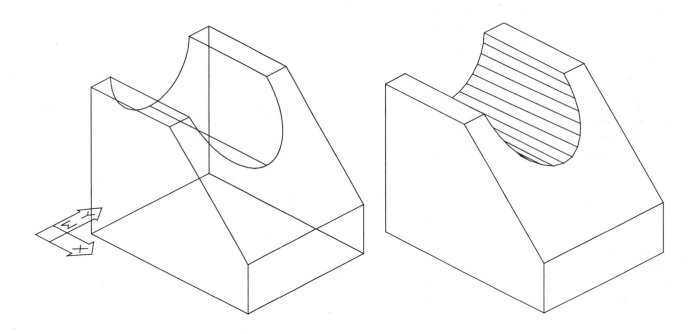

PURPOSE:
This tutorial is designed to use the Project.LSP routine to extract the Front, Top, Right Side, Auxiliary, and Isometric views from an existing wireframe drawing called "Incline.Dwg". It is important that this drawing is wireframe and not consist of surfaces since mesh entities will not project. Refer to the following for special system settings and suggested command sequences.

SYSTEM SETTINGS:
Since this drawing is provided on diskette, edit an existing drawing called "Incline." Follow the steps in this tutorial for creating orthographic views from a wireframe model.

LAYERS:
No layers needed to be created at this time.

SUGGESTED COMMANDS:
Begin this tutorial by defining five UCS'; Front, Top, R_side, Aux, and Iso. When each is created using the UCS-3point option, save each coordinate system using the UCS-Save option. Load the Project.LSP routine, restore each coordinate system, and project entities to the current coordinate system. Make the projected entities into blocks corresponding to the coordinate systems and insert each block into a flat 2D drawing sheet.

DIMENSIONING:
This drawing may be dimensioned at a later date. Consult your instructor before continuing.

PLOTTING:
This tutorial exercise may be plotted on "B"-size paper (11" x 17"). Plot Incline at a scale of full size, or 0.50=1.

VERSION OF AUTOCAD:
This tutorial exercise may be completed using either AutoCAD Release 10 or Release 11.

Step #1

Before projecting any entities into primary viewing planes, multiple UCS' need to be established to project into the current UCS. Even though the model was constructed using numerous UCS', set up a new UCS for each viewing plane as review. The UCS-3point option will be used for most cases. When finished defining the new UCS, save it under a unique name (relative to the viewing position).

Command: **UCS**
Origin/ZAxis/3point/Entity/View/X/Y/Z/Prev/Restore/Save/
    Del/?/<World>: **3point**
Origin point <0,0,0>: **Endp**
of *(Select the endpoint at "A")*
Point on positive portion of the X axis<0.000>: **Endp**
of *(Select the endpoint at "B")*
Point on positive-Y portion of the UCS X-Y plane <0.000>:
    Endp
of *(Select the endpoint at "C")*

Command: **UCS**
Origin/ZAxis/3point/Entity/View/X/Y/Z/Prev/Restore/Save/
    Del/?/<World>: **S**
?/Name of UCS: **Front**

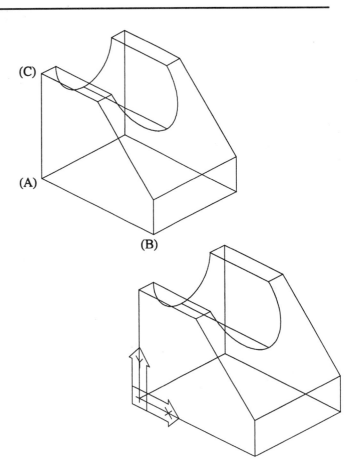

Step #2

Use the UCS command to define and save a new coordinate system called Top.

Command: **UCS**
Origin/ZAxis/3point/Entity/View/X/Y/Z/Prev/Restore/Save/
    Del/?/<World>: **3point**
Origin point <0,0,0>: **Endp**
of *(Select the endpoint at "A")*
Point on positive portion of the X axis <0.000>: **Endp**
of *(Select the endpoint at "B")*
Point on positive-Y portion of the UCS X-Y plane <0.000>:
    Endp
of *(Select the endpoint at "C")*

Command: **UCS**
Origin/ZAxis/3point/Entity/View/X/Y/Z/Prev/Restore/Save/
    Del/?/<World>: **S**
?/Name of UCS: **Top**

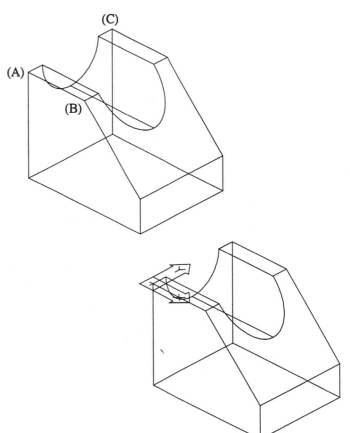

Step #3

Use the UCS command to define and save a new coordinate system called R_side.

Command: **UCS**
Origin/ZAxis/3point/Entity/View/X/Y/Z/Prev/Restore/Save/
     Del/?/<World>: **3point**
Origin point <0,0,0>: **Endp**
of *(Select the endpoint at "A")*
Point on positive portion of the X axis <0.000>: **Endp**
of *(Select the endpoint at "B")*
Point on positive-Y portion of the UCS X-Y plane <0.000>:
     Endp
of *(Select the endpoint at "C")*

Command: **UCS**
Origin/ZAxis/3point/Entity/View/X/Y/Z/Prev/Restore/Save/
     Del/?/<World>: **S**
?/Name of UCS: **R_side**

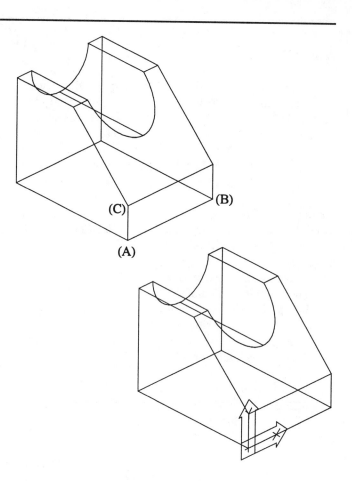

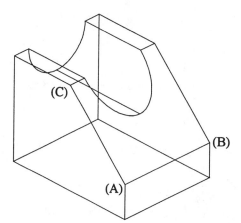

Step #4

Use the UCS command to define and save a new coordinate system called Aux.

Command: **UCS**
Origin/ZAxis/3point/Entity/View/X/Y/Z/Prev/Restore/Save/
     Del/?/<World>: **3point**
Origin point <0,0,0>: **Endp**
of *(Select the endpoint at "A")*
Point on positive portion of the X axis <0.000>: **Endp**
of *(Select the endpoint at "B")*
Point on positive-Y portion of the UCS X-Y plane <0.000>:
     Endp
of *(Select the endpoint at "C")*

Command: **UCS**
Origin/ZAxis/3point/Entity/View/X/Y/Z/Prev/Restore/Save/
     Del/?/<World>: **S**
?/Name of UCS: **Aux**

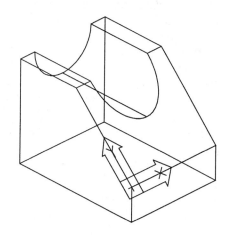

Step #5

Use the UCS-View option to define and save a new coordinate system called Iso. Use the UCS-Origin option to move the origin to the endpoint at "A".

Command: **UCS**
Origin/ZAxis/3point/Entity/View/X/Y/Z/Prev/Restore/Save/
     Del/?/<World>: **V**

Command: **UCS**
Origin/ZAxis/3point/Entity/View/X/Y/Z/Prev/Restore/Save/
     Del/?/<World>: **O**
Origin point <0,0,0>: **Endp**
of *(Select the endpoint at "A")*

Command: **UCS**
Origin/ZAxis/3point/Entity/View/X/Y/Z/Prev/Restore/Save/
     Del/?/<World>: **Save**
?/Name of UCS: **Iso**

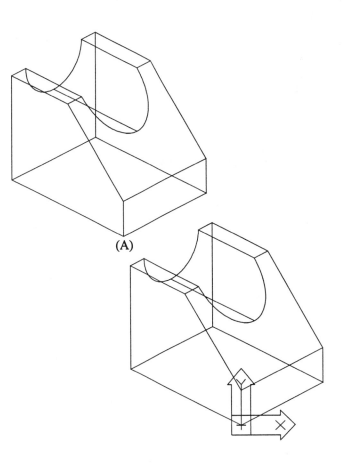

(A)

Step #6

Use the UCS command to inquire that the following coordinate systems have been created and saved:
               Front (Origin at "A")
               Top (Origin at "B")
               R_side (Origin at "C")
               Aux (Origin at "D")
               Iso (Origin at "C")

Command: **UCS**
Origin/ZAxis/3point/Entity/View/X/Y/Z/Prev/Restore/Save/
     Del/?/<World>: **?**

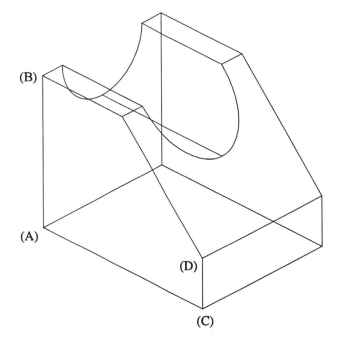

(B)

(A)

(D)

(C)

Step #7

Begin restoring each individual UCS. Load the Project.LSP routine and begin projecting the selected entities perpendicular to the UCS. Make the projected entities into a block for later insertion into an orthographic drawing. (Be sure the file Project.LSP is located in the same subdirectory as all AutoCAD files.)

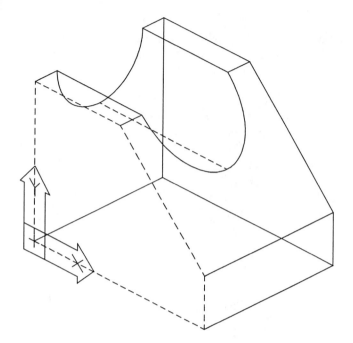

Command: **UCS**
Origin/ZAxis/3point/Entity/View/X/Y/Z/Prev/Restore/Save/
    Del/?/<World>: **R**
?/Name of UCS to restore: **Front**

Command: **(load "project")**
C: Project

Command: **Project**
Extrusion and mesh information will not be projected.
Select objects: *(Select all dashed entities illustrated at the right)*
Select objects: *(Strike Enter to begin the projection process)*
0 entities not projected
Project more entities? <N>: *(Strike Enter to continue)*
Make projected entitie(s) into block? <N>: **Yes**
Block name: **Front**
Insertion point <UCS 0,0,0>: *(Strike Enter to accept this default)*

Command: **Redraw**

Step #8

Restore the Top UCS. Use the Project command and select the dashed entities at the right to form the top view. Once the Project.LSP routine is loaded into a drawing, it remains loaded for the duration of this drawing and can be used as a regular AutoCAD command. Create a block called Top with the current location of the UCS as the insertion point. Redraw the drawing when the entities disappear from the screen and continue.

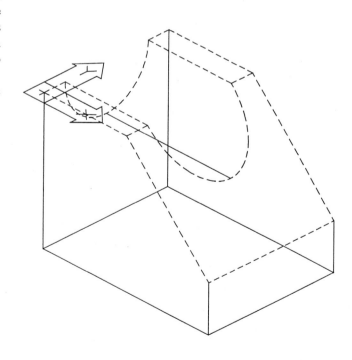

Command: **UCS**
Origin/ZAxis/3point/Entity/View/X/Y/Z/Prev/Restore/Save/
    Del/?/<World>: **R**
?/Name of UCS to restore: **Top**

Command: **Project**
Extrusion and mesh information will not be projected.
Select objects: *(Select all dashed entities illustrated at the right)*
Select objects: *(Strike Enter to begin the projection process)*
0 entities not projected
Project more entities? <N>: *(Strike Enter to continue)*
Make projected entitie(s) into block? <N>: **Yes**
Block name: **Top**
Insertion point <UCS 0,0,0>: *(Strike Enter to accept this default)*

Command: **Redraw**

Step #9

Restore the R_side UCS. Use the Project command and select the dashed entities at the right to form the right side view. Create a block called R_side with the current location of the UCS as the insertion point. Redraw the drawing when the entities disappear from the screen and continue.

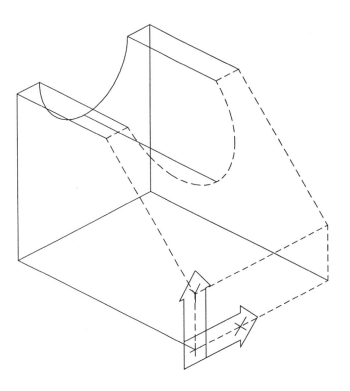

Command: **UCS**
Origin/ZAxis/3point/Entity/View/X/Y/Z/Prev/Restore/Save/
    Del/?/<World>: **R**
?/Name of UCS to restore: **R_side**

Command: **Project**
Extrusion and mesh information will not be projected.
Select objects: *(Select all dashed entities illustrated at the*
    right)
Select objects: *(Strike Enter to begin the projection process)*
0 entities not projected
Project more entities? <N>: *(Strike Enter to continue)*
Make projected entitie(s) into block? <N>: **Yes**
Block name: **R_side**
Insertion point <UCS 0,0,0>: *(Strike Enter to accept this*
    default)

Command: **Redraw**

Step #10

Restore the Aux UCS. Use the Project command and select the dashed entities at the right to form the auxiliary view. Create a block called Aux with the current location of the UCS as the insertion point. Redraw the drawing when the entities disappear from the screen and continue.

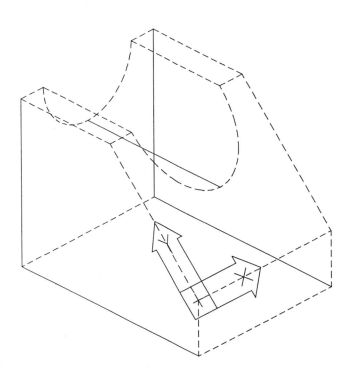

Command: **UCS**
Origin/ZAxis/3point/Entity/View/X/Y/Z/Prev/Restore/Save/
    Del/?/<World>: **R**
?/Name of UCS to restore: **Aux**

Command: **Project**
Extrusion and mesh information will not be projected.
Select objects: *(Select all dashed entities illustrated at the*
    right)
Select objects: *(Strike Enter to begin the projection process)*
0 entities not projected
Project more entities? <N>: *(Strike Enter to continue)*
Make projected entitie(s) into block? <N>: **Yes**
Block name: **Aux**
Insertion point <UCS 0,0,0>: *(Strike Enter to accept this*
    default)

Command: **Redraw**

Step #11

Restore the Iso UCS. Use the Project command and select the dashed entities at the right to form the isometric view. Remember this UCS was originally defined using the UCS-View option. All selected entities will be projected directly in the same direction as the display screen on the monitor. Create a block called Iso with the current location of the UCS as the insertion point. Redraw the drawing when the entities disappear from the screen and continue.

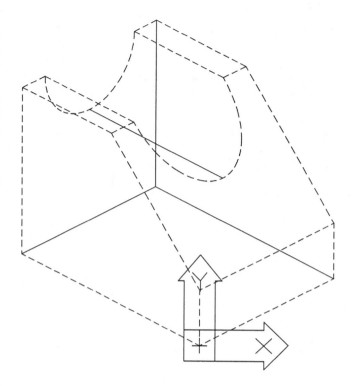

Command: **UCS**
Origin/ZAxis/3point/Entity/View/X/Y/Z/Prev/Restore/Save/
    Del/?/<World>: **R**
?/Name of UCS to restore: **Iso**

Command: **Project**
Extrusion and mesh information will not be projected.
Select objects: *(Select all dashed entities illustrated at the right)*
Select objects: *(Strike Enter to begin the projection process)*
0 entities not projected
Project more entities? <N>: *(Strike Enter to continue)*
Make projected entitie(s) into block? <N>: **Yes**
Block name: **Iso**
Insertion point <UCS 0,0,0>: *(Strike Enter to accept this default)*

Command: **Redraw**

Step #12

Change to the World coordinate system using the UCS-World option.

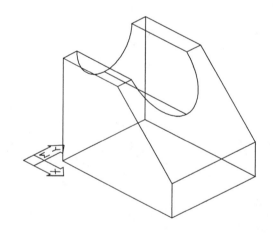

Command: **UCS**
Origin/ZAxis/3point/Entity/View/X/Y/Z/Prev/Restore/Save/
    Del/?/<World>: *(Strike Enter to accept the default)*

Step #13

Use the Vpoint command to return to plan view.

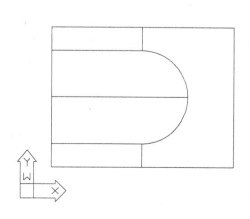

Command: **Vpoint**
Rotate/<Vpoint> <2.49,-2.49,2.03>: **0,0,1**

Step #14

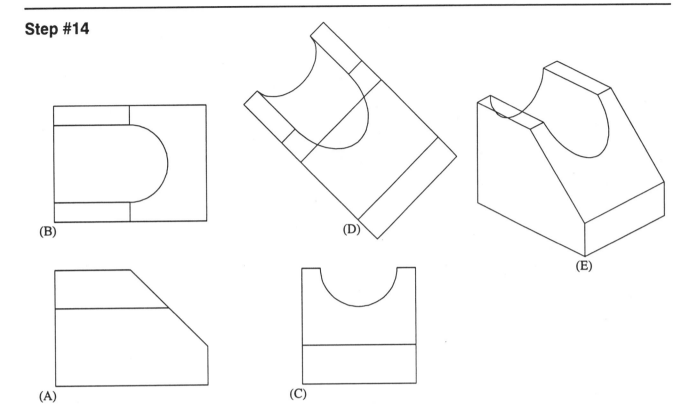

(B)

(D)

(A)

(C)

(E)

To prepare the display for the insertion of the five blocks, increase the limits of the screen to an upper right corner of 33,21. Change the snap value to 0.5 units. When completed with these two operations, immediately perform a Zoom-All to force a system regeneration which is needed to update the drawing to the new limits. The wireframe construction of the incline is no longer needed; however, we wish to keep it for future reference. Move the model from a base point of 0,0 to a second point of displacement at 36,24. In other words, the model is moved off the limits of the screen to make room for the five inserted views. Create the following layers using the Layer command: "Views" and "Hidden." Assign the color red to layer "Hidden," along with the hidden linetype. Set the new current layer to "Views." Proceed by inserting the five views.

Command: **Limits**
ON/OFF/<Lower left corner> <0.00,0.00>: *(Strike Enter)*
Upper right corner<12.00,9.00>: **33,21**

Command: **Snap**
Snap spacing or ON/OFF/Aspect/Rotate/Style<1.00>: **0.50**

Command: **Zoom**
All/Center/Dynamic/Extents/Left/Previous/Window/
    <Scale(X)>: **All**

Command: **Move**
Select objects: *(Select the entire wireframe by using "Window")*
Select objects: *(Strike Enter to continue with this command)*
Base point or displacement: **0,0**
Second point of displacement: **36,24**

Command: **Layer**
?/Make/Set/New/ON/OFF/Color/Ltype/Freeze/Thaw: **N**
New layer name(s): **Views, Hidden**
?/Make/Set/New/ON/OFF/Color/Ltype/Freeze/Thaw: **C**
Color: **Red**
Layer name(s) for color Red <0>: **Hidden**
?/Make/Set/New/ON/OFF/Color/Ltype/Freeze/Thaw: **Lt**
Linetype (or?) <Continuous>: **Hidden**
Layer name(s) for linetype Hidden <0>: **Hidden**
?/Make/Set/New/ON/OFF/Color/Ltype/Freeze/Thaw: **S**
New current layer <0>: **Views**
?/Make/Set/New/ON/OFF/Color/Ltype/Freeze/Thaw:
    (Strike Enter to exit this command)

Use the Insert command to place the five blocks on the new screen limits using the following insertion points:

(A)	Front	(1.50,1.50)
(B)	Top	(1.50, 10.00)
(C)	R_side	(14.50,1.50)
(D)	Aux	(17.50,10.00)
(E)	Iso	(28.00,8.00)

When inserting the block "Aux", use a rotation angle of 45 degrees to make the view parallel to the incline located in the front view. When inserting the block "Iso," use a scale factor of 0.75 to reduce the isometric view in size. The completed operation should appear similar to the illustration above.

Step #15

The auxiliary view should line up exactly with the view it is being projected from. To align the auxiliary view with the front view, construct a line using the Osnap-Endpoint and Perpend options to place the line segment exactly perpendicular from the front to the auxiliary view.

Command: **Line**
From point: **Endp**
of *(Select the endpoint of the front view at "A")*
To point: **Per**
to *(Select anywhere along the line at "B")*
To point: *(Strike Enter to exit this command)*

Selecting anywhere along the line in the auxiliary view will allow AutoCAD to search for the exact perpendicular line from the front view.

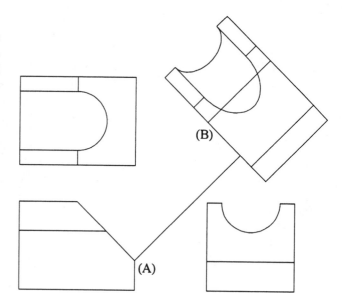

Step #16

Use the Move command to move the auxiliary view from the endpoint of the view at "A" to the endpoint of the construction line segment at "B".

Command: **Move**
Select objects: *(Select the auxiliary view anywhere on the view)*
Select objects: *(Strike Enter to continue)*
Base point or displacement: **Endp**
of *(Select the endpoint of the line on the auxiliary at "A")*
Second point of displacement: **Endp**
of *(Select the endpoint of the construction line at "B")*

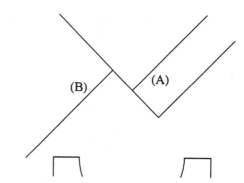

Step #17

The complete move should appear similar to the illustration at the right. Since there is no longer a need for the construction line, use the Erase command to delete it.

Command: **Erase**
Select objects: *(Select the construction line)*
Select objects: *(Strike Enter to execute this command)*

Before continuing, use the explode command to break up the Front, Aux, and Iso blocks into individual entities.

Command: **Explode**
Select block reference, polyline, dimension, or mesh: *(Pick any entity in the Front view. Repeat the same procedure for Aux and Iso)*

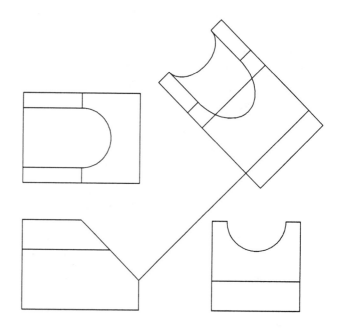

Step #18

Magnify the front view using the Zoom-Window option. The purpose of projecting the solid line at "A" in the front view is to show the depth of the cove cut. Unfortunately, this line needs to be dashed since this feature is hidden in the front view. The Chprop command is used to convert the line from its current layer to the new layer of "Hidden".

Command: **Chprop**
Select objects: *(Select the line at "A")*
Select objects: *(Strike Enter to continue)*
Change what property(Color/LAyer/LType/Thickness)? **LA**
New layer <0>: **Hidden**
Change what property(Color/LAyer/LType/Thickness)?
    (Strike Enter to exit this command)

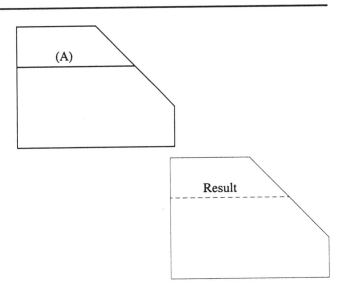

Step #19

Display the overall drawing using the Zoom-Previous option. Then use Zoom-Window to magnify the auxiliary view. The solid line in the auxiliary view also represents an invisible edge requiring a hidden line. As in the previous step, use the Chprop command to convert the line into a hidden line.

Command: **Chprop**
Select objects: *(Select the line at "A")*
Select objects: *(Strike Enter to continue)*
Change what property(Color/LAyer/LType/Thickness)? **LA**
New layer <0>: **Hidden**
Change what property(Color/LAyer/LType/Thickness)?
    (Strike Enter to exit this command)

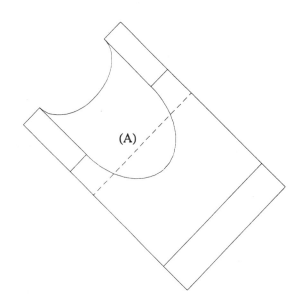

Step #20

Use Zoom-Previous to display the overall drawing and use Zoom-Window to magnify the isometric view. Hidden lines are used in the multi-view drawing and not in isometric drawings. Only visible surfaces are displayed. The rear semi-circle needs to be trimmed to size using the Trim command. Select the dashed line as the cutting edge and select the curve at "A" as the entity to trim.

Command: **Trim**
Select cutting edges...
Select objects: *(Select the dashed line at the right)*
Select objects: *(Strike Enter to continue)*
Select object to trim: *(Select the curve at "A")*
Select object to trim: *(Strike Enter to exit this command)*

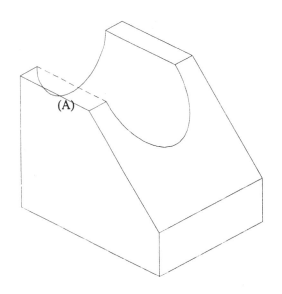

Step #21

The resulting isometric is illustrated at the right. Use the Zoom-Previous option to return to the overall screen displaying all five views similar to the illustration below. At this time, center lines are added in addition to proper dimensions. The Project.LSP routine is not the answer to everyone's needs; it does, however, make it possible to extract from a 3D wireframe model the views necessary to describe an object.

This lisp routine is also available to AutoCAD Release 11 users. The command has been enhanced to place projected entities on a predefined layer. This would allow hidden entities to be placed on a hidden layer without exploding the views and using the Chprop command. As the projected entities are made into a block or wblock, the screen is automatically redrawn. To activate this command in Release 11, type "Project1" instead of "Project".

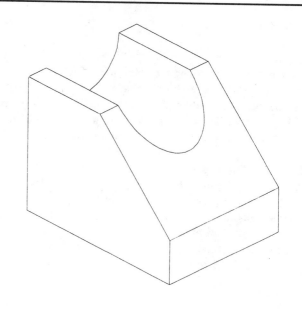

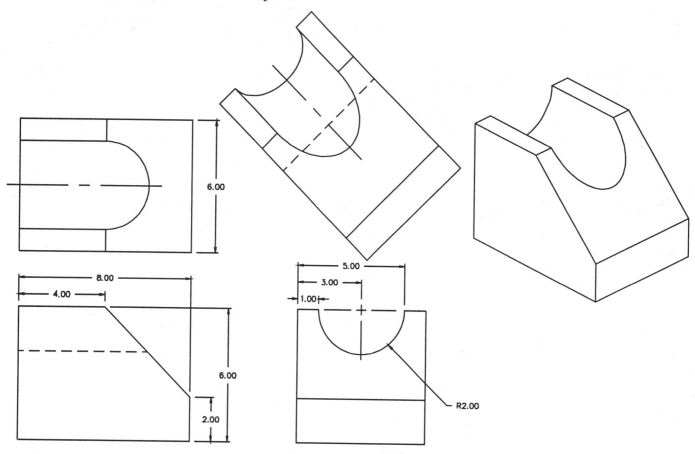

Problems for Unit 9

Directions for Problems 9–1 through 9–20:

Each drawing is complete as a wireframe model. Use the small isometric drawing in the upper right corner to better visualize the wireframe. Four user coordinate systems have been assigned to each model, namely Front, Top, R_side, and Iso. Use the UCS-Restore command to bring up each individual coordinate system and use the Project.Lsp routine to project selected entities to the current coordinate system.

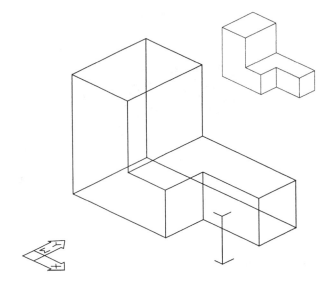

Create blocks of each group of entities corresponding to the front, top, and right side views. Insert these views onto a flat display screen. It is helpful to have Snap turned on in order for the orthographic views to line up. If necessary, use the Chprop command to convert continuous lines into hidden lines. Where circles apply, add proper center lines using conventional methods.

Add all dimensions to the multi-view drawing. When the dimensioning is complete, move views to better locations without disrupting the line-up of the orthographic views.

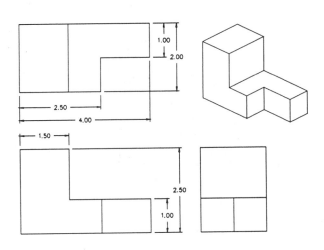

Problem 9-2

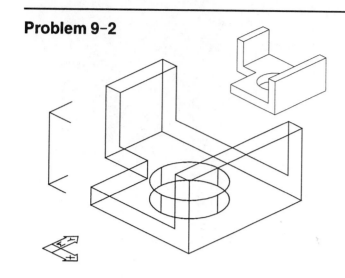

Problem 9-3

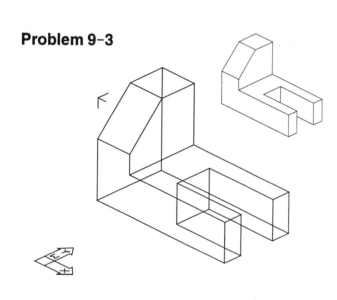

Problem 9-4

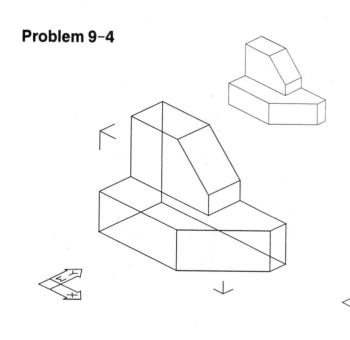

Problem 9-5

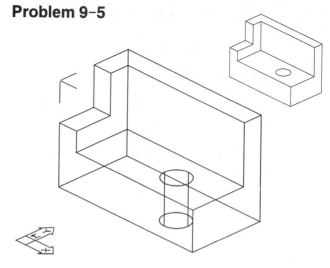

Problem 9-6

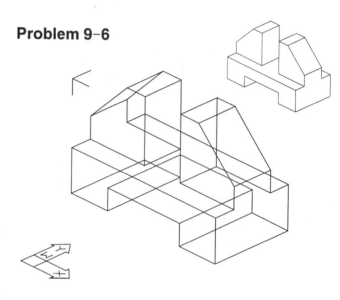

Problem 9-7

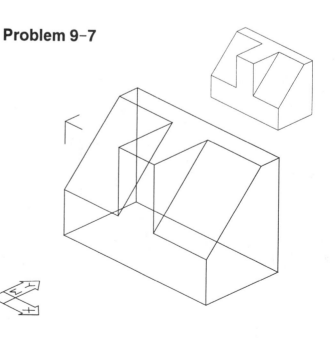

Problem 9-8

Problem 9-11

Problem 9-9

Problem 9-12

Problem 9-10

Problem 9-13

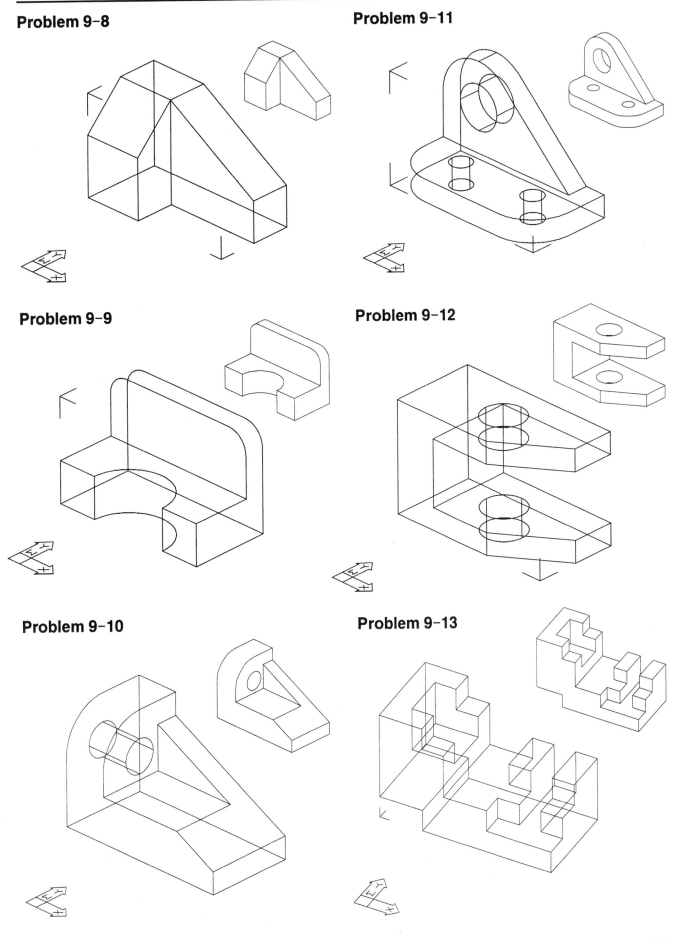

Problem 9-14

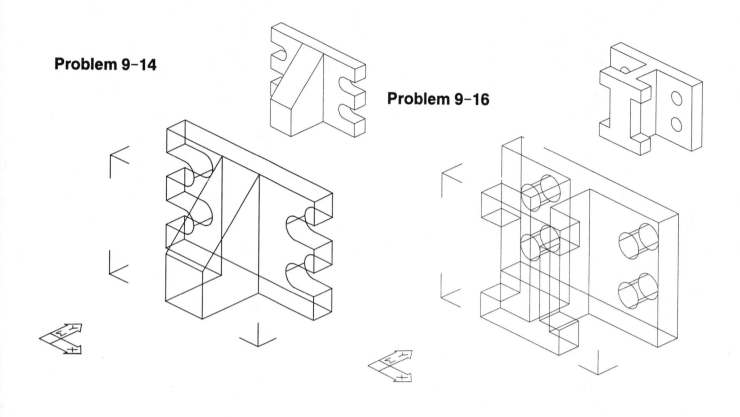

Problem 9-16

Problem 9-15

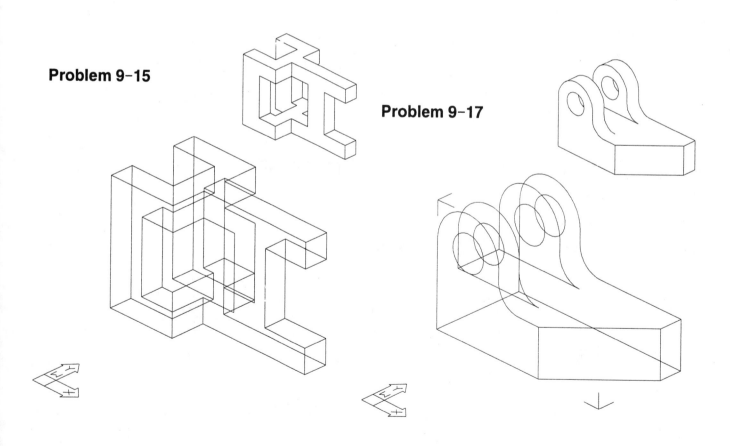

Problem 9-17

Problem 9–18

Problem 9–20

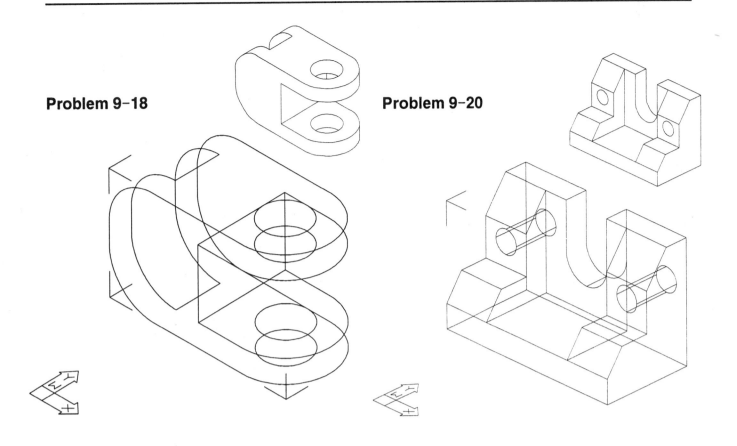

Problem 9–19

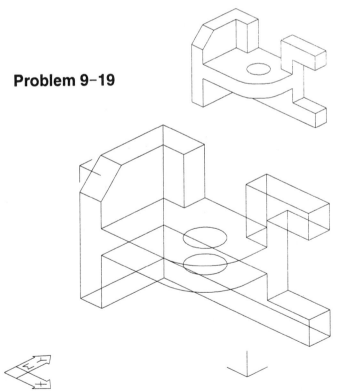

Note: Wireframe models created in Unit 8 may also be projected to lay out their orthographic views in addition to being dimensioned. Consult your instructor if any of these problems are to be used.

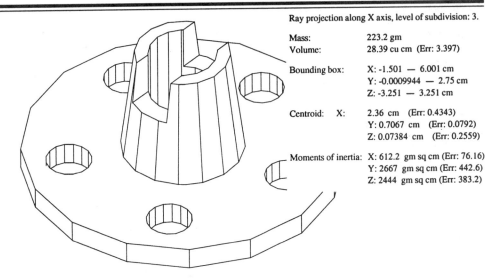

Ray projection along X axis, level of subdivision: 3.

Mass:	223.2 gm
Volume:	28.39 cu cm (Err: 3.397)
Bounding box:	X: -1.501 — 6.001 cm
	Y: -0.0009944 — 2.75 cm
	Z: -3.251 — 3.251 cm
Centroid: X:	2.36 cm (Err: 0.4343)
	Y: 0.7067 cm (Err: 0.0792)
	Z: 0.07384 cm (Err: 0.2559)
Moments of inertia:	X: 612.2 gm sq cm (Err: 76.16)
	Y: 2667 gm sq cm (Err: 442.6)
	Z: 2444 gm sq cm (Err: 383.2)

Solids Modeling

Contents

Thus far, we have seen how wireframe models can be surfaced for hidden line removal or left unsurfaced for the purpose of projecting primary views onto a flat 2D sheet of paper. One limiting factor of wireframe models is they are basically hollow, even when surfaced. This is where solid modeling picks up. These types of models are more informationally correct than wireframe models since a solid object can be analyzed by calculating such items as mass properties, center of gravity, surface area, moments of inertia, and much more. The solid model starts the true design process by defining objects as a series of primitives; boxes, cubes, cylinders, spheres, and wedges are all examples of primitives. These building blocks are then joined together or subtracted from each other using certain modifying commands. Fillets and chamfers may be created to give the solid model a more realistic appearance. 2D views may be extracted from the solid model along with a cross-section of the model. Follow the next series of pages that explain basic solid modeling concepts before completing the three tutorial exercises at the end of this unit.

Solid Modeling Basics

All objects, no matter how simple or complex, are composed of simple geometric shapes or primitives. The shapes range from boxes to cylinders to cones and so on. Solids modeling allows for the creation of these primitives. Once created, the shapes are either merged or subtracted to form the final object. Follow the next series of steps to form the object at the right.

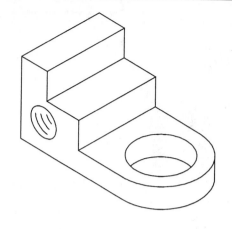

Begin the process of solids modeling by constructing a solid slab that will represent the base of the object. This is accomplished using the AutoCAD Advanced Modeling Extension (AME) command Solbox, short for Solid Box. The length, width, and height of the box are supplied. The result is a solid slab.

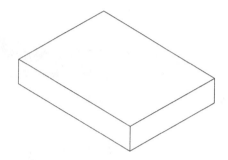

Next, a cylinder is constructed using the Solcyl, or Solid Cylinder command. This cylinder will eventually form the curved end of the object. One of the advantages of constructing a solid model is the ability of merging primitives together to form composite solids. Using Constructive Solids Geometry (CSG) commands such as Solunion, the cylinder and box are combined to form the complete base of the object.

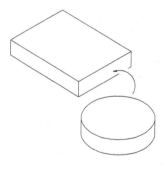

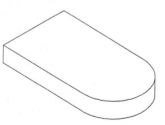

As the object progresses, another solid box is created and moved into position on top of the base. There, the Solunion command is used to join this new block with the base. As new shapes are added during this process, they all become part of the same solid.

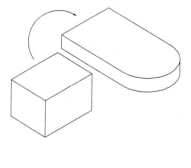

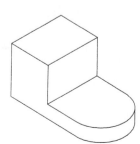

Solid Modeling Basics Continued

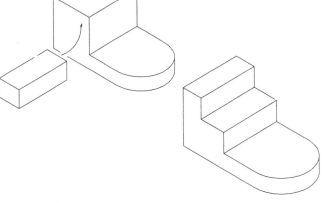

Yet another box is created and moved into position. However, instead of combining the blocks, this new box is removed from the solid. This process is called subtraction and when complete, creates the step illustrated at the right. The AutoCAD command used during this process is Solsub.

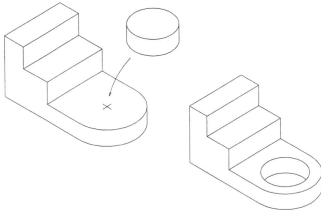

Holes are formed in a similar fashion. First the Solcyl command is used to create a cylinder the diameter and depth of the desired hole. The cylinder is moved into the solid and there subtracted using the Solsub command. Again, the object at the right represents a solid object.

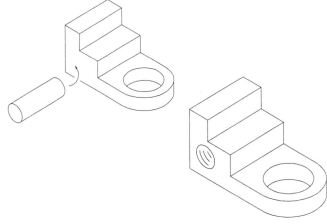

Using existing AutoCAD tools such as UCS, another cylinder is created using the Solcyl command. It too is moved into position where it is subtracted using the Solsub command. The complete solid model of the object is illustrated at the far right.

Rewards from constructing a solid model out of an object come in many forms. Profiles of different surfaces of the solid model may be taken. Section views of solids may automatically be formed and cross-hatched through solid models. A very important analysis tool is the Solmassp commands which is short for mass property extraction. Information such as the mass and volume of the solid object may be calculated along with centroids and moments of inertia, components that are used in Computer Aided Engineering (CAE) and in Finite Element Analysis (FEA) of the model.

Mass:	223.2 gm
Volume:	28.39 cu cm (Err: 3.397)
Bounding box:	X: -1.501 — 6.001 cm
	Y: -0.0009944 — 2.75 cm
	Z: -3.251 — 3.251 cm
Centroid: X:	2.36 cm (Err: 0.4343)
	Y: 0.7067 cm (Err: 0.0792)
	Z: 0.07384 cm (Err: 0.2559)
Moments of inertia:	X: 612.2 gm sq cm (Err: 76.16)
	Y: 2667 gm sq cm (Err: 442.6)
	Z: 2444 gm sq cm (Err: 383.2)

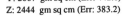

CSG, B-Rep, and Tree Structures

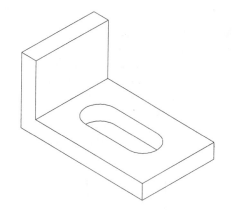

The construction of the object on the previous pages was possible through the use of Constructive Solids Geometry or CSG. An object is said to be constructed of a series of primitive shapes such as boxes, cylinders, and cones to name a few. These geometric shapes are then merged together or subtracted to create the solid model. In the object at the right, two solid primitives in the form of boxes are constructed and then joined using the Boolean operation of union. A solid slot is constructed, moved into postition, and subtracted from the main object to complete the solid model. The CSG method allows the solid to retain information regarding the structure and dimensions of the solid model.

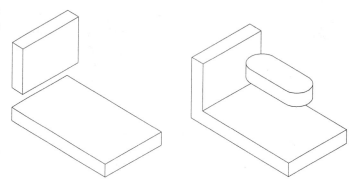

Another method used to define solid models is Boundary Representation or B-Rep. This method holds information regarding the surfaces, edges, and vertices that make up the boundary of the solid; hence, B-Rep. The AutoCAD AME utilizes both CSG and B-Rep information to create the most accurate solid model.

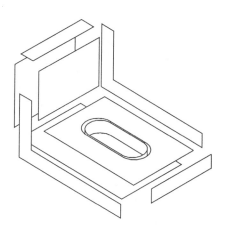

As multiple Boolean operations are performed, a hierarchy or tree structure forms to keep track of the steps used to construct the solid model. It is important that the solid model be broken down into individual primitives. A typical tree structure is illustrated at the right representing construction of the object above. Read this tree structure from bottom up. First a union of the two solid boxes was performed, represented by "A" and "B". Then the solid slot was subtracted from the main object to form the slot represented by "C". The results leave the final object shown at the top of the tree. Multiple Boolean operations may be performed at one time; however, these operations are organized in pairs and these pairs form the tree.

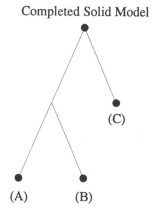

Completed Solid Model

(C)

(A)   (B)

Solid Primitives

SOLBOX

The Solbox command constructs a solid primitive of a 3D box. One corner of the box is located along with its other diagonal corner. A height is assigned to complete the definition of the box. A cube may also be constructed by selecting the appropriate option in the command prompt below. If all three dimensions of a box are known, the solid box may be constructed by entering values for its length, width, and height. Illustrated at the right are a few examples of solid boxes.

Command: **Solbox**
Corner of box: **4.00,5.50** *(At "A")*
Cube/Length/<Other corner>: **L**
Length: **4.00** *("A" to "B")*
Width: **4.00** *("B" to "C")*
Height: **8.00** *("B" to "D")*

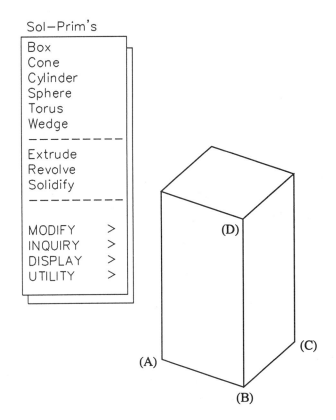

SOLCONE

The Solcone command constructs a solid primitive of a cone with a circular or elliptical base connecting at a central point of specified height. Examples are illustrated at the right with sample prompt sequences below.

The following is a sample prompt sequence for the cone illustrated at "A":

Command: **Solcone**
Elliptical/<Center point>: *(Identify a point)*
Diameter/<Radius>: *(Enter a radius value)*
Height of cone: *(Enter a height for the cone)*

The following is a sample prompt sequence for the cone illustrated at "B":

Command: **Solcone**
Elliptical/<Center point>: **Elliptical**
<Axis endpoint 1>/Center: *(Identify an axis endpoint)*
Axis endpoint 2: *(Identify the second axis endpoint)*
Other axis distance: *(Identify the other axis distance)*
Height of cone: *(Enter a height for the cone)*

Cone "A"

Cone "B"

Solid Primitives Continued

SOLWEDGE

Yet another solid primitive is the wedge which consists of a box that has been diagonally cut. The base of the wedge is drawn parallel to the current UCS. The sloped surface tapers along the X axis. The prompts for this command are very similar to the Solbox command.

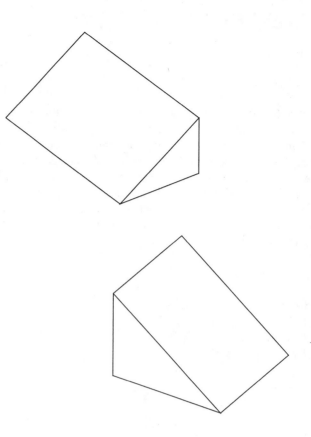

Command: **Solwedge**
Corner of wedge: *(Locate one corner of the wedge)*
Length:<Opposite corner>: *(Locate opposite corner of the wedge)*
Height: *(Enter a nonzero value for the height of the wedge)*

The following prompts illustrate the construction of a wedge by providing the length, width, and height dimensions.

Command: **Solwedge**
Corner of wedge: *(Locate one corner of the wedge)*
Length:<Opposite corner>: **Length**
Length: *(Enter the length of the wedge)*
Width: *(Enter the width of the wedge)*
Height: *(Enter the height of the wedge)*

SOLCYL

The Solcyl command is similar to the Solcone command except that a cylinder without taper is drawn. The central axis of a cylinder is along the Z axis of the current UCS. The following prompts illustrate construction of a cylinder by radius and diameter. A cylinder may also be elliptical in shape.

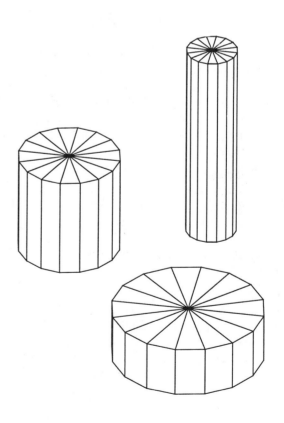

Command: **Solcyl**
Elliptical/<Center point>: *(Enter coordinates for the center of the cylinder)*
Diameter/<Radius>: *(Enter a value for the cylinder radius)*
Height of cylinder: *(Enter a value for the cylinder height)*

Command: **Solcyl**
Elliptical/<Center point>: *(Enter coordinates for the center of the cylinder)*
Diameter/<Radius>: **Diameter**
Diameter: *(Enter a value for the diameter of the cylinder)*
Height of cylinder: *(Enter a value for the cylinder height)*

Solid Primitives Continued

SOLSPHERE

The Solshpere command constructs a type of ball with all points along its surface equal in distance from a central center. As in the cylinder, the central axis of a sphere is along the Z axis of the current UCS. Spheres may be drawn using either diameter or radius dimensions.

Command: **Solsphere**
Center point: *(Enter the center of the sphere)*
Diameter/<Radius>: *(Enter the radius of the sphere or "D" to be prompted for the diameter of the sphere)*

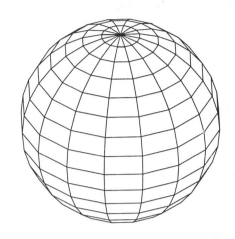

SOLTORUS

A torus is formed when a circle is revolved about a line in the same plane as the circle. In other words, a torus is similar to a 3D donut. A torus may be constructed using either a radius or diameter method. When using the radius method, two radius values must be used to define the torus; one for the radius of the tube and the other for the radius from the center of the torus to the center of the tube. Two diameter values would be used when specifying a torus by diameter. Once the torus is constructed, it lies parallel to the current UCS. Follow the prompts below to define a torus by radius values.

Command: **Soltorus**
Center of torus: *(Identify the center of the torus through a coordinate or by picking)*
Diameter/<Radius> of torus: *(Enter a value for the radius of the torus at "A")*
Diameter/<Radius> of tube: *(Enter a value for the radius of the tube of the torus at "B")*

Follow the prompts below to define a torus through diameter values.

Command: **Soltorus**
Center of torus: *(Identify the center of the torus through a coordinate or by picking)*
Diameter/<Radius> of torus: **Diameter**
Diameter: *(Enter a value for the diameter of the torus at "C")*
Diameter/<Radius> of tube: **Diameter**
Diameter: *(Enter a value for the diameter of the tube of the torus at "D")*

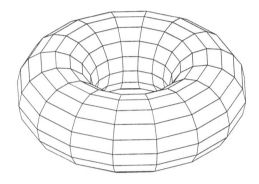

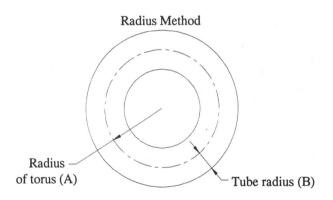

Radius Method

Radius of torus (A)

Tube radius (B)

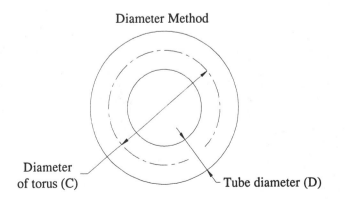

Diameter Method

Diameter of torus (C)

Tube diameter (D)

Creating Solid Extrusions

The Solext command creates a solid by extrusion. Only polylines and circles may be extruded. Once these entities are selected, an extrusion height is asked for followed by an extrusion taper angle. If the entities selected are not polylines, use the Pedit command and convert them to polylines.

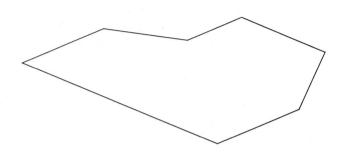

Command: **Pedit**
Select polyline: *(Select a line at the right)*
Entity selected is not a polyline.
Do you want it to turn into one? <N>: **Yes**
Close/Join/Width/Edit vertex/Fit curve/Spline curve/
    Decurve/Undo/eXit <X>: **J**
Select objects: *(Select the remaining lines of the object)*
6 segments added to polyline.
Close/Join/Width/Edit vertex/Fit curve/Spline curve/
    Decurve/Undo/eXit <X>: *(Strike Enter to exit this
    command)*

Once entities are polylines, use the prompts below to construct a solid extrusion of the object below. For the height of the extrusion, a positive numeric value may be entered or the distance may be determined by picking two points on the display screen.

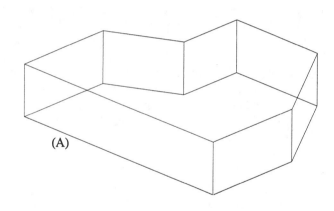

(A)

Command: **Solext**
Select polylines and circles for extrusion...
Select objects: *(Select the polyline at "A")*
Select objects: *(Strike Enter to continue)*
Height of extrusion: **1.00**
Extrusion taper angle from Z <0>: *(Strike Enter to accept
    default)*

An optional taper may be created along with the extrusion by entering an angle value for the prompt, "Extrusion taper angle from Z".

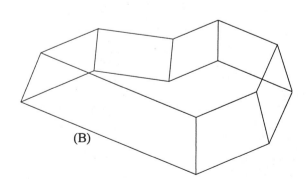

(B)

Command: **Solext**
Select polylines and circles for extrusion...
Select objects: *(Select the polyline at "B")*
Select objects: *(Strike Enter to continue)*
Height of extrusion: **1.00**
Extrusion taper angle from Z <0>: **15**

Creating Revolved Solids

The Solrev command creates a solid by revolving an entity about an axis of revolution. Only polylines, polygons, circles, ellipses, and 3D polylines may be revolved. If a group of entities are not in the form of polylines, group them together using the Pedit command. The resulting image at the right represents a solid entity.

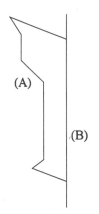
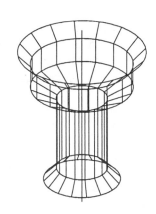

Command: **Solrev**
Select polyline or circle for revolution...
Select objects: *(Select the polyline at "A")*
Select objects: *(Strike Enter to continue)*
Axis of revolution - Entity/X/Y/<Start point of axis>: **Entity**
Entity to revolve about: *(Select the line at "B")*
Included angle<full circle>: *(Strike Enter to accept default)*

A practical application of this type of solid would be to first construct an additional solid consisting of a cylinder using the Solcyl command. Be sure this solid is larger in diameter than the revolved solid. Existing Osnap options are fully supported in solid modeling. Use the Center option of Osnap along with the Move command to position the revolved solid inside of the cylinder.

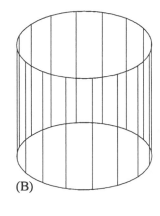
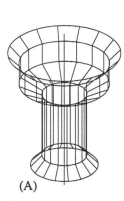

Command: **Move**
Select objects: *(Select the revolved solid at the right)*
Select objects: *(Strike Enter to continue)*
Base point or displacement: **Cen**
of *(Select the center of the revolved solid at "A")*
Second point of displacement: **Cen**
of *(Select the center of the cylinder at "B")*

Once the revolved solid is positioned inside of the cylinder, the Solsub command is used to subtract the revolved from the cylinder. Next, the Solmesh command is used to surface the solid at "A". Finally, the Hide command is used to perform a hidden line removal at "B" to check that the solid is correct, which would be difficult to interpret in wireframe mode. All of these commands will be discussed in detail in the pages to follow.

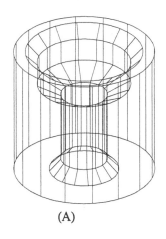
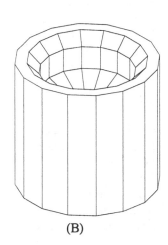

Modifying Solid Models

To create details such as cuts and holes from solid primitives, Boolean operations are used to add or subtract geometry from each other. In the example at "A", a cylinder has been constructed along with a square block. Depending on which Boolean operation is used, the results could be quite different. In example "B", both the square slab and cylinder are one entity and considered one solid object. This is the purpose of the Solunion command; to join or unite two solid primitives into one. Example "C" illustrates the intersection of the two solid primitives or the area that both solids have in common. This solid is obtained by using the Solint command. Example "D" shows the results of removing or subtracting the cylinder from the square slab; a hole is formed inside the square slab as a result of using the Solsub command. All Boolean operation commands may work on numerous solid primitives; that is, if subtracting numerous cylinders from a slab, all cylinders at the same time may be subtracted. These commands may be entered in from the keyboard or may be selected from the pulldown menu area under Sol-Modify.

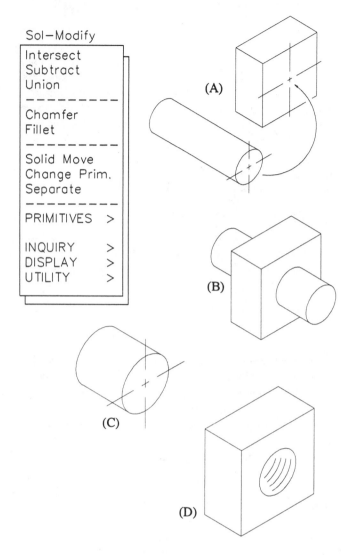

Merging Solids

Illustrated at the right is an object consisting of one horizontal solid box, two vertical solid boxes, and two extruded semi-circular shapes. All solids have been positioned either with the Move or Solmove commands. The problem is to join all of these shapes into one using the Solunion command. The order of selection of these solids for this command is not important. Follow the prompts below and the example at the right.

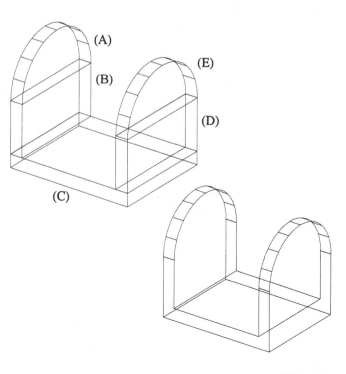

Command: **Solunion**
Select objects: *(Select the solid extrusion at "A")*
Select objects: *(Select the vertical solid box at "B")*
Select objects: *(Select the horizontal solid box at "C")*
Select objects: *(Select the vertical solid box at "D")*
Select objects: *(Select the solid extrusion at "E")*
Select objects: *(Strike Enter to perform the union of the solids)*

Subtracting Solids from Each Other

Using the same problem from the previous page, let's now add a hole in the center of the base. The cylinder was already created using the Solcyl command. It now needs to be moved to the exact center of the base. The Move command along with XYZ filters will be used to accomplish this.

Command: **Move**
Select objects: *(Select the cylinder at "A")*
Select objects: *(Strike Enter to continue)*
Base point or displacement: **Cen**
of *(Select the bottom center of the cylinder at "A")*
Second point of displacement: **.X**
of **Mid**
of *(Select the midpoint of the bottom of the base at "B")*
(Need YZ): **.Y**
of **Mid**
of *(Select the midpoint of the bottom of the base at "C")*
(Need Z): **0**

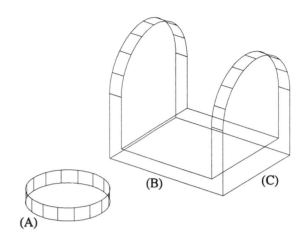

Now that the solids are in position, use the Solsub command to subtract the cylinder from the base of the main solid.

Command: **Solsub**
Source objects...
Select objects: *(Select the main solid as source at "A")*
Select objects: *(Strike Enter to continue)*
Objects to subtract from them...
Select objects: *(Select the cylinder at "B")*
Select objects: *(Strike Enter to perform the subtraction operation)*

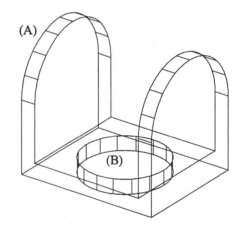

Two more holes need to be added to the vertical sides of the object. A cylinder was already constructed, however, it is in the vertical position. This entity needs to be rotated along the Y axis using the Solmove command. When issuing this command and selecting the cylinder, a special icon appears that shows the X, Y, and Z axes. For motion description, enter a value of "RY90" which will rotate the cylinder 90 degrees in the counterclockwise direction using the Y axis as the pivot.

Command: **Solmove**
Select objects: *(Select the vertical cylinder at "A")*
Select objects: *(Strike Enter to continue)*
<Motion description>/?: **RY90**

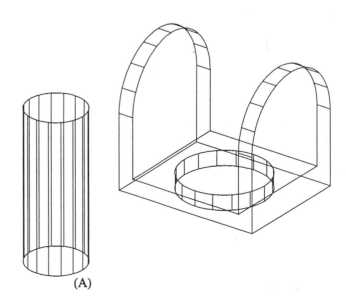

Subtracting Solids Continued

Use the Move command to move the long cylinder to the vertical sides of the main solid using the Osnap-Center option as an aid.

Command: **Move**
Select objects: *(Select the cylinder at "A")*
Select objects: *(Strike Enter to continue)*
Base point or displacement: **Cen**
of *(Select the center of the cylinder at "A")*
Second point of displacement: **Cen**
of *(Select the center of the semi-circular solid at "B")*

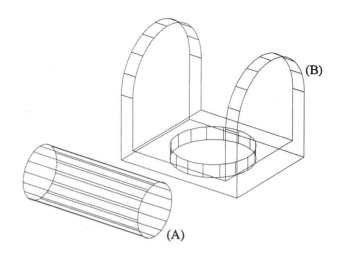

Use the Solsub command to subtract the long cylinder from the two vertical sides of the object.

Command: **Solsub**
Source objects...
Select objects: *(Select the main solid as source at "A")*
Select objects: *(Strike Enter to continue)*
Objects to subtract from them...
Select objects: *(Select the cylinder at "B")*
Select objects: *(Strike Enter to perform the subtraction operation)*

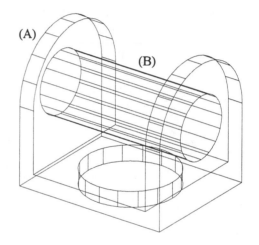

The results of both union and subtraction operations are illustrated at the right. Because the AME designs solid images in wireframe mode, use the Solmesh command to surface the solid model before using the Hide command to perform a hidden line removal of the object.

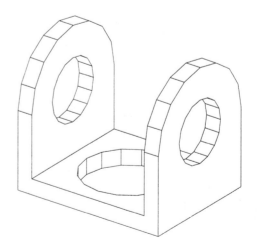

Curve Tessellation

Tessellation refers to the lines that are displayed on any curved surface to help visualize the surface. Tessellation lines are automatically formed when constructing solid primitives such as cylinders and cones. These lines are also calculated when performing such solid modeling operations as Solsub and Solunion to name a few.

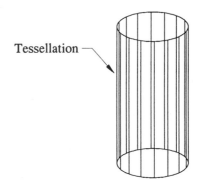

Tessellation

The number of tessellation lines per curved entity is controlled by a system variable called Solwdens. By default, this variable is set to a value of 4. Illustrated at the right are the results of setting this variable to other values such as 1 and 8. The more lines used to describe a curved surface, the more accurate the surface will look; however, the longer the surface will take to process hidden line removals using the Hide command and Boolean operation commands such as Solunion and Solsub.

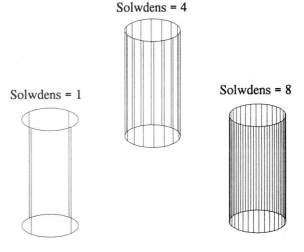

Solwdens = 4

Solwdens = 1

Solwdens = 8

The above example of tessellation lines on cylinders was illustrated in wireframe mode. When surfacing these entities, the results are displayed in the illustrations at the right. The cylinder with Solwdens = 1 processes much quicker that the cylinder with Solwdens = 8 since there are less surfaces to process in such operations as hidden line removals. An average value for Solwdens is 4 which seems adequate for most applications.

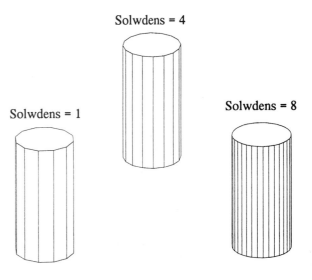

Solwdens = 4

Solwdens = 1

Solwdens = 8

Chamfering Solid Models

The Solcham command creates a chamfer and automatically performs a union or subtraction operation to satisfy the chamfer specifications for a particular solid. This command requires a base surface to begin the chamfer calculations. As a surface is selected, sometimes it highlights, sometimes it does not. During this selection process, the prompt "<OK>/Next" appears which allows the user to accept the current surface, or move to the next surface to see if it highlights. Keep repeating this procedure until the desired edge is selected. Once the base surface is identified, the edges to chamfer are selected. Two chamfer distances are asked for; if both values are equal, the chamfer will form a 45-degree angle; if both values are not equal, a beveled edge is formed at an angle other than 45 degrees.

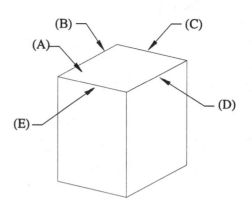

Command: **Solcham**
Select base surface: *(Select top surface "A")*
<OK>/Next: *(If the top surface highlights, strike Enter; if not, enter "N" for the next surface and continue until the top surface highlights)*
Select edges to be chamfered (Press Enter when done): *(Select the edges at "B", "C", and "D")*
Enter distance along first surface<0.000>: **0.125**
Enter distance along second surface<0.125>: *(Strike Enter)*

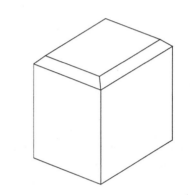

Filleting Solid Models

Solid models may be filletted using the Solfill command. This command is similar to the Solcham command except that edges are rounded instead of beveled. Multiple edges may be selected using this command. Enter the desired radius and the selected edges are filletted. A solid fillet primitive is automatically created and added to or subtracted from the existing solid model.

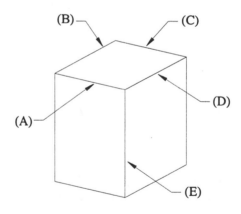

Command: **Solfill**
Select edges to be filleted (Press Enter when done): *(Select edges "A", "B", "C", and "D")*
Diameter/<Radius> of fillet <0.000>: **0.125**

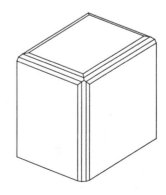

Using the Solmove Command

As an aid in moving and positioning solid models, the Solmove command can be used. This is not to be confused with the regular AutoCAD Move comamnd. When issuing this command and selecting a solid model to position, a new icon appears at the current location of the UCS. Because this icon is specific to the Solmove command, it is called the MCS or Motion Coordinate System. Three axes are provided to assist in an understanding of where to move the solid model. These axes are labeled with a series of cones; the X axis is identified with one cone; the Y axis identified with two cones; the Z axis with three cones. The next prompt asks for a motion description. These key-ins to describe motion are quite different and are listed in the command when entering the "?" at the motion description prompt.

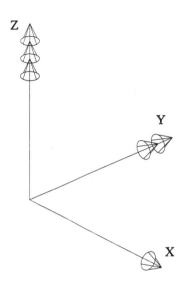

At the right is a cylinder 1 unit in diameter and 2 units tall. The user coordinate and motion coordinate system icons are located at the center base of the cylinder. The next four examples will reference this illustration. Please refer back to this example. The prompts for the Solmove command are:

Command: **Solmove**
Select objects: *(Select the cylinder at the right)*
Select objects: *(Strike Enter to continue)*
<Motion description>/?: *(Enter a motion description definition)*

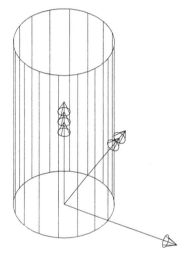

To move the cylinder 1 unit in the X direction and 1 unit in the Y direction, the key phrase is "Translate". Use the motion coordinate system to determine the direction of the move. The key-in to move the solid is "TX1.00,TY1.00".

Command: **Solmove**
Select objects: *(Select the cylinder at the right)*
Select objects: *(Strike Enter to continue)*
<Motion description>/?: **TX1.00,TY1.00**

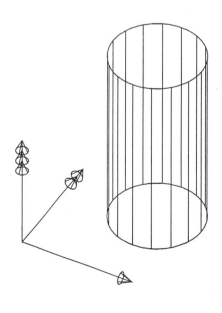

Using the Solmove Command Continued

At the right, the cylinder is moved 1.50 units in the negative X direction by 1.00 units in the positive Z direction. Again, use the motion coordinate system to determine the direction of motion.

Command: **Solmove**
Select objects: *(Select the cylinder at the right)*
Select objects: *(Strike Enter to continue)*
<Motion description>/?: **TX-1.50,TZ1.00**

The Solmove command does more than just move or translate; solid models may also be rotated using this motion coordinate system. In the example at the right, the cylinder was rotated 90 degrees clockwise using the X axis as the pivot. The clockwise rotation demands a negative value.

Command: **Solmove**
Select objects: *(Select the cylinder at the right)*
Select objects: *(Strike Enter to continue)*
<Motion description>/?: **RX-90**

This last example illustrates how numerous motion directions can be used together to translate the solid model around. At the right, the cylinder was first rotated about the Y axis at 90 degrees counterclockwise. Then it was moved 1.00 units in the X direction. Instead of using the Solmove command twice, both operations are entered at the motion description prompt separated by a comma.

Command: **Solmove**
Select objects: *(Select the cylinder at the right)*
Select objects: *(Strike Enter to continue)*
<Motion description>/?: **RY90,TX1.00**

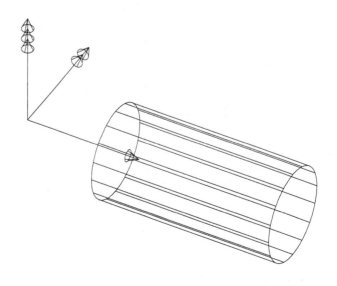

Displaying Solid Models

The pull-down menu bar illustrated at the right lists the numerous ways to display the image of a solid model. By default, all solid models are construced in wireframe mode. Should the need arise to perform a hidden line removal, the Solmesh command places faces along all surfaces of the model. This enables the model to display visible surfaces when using the Hide command. Most AME commands function in wireframe mode; so if rendering a solid, all surfaces need to be converted back from mesh patterns to wireframe mode using the Solwire command. The next series of pages discuss some of the more advanced display commands of solids modeling such as Cutting Sections and Creating Profiles.

```
Sol-Display
┌─────────────────┐
│ Mesh            │
│ Wireframe       │
│ ─ ─ ─ ─ ─ ─ ─   │
│ Copy Feature    │
│ Cut Section     │
│ Profile         │
│ ─ ─ ─ ─ ─ ─ ─   │
│ Set Decomp.     │
│ Set Subdiv.     │
│ Set Wire Dens.  │
│ ─ ─ ─ ─ ─ ─ ─   │
│ PRIMITIVES  >   │
│ MODIFY      >   │
│ INQUIRY     >   │
│                 │
│ UTILITY     >   │
└─────────────────┘
```

Cutting Sections from Solid Models

The image at the right is the familiar cylinder with a revolved solid removed from its interior. At times it is advantageous to view interior details through section views. This requires some preparation before performing this operation. First, the section is cut in relation to the position of the current UCS. At the right, the UCS icon has been revolved 90 degrees by the X axis and positioned at the center base of the cylinder. This determines the cutting plane angle that creates the section view out of the solid model.

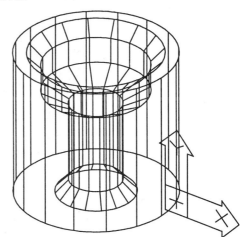

Before issuing the section view extraction command, a pattern needs to be identified for the section to represent. Without it, the section will be created; however, the surfaces making up the section will be void of any cross-hatch lines. This is the purpose of the Solhpat command.

Command: **Solhpat**
Hatch pattern <None>: **Ansi31**

Command: **Solsect**
Select objects: *(Select the solid anywhere)*

As the section constructs itself inside the solid model, use the Move command to move the section view to a better location. Also, lines not cut by the cutting plane line are omitted from the section view resulting in two halves of the section view that are not connected. The Line command is used to connect edges not cut, as in the example at the far right.

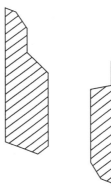

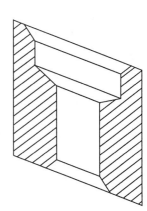

Shading Solid Models

For a quick view of what the object would look like as a
shaded image, use the Shade command to accomplish this.
This command may be selected from the Display menu of
the pull-down menu bar; another way to execute this
command is by entering it through the keyboard.

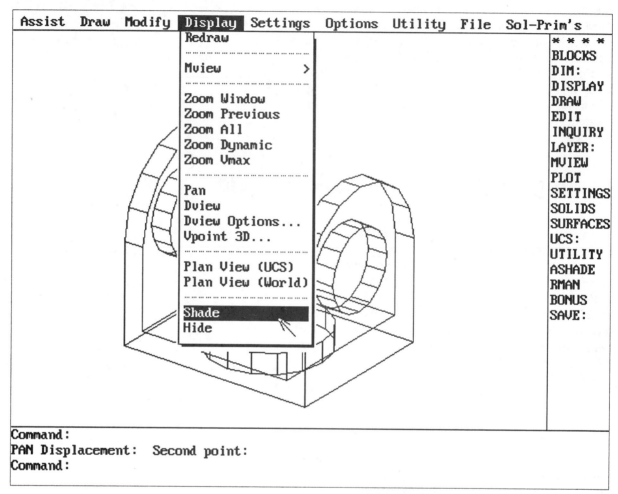

This command automatically produces a hidden line removal in addition to displaying the shaded image in the
current color. When executing this command, the screen
temporarily will go blank while the system calculates the
surfaces to shade. The length of time to accomplish this is
of course dependent on the complexity of the object being
shaded. A percentage of completion is displayed at the
bottom of the prompt line. Once the shaded image is
displayed on the screen, it cannot be plotted; it can, however, be made into a slide using the Mslide command.
This image consists of a colorization of the solid model
and is not meant for different surface tones or casting of
shadows.

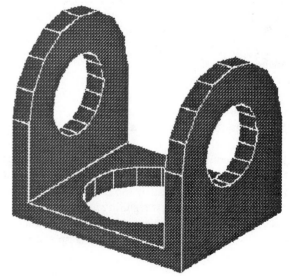

Listing Solids

The commands listed under the Sol-Inquiry pull-down menu allow the user to list various characteristics associated with a selected solid model. The Sollist command displays information used to define the solid model. The Solmassp command lists the mass property information associated with the solid model. The Solarea command displays the area of the solid. The three types of units affect the following solid modeling variables: Sollength, Solareav, Solvolume, and Solmass. British units assign the following values: Sollength=feet, Solareav=square feet, Solvolume=cubic feet, and Solmass=pounds. CGS units assign the following values; Sollength=centimeters, Solareav=Square centimeters, Solvolume=cubic centimeters, and Solmass=grams. SI units assign the following values: Sollength=meters, Solareav=square meters, Solvolume=cubic meters, and solmass=kilograms.

```
Sol-Inquiry
List Solid
Mass Property
Solid Area
- - - - - - -
British units
CGS units
SI units
- - - - - - -


PRIMITIVES    >
MODIFY        >

DISPLAY       >
UTILITY       >
```

(A)

Using the Sollist command displays the following prompts:
Edge/Face/Tree/<Solid>:

The results of keeping the default value are displayed at the right. This displays the top-most level of the tree structure that went into the creation of the solid model shown above at "A". Each line is briefly described below:

Solid type = SUBTRACTION – This represents the highest level on the CSG tree and happens to be the last Boolean operation performed to create the solid model. An entity handle is assigned to the resulting composite solid.

Component handles: 9D and 5C – These handles correspond to the objects affected by the Boolean operation used to create the composite solid.

Area and Material – These list the area and material of the solid.

Representation – indicated whether the solid model is displayed in wireframe mode or mesh mode. Both types of representation are controlled by the Solwire and Solmesh commands.

Rigid motion – These are values that keep track of any rotations, translations, and scaling applied to the solid model.

Using the Sollist command and selecting the Tree option displays a listing of the solid primitives and Boolean operations involved in the creation of the solid model. This listing is illustrated at the right.

```
Solid type = SUBTRACTION    Handle = 14F
   Component handles:  9D and 5C
   Area = 35.414214   Material = MILD_STEEL
   Representation = WIREFRAME   Shade type = CSG
Rigid motion:
      +1.000000      +0.000000      +0.000000      +0.000000
      +0.000000      +1.000000      +0.000000      +0.000000
      +0.000000      +0.000000      +1.000000      +0.000000
      +0.000000      +0.000000      +0.000000      +1.000000
```

```
Solid type = SUBTRACTION    Handle = 14F
   Component handles:  9D and 5C
   Area = 35.414214   Material = MILD_STEEL
   Representation = WIREFRAME   Shade type = CSG

.... Solid type = SUBTRACTION    Handle = 9D
.... Component handles:  78 and 4D
.... Area not computed   Material = MILD_STEEL
.... Representation = WIREFRAME   Shade type = CSG
.... Node level = 1

.... Solid type = UNION    Handle = 78
.... Component handles:  29 and 3B
.... Area not computed   Material = MILD_STEEL
.... Representation = WIREFRAME   Shade type = CSG
.... Node level = 2

.... Solid type = BOX (4.000000, 2.000000, 1.000000)   Handle = 29
.... Area not computed   Material = MILD_STEEL
.... Representation = WIREFRAME   Shade type = CSG
.... Node level = 3

.... Solid type = BOX (1.000000, 2.000000, 1.500000)   Handle = 3B
.... Area not computed   Material = MILD_STEEL
.... Representation = WIREFRAME   Shade type = CSG
.... Node level = 3

.... Solid type = BOX (2.000000, 1.000000, 1.000000)   Handle = 4D
.... Area not computed   Material = MILD_STEEL
.... Representation = WIREFRAME   Shade type = CSG
.... Node level = 2

.... Solid type = WEDGE (1.000000, 1.000000, 1.000000)   Handle = 5C
.... Area not computed   Material = MILD_STEEL
.... Representation = WIREFRAME   Shade type = CSG
.... Node level = 1
```

Listing Solids Continued

SOLMASSP

Because the solid model represents the most informationally correct solution to a part, a major advantage of the model is its ability to be analyzed for design purposes. The following properties of a solid model are calculated by this command: Mass, Volume, Bounding Box, Centroid, Moments of Inertia, Products of Inertia, Radii of Gyration, and Principal Moments. All calculations are based on the position of the current UCS. If errors are detected, they are enclosed in parentheses. When identifying the centroid of a solid model, a point is placed on the object in the current layer. The type of point displayed is controlled by the Pdmode system variable; the size of this point is controlled by the system variable Pdsize. When displaying the results of a typical mass property listing, the Solmassp command will prompt the user to write the results to a file. If this prompt is answered "Yes" a text file is created from a name supplied by the user or the default name of the solid model. A file extension of .Mpr will be used to separate the mass property text file from the standard drawing file. Illustrated at the right is a sample set of mass property calculations.

Ray projection along X axis, level of subdivision: 3.
Mass: 67.55 gm
Volume: 8.594 cu cm (Err: 0.7242)

Bounding box:
X: 4 — 8 cm
Y: 5.5 — 7.5 cm
Z: 0 — 2.5 cm

Centroid:
X: 5.373 cm   (Err: 0.548)
Y: 6.532 cm   (Err: 0.5621)
Z: 0.8411 cm   (Err: 0.05665)

Moments of inertia:
X: 2979 gm sq cm (Err: 262.1)
Y: 2104 gm sq cm (Err: 243.3)
Z: 4938 gm sq cm (Err: 501.8)

Products of inertia:
XY: 2371 gm sq cm (Err: 246.3)
YZ: 373.3 gm sq cm (Err: 26.13)
ZX: 287.2 gm sq cm (Err: 23.15)

Radii of gyration:
X: 6.64 cm
Y: 5.582 cm
Z: 8.55 cm

Principal moments (gm sq cm) and X-Y-Z directions about centroid:
I: 43.81 along [0.9607 -0.000183 -0.2778]
J: 112.2 along [0.2559 -0.3882 0.8853]
K: 105.8 along [-0.108 -0.9216 -0.3729]

Solid Modeling Utility Commands

The following utility commands allow the user to control items such as the type of material the solid model consists of (Solmat). The Solvars command controls system variables specific to solid modeling commands. A new current UCS may be set to a face or edge of a solid using the Solucs command. The Solin and Solout commands were primarily designed to import models to or export models from AutoSolid. Unused solid modeling information is easily removed from a drawing using the Solpurge command. To free-up precious memory, both AME and AMElite may be removed using the Unload option from the pull-down menu area. A few of the above commands are described in detail as follow.

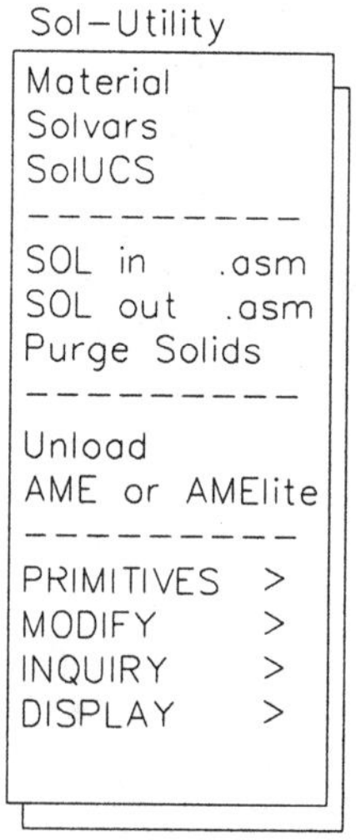

```
Sol—Utility
┌─────────────────┐
│ Material        │
│ Solvars         │
│ SolUCS          │
│ ─────────       │
│ SOL in    .asm  │
│ SOL out   .asm  │
│ Purge Solids    │
│ ─────────       │
│ Unload          │
│ AME or AMElite  │
│ ─────────       │
│ PRIMITIVES  >   │
│ MODIFY      >   │
│ INQUIRY     >   │
│ DISPLAY     >   │
└─────────────────┘
```

SOLMAT

This command assigns a default material to the solid model. An existing solid model may have its current material changed to reflect a new material from the list at the right. New materials may be added to this list. These material definitions are very important to the solid model since this information is used when performing a mass property calculation when executing the Solmassp command. By default, all solid models are assigned the material "Mild Steel". The following prompt sequence may be used to list or change this material:

Command: **Solmat**
Change/Edit/<eXit>/LIst/LOad/New/Remove/SAve/SEt/?:
    ?

```
Defined in drawing:
  MILD_STEEL
Defined in file:
  ALUMINUM        --    Aluminum
  BRASS           --    Soft Yellow Brass
  BRONZE          --    Soft Tin Bronze
  COPPER          --    Copper
  GLASS           --    Glass
  HSLA_STL        --    High Strength Low Alloy Steel
  LEAD            --    Lead
  MILD_STEEL      --    Mild Steel
  NICU            --    Monel 400
  STAINLESS_STL   --    Austenic Stainless Steel

Change/Edit/<eXit>/LIst/LOad/New/Remove/SAve/SEt/?:
```

SOLVARS

This command controls specific system variables that affect solid modeling commands and is similar to the AutoCAD Setvar command. The illustration at the right lists these variables, what their default values are, and displays a short description of the variable.

```
SOLAMEVER    1.02        AME version        (read only)
SOLAREAU     sq cm       Area units
SOLAXCOL     3           Solid axes color
SOLDECOMP    X           Mass property decomposition direction
SOLDELENT    3           Entity deletion
SOLDISPLAY   wire        Display type
SOLHANGLE    0.000000    Hatch angle
SOLHPAT      NONE        Hatch pattern
SOLHSIZE     1.000000    Hatch size
SOLLENGTH    cm          Length units
SOLMASS      gm          Mass units
SOLMATCURR   MILD_STEEL  Current material (read only)
SOLPAGELEN   25          Length of Text Page
SOLRENDER    CSG         Rendering type
SOLSERVMSG   3           Solid Server message display level
SOLSOLIDIFY  2           Automatic solidification
SOLSUBDIV    3           Mass property subdivision level
SOLVOLUME    cu cm       Volume units
SOLWDENS     4           Mesh wireframe density
```

SOLUCS

This command will align the UCS with a selected face or edge. The prompts for this command are as follow:

Command: **Solucs**
Edge/<Face>: *(Select a face or enter "E" to select an edge)*
<OK>/Next: *(Accept the highlighted face or enter "N" to highlight the next face)*

As with similar AME commands displaying the prompt "<OK>/Next", the user has the option to accept the selected face with "OK" or choosing another face through "Next".

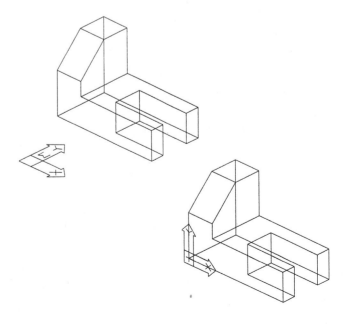

AME Command Aliasing

As part of the ACAD.PGP file, a majority of the AME commands may be typed in at the keyboard in abbreviated style. As an example, to draw a solid box, the AME command is Solbox. In the ACAD.PGP file, this command has been renamed to "Box". If typing is to be kept at a minimum, study the complete listing of AME commands on this page to see what the name of the command has been renamed to. Then enter that command from the keyboard at the command prompt. These commands are automatically supplied with the AutoCAD software package.

; Primitives.

BOX,	*SOLBOX
WED,	*SOLWEDGE
WEDGE,	*SOLWEDGE
CON,	*SOLCONE
CONE,	*SOLCONE
CYL,	*SOLCYL
CYLINDER,	*SOLCYL
SPH,	*SOLSPHERE
SPHERE,	*SOLSPHERE
TOR,	*SOLTORUS
TORUS,	*SOLTORUS

; Complex Solids.

FIL,	*SOLFILL
SOLF,	*SOLFILL
CHAM,	*SOLCHAM
SOLC,	*SOLCHAM
EXT,	*SOLEXT
EXTRUDE,	*SOLEXT
REV,	*SOLREV
REVOLVE,	*SOLREV
SOL,	*SOLIDIFY

; Boolean operations.

UNI,	*SOLUNION
UNION,	*SOLUNION
INT,	*SOLINT

INTERSECT,	*SOLINT
SUB,	*SOLSUB
SUBTRACT,	*SOLSUB
DIF,	*SOLSUB
DIFF,	*SOLSUB
DIFFERENCE,	*SOLSUB
SEP,	*SOLSEP
SEPARATE,	*SOLSEP

; Modification and Query commands.

SCHP,	*SOLCHP
CHPRIM,	*SOLCHP
MAT,	*SOLMAT
MATERIAL,	*SOLMAT
MOV,	*SOLMOVE
SL,	*SOLLIST
SLIST,	*SOLLIST
MP,	*SOLMASSP
MASSP,	*SOLMASSP
SA,	*SOLAREA
SAREA,	*SOLAREA
SSV,	*SOLVAR

; Documentation commands.

FEAT,	*SOLFEAT
PROF,	*SOLPROF
PROFILE,	*SOLPROF
SU,	*SOLUCS
SUCS,	*SOLUCS

; Model representation commands.

SW,	*SOLWIRE
WIRE,	*SOLWIRE
SM,	*SOLMESH
MESH,	*SOLMESH

Tutorial Exercise #24
Bplate.Dwg

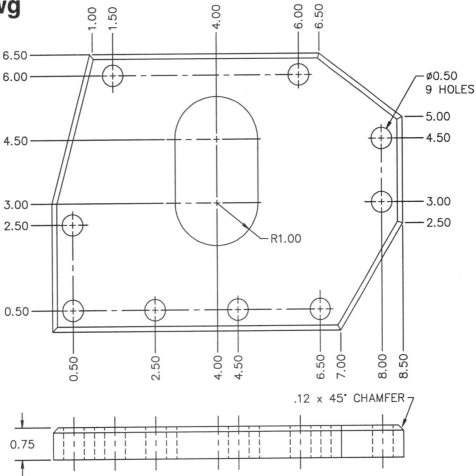

PURPOSE:
The purpose of this tutorial is to produce a solid model of the Plate using AutoCAD AME commands.

SYSTEM SETTINGS:
Begin a new drawing called "Bplate." Use the Units command to change the number of decimal places past the zero from 4 to 2. Keep the remaining default unit values. Using the Limits command, keep 0,0 for the lower left corner and change the upper right corner from 12,9 to 15.50,9.50. Use the Grid command and change the grid spacing from 1.00 to 0.50 units. Do not turn the snap or ortho On.

LAYERS:
Special layers do not have to be created for this tutorial exercise.

SUGGESTED COMMANDS:
Begin this tutorial by constructing the profile of the Plate using polylines. Add all circles and enter the AME. Use the Solext command to extrude all entities the thickness of the base at 0.75 units. Use the Solsub command to subtract all cylinders from the Plate forming the holes in the Plate. Use the Solmesh command to surface the model before performing a hidden line removal using the Hide command.

DIMENSIONING:
This tutorial does not need to be dimensioned.

PLOTTING:
This tutorial exercise may be plotted on "B"-size paper (11" x 17"). Use a plotting scale of 1=1 to produce a full size plot.

VERSION OF AUTOCAD:
This tutorial exercise must be completed using AutoCAD Release 11 with the Advanced Modeling Extension (AME).

Step #1

Begin the plate by establishing a new coordinate system using the UCS command. Define the origin at 2.00,1.50. Use the Ucsicon command to update the UCS icon to the new coordinate system location on the display screen. Use the Pline command to draw the profile of the plate.

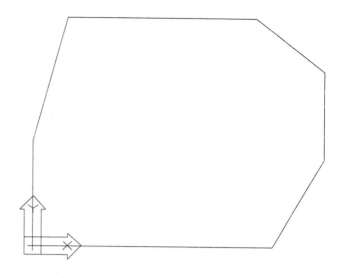

Command: **UCS**
Origin/ZAxis/3point/Entity/View/X/Y/Z/Prev/Restore/Save/
    Del/?/<World>: **O**
Origin point: <0,0,0>: **2.00,1.50**

Command: **Ucsicon**
ON/OFF/All/Noorigin/ORigin <ON>: **OR**

Command: **Pline**
From point: **0,0**
Current line-width is 0.00
<Endpoint of line>: **@7.00<0**
<Endpoint of line>: **@1.50,2.50**
<Endpoint of line>: **@2.50<90**
<Endpoint of line>: **@-2.00,1.50**
<Endpoint of line>: **@5.50<180**
<Endpoint of line>: **@-1.00,-3.50**
<Endpoint of line>: **Close**

Step #2

Draw the 9 circles of 0.50 diameter by placing one circle at "A" and copying the remaining circles to their desired locations. Use the Copy-Multiple option to accomplish this.

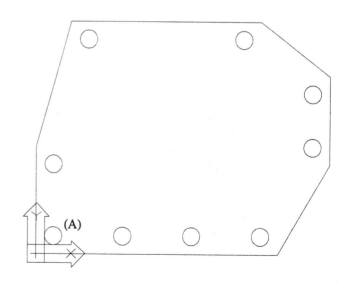

Command: **Circle**
3P/2P/TTR/<Center point>: **0.50,0.50**
Diameter/<Radius>: **D**
Diameter: **0.50**

Command: **Copy**
Select objects: **L** *(for the last circle)*
Select objects: *(Strike Enter to continue)*
<Base point or displacement>/Multiple: **M**
Base point: **@** *(References the center of the 0.50 circle)*
Second point of displacement: **2.50,0.50**
Second point of displacement: **4.50,0.50**
Second point of displacement: **6.50,0.50**
Second point of displacement: **8.00,3.00**
Second point of displacement: **8.00,4.50**
Second point of displacement: **0.50,2.50**
Second point of displacement: **1.50,6.00**
Second point of displacement: **6.00,6.00**
Second point of displacement: *(Strike Enter to exit this command)*

Step #3

Form the slot by placing two circles using the Circle command followed by two lines drawn from the quadrants of the circles using the Osnap-Quadrant option.

Command: **Circle**
3P/2P/TTR/<Center point>: **4.00,3.00**
Diameter/<Radius>: **1.00**

Command: **Circle**
3P/2P/TTR/<Center point>: **4.00,4.50**
Diameter/<Radius>: **1.00**

Command: **Line**
From point: **Qua**
of *(Select the quadrant of the circle at "A")*
To point: **Qua**
of *(Select the quadrant of the circle at "B")*
To point: *(Strike Enter to exit this command)*

Command: **Line**
From point: **Qua**
of *(Select the quadrant of the circle at "C")*
To point: **Qua**
of *(Select the quadrant of the circle at "D")*
To point: *(Strike Enter to exit this command)*

Command: **Redraw**

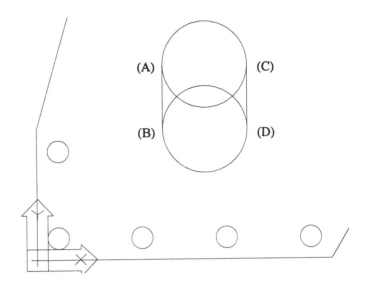

Step #4

Use the Trim command to trim away any unnecessary arcs to form the slot.

Command: **Trim**
Select cutting edges...
Select objects: *(Select both vertical lines at "A" and "B")*
Select objects: *(Strike Enter to continue)*
Select object to trim: *(Select the circle at "C")*
Select object to trim: *(Select the circle at "D")*
Select object to trim: *(Strike Enter to exit this command)*

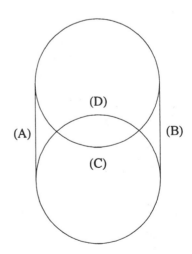

Step #5

In a few moments, the AME Solext command will be used to extrude all entities to the thickenss of the plate which is 0.75. This command, however, only operates on polylines and circles. Currently, all entities may be extruded except for the two arcs and lines representing the slot. Use the Pedit command to convert these entities into a single polyline.

Command: **Pedit**
Select polyline: *(Select the bottom arc "A")*
Entity selected is not a polyline
Do you want to turn it into one?<Y>: **Yes**
Open/Join/Width/Edit vertex/Fit curve/Spline curve/
    Decurve/Undo/eXit <X>: **J**
Select objects: *(Select the entities labeled "B", "C" and "D")*
Select objects: *(Strike Enter to continue)*
3 segments added to polyline.
Open/Join/Width/Edit vertex/Fit curve/Spline curve/
    Decurve/Undo/eXit <X>: **X**

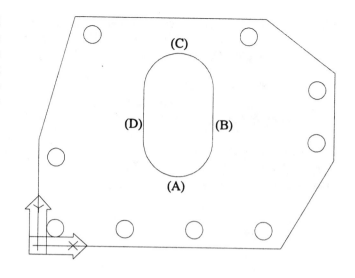

Step #6

From the pull-down menu area, load the AME by selecting the appropriate area. Next, under the Sol-Prim's menu, select Extrude, which really is the Solext command (Solid Extrude). The Advanced Modeling Extension may also be loaded by entering in the following command sequence.

Command: **(Xload "AME")**

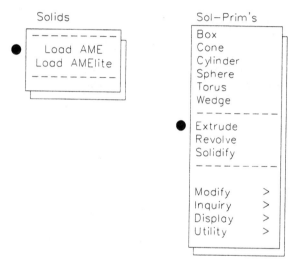

Step #7

Selecting Extrude prompts the user to select the objects to extrude. Select all polylines and circles that make up the plate. Enter a value of 0.75 as the height of the extrusion. This process may take a few minutes depending on the speed and amount of memory for your machine. Turn off the UCS icon.

Command: **Solext** *(Already selected from the pull-down menu)*
Initializing Advanced Modeling Extension.
Select polylines and circles for extrusion...
Select objects: **W**
First corner: **-1.00,-1.00**
Other corner: **9.00,7.00**
Select objects: *(Strike Enter to continue)*
Height of extrusion: **0.75**
Extrusion taper angle from Z <0>: *(Strike Enter to accept default)*

Command: **Ucsicon**
ON/OFF/All/Noorigin/ORigin <ON>: **OFF**

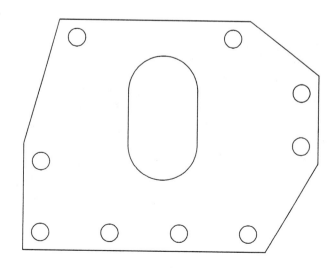

Step #8

Use the Vpoint command to view the plate in 3D using the prompts and settings below. Use the Solcham command to place a chamfer along the top edge of the plate. When prompted to select the base surface, select the entire top surface of the plate. This surface may not be selected the first time. Use the "Next" prompt until the top surface is selected. Select all individual edges of the top of the plate as the edges to be chamfered. Enter 0.12 as the two chamfer distances to place a 45-degree chamfer. Wait a moment while the AME updates the model to include the chamfer.

Command: **Vpoint**
Rotate/<View point> <0.00,0.00,1.00>: **1,-1,1**

Command: **Solcham** *(Or select from the Sol-Modify menu)*
Select base surface: *(Select the top surface at "A")*
<OK>/Next: **N** *(May be required if the top surface did not select)*
<OK>/Next: *(Strike Enter only if top surface is selected)*
Select edges to be chamfered (Press ENTER when done):
    (Select all individual edges of the top of the plate, edges "A" to "G", then strike Enter)
7 edges selected
Enter distance along first surface <0.00>: **0.12**
Enter distance along second surface <0.12>: *(Strike Enter)*
Phase I - Boundary evaluation begins.
Phase II - Tessellation computation begins.
Updating the Advanced Modeling Extension database.

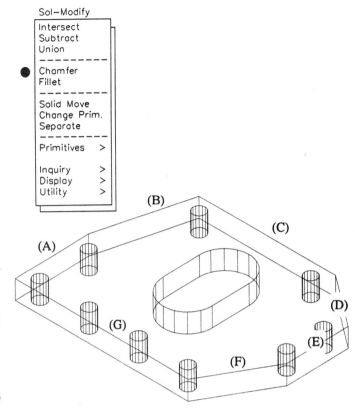

Step #9

With all entities extruded the distance 0.75, the plate is now a solid object complete with .12 x 45-degree chamfer. The holes and slot however are also solid. The Solsub command is used to subtract the holes and slot from the base of the plate. This resembles actually drilling holes and milling the slot to size. This command may be selected from the Sol-Modify pull-down menu area. As with all AME commands, this process may take some time.

Command: **Solsub**
Source objects...
Select objects: *(Select the base of the plate along any edge)*
Select objects: *(Strike Enter to continue)*
1 solid selected.
Objects to subtract from them...
Select objects: *(Carefully select every hole and slot)*
Select objects: *(Strike Enter to begin the subtraction process)*
10 solids selected.
Phase I - Boundary evaluation begins.
Phase II - Tessellation computation begins.
Updating Advanced Modeling Extension database.

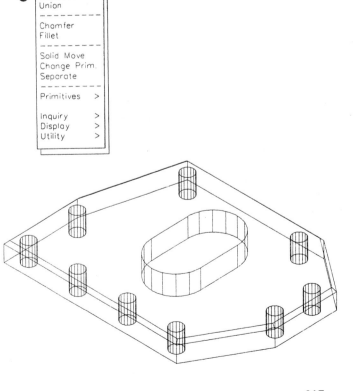

Step #10

The correct solid model of the plate is illustrated at the right. By default, the current display is that of a wireframe view. To surface the model, the Solmesh command is used. This command may be selected from the Sol-Display pull-down menu area. This command will take some time to process depending on the speed of your computer.

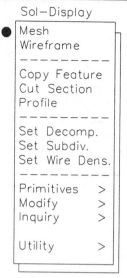

Command: **Solmesh**
Select solids to be meshed. . .
Select objects: *(Select an entity on the plate; all should highlight)*
Select objects: *(Strike Enter to continue with this command)*
1 solid selected.
Phase I of surface meshing.
Phase II of surface meshing.
Surface meshing of current solid is complete.
Creating block for mesh representation. . .
Done.

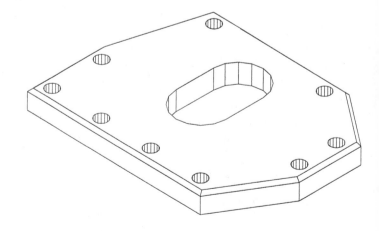

Step #11

View the completed object by issuing the Hide command to preform a hidden line removal. Since this command takes the most time to process, it may be a good idea to save the view as a slide which may be viewed much quicker at a later date.

Command: **Hide**
Regenerating drawing.
Removing hidden lines: **XXX**

Command: **Mslide**
Slide file <Bplate>: **Bplate-h**

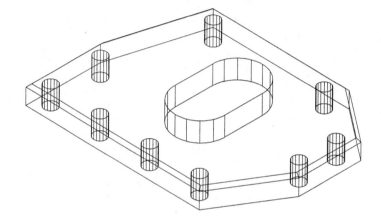

Step #12

To appreciate the automatic surfacing performed by the Solmesh command, set the system variable Splframe to a value of 1. This will turn on all of the 3D faces that went into the construction of the object. The faces will not appear until a screen regeneration is forced using the Regen command. This effect is illustrated at the right. To remove the display of the surfaces, set Splframe to a value of 0 and perform another screen regeneration.

Command: **Splframe**
New value for SPLFRAME <0>: **1**

Command: **Regen**
Regenerating drawing.

Command: **Splframe**
New value for SPLFRAME <1>: **0**

Command: **Regen**
Regenerating drawing.

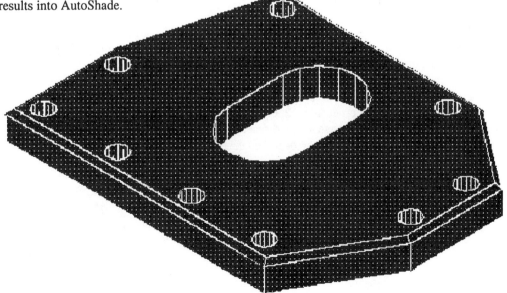

Step #13

Illustrated below is a display of the plate that has been shaded using the Shade command. This will produce a shaded image consisting of hidden line removal in addition to the shading being performed in the original color of the model. There are no areas of shadows using the Shade command. For best results, add lights and cameras, create a scene, and finally create a filmroll file of this or any surfaced object and import the results into AutoShade.

Command: **Shade**
Regenerating drawing.
Shading xx% done.

Command: **Mslide**
Slide file <Bplate>: **Bplate-s**

Tutorial Exercise #25
Guide.Dwg

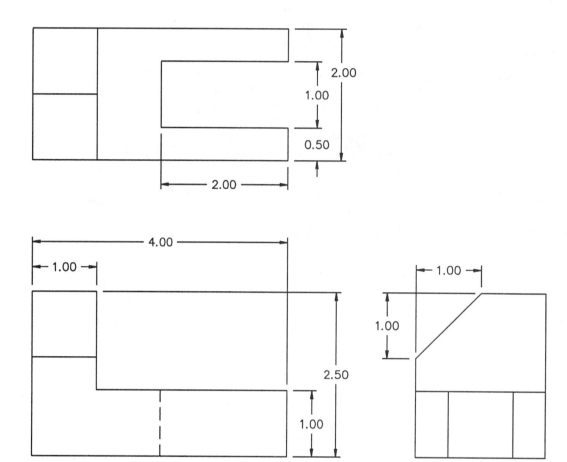

PURPOSE:
This tutorial exercise is designed to produce a 3D solid model of the Guide from the information supplied in the orthographic drawing above.

SYSTEM SETTINGS:
Begin a new drawing called "Guide." Use the Units command to change the number of decimal places past the zero from 4 to 2. Keep the remaining default unit values. Using the Limits command, keep 0,0 for the lower left corner and change the upper right corner from 12,9 to 10.50,8.00. Use the Grid command and change the grid spacing from 1.00 to 0.25 units. Do not turn the snap or ortho On.

LAYERS:
Special layers do not have to be created for this tutorial exercise.

SUGGESTED COMMANDS:
Begin this drawing by constructing solid primitives of all components of the Guide using the Solbox and Solwedge commands. Move the components into position and begin merging solids using Solunion. To form the rectangular hole, move the solid box into position and use the Solsub command to subtract the rectangle from the solid, thus forming the hole. Do the same procedure for the wedge. Use the Solmesh command to surface the solid. Perform a hidden line removal and view the solid.

DIMENSIONING:
This tutorial exercise does not require dimensioning.

PLOTTING:
This tutorial exercise may be plotted on "B"-size paper (11" x 17"). Use a plotting scale of 1=1 to produce a full size plot.

VERSION OF AUTOCAD:
This tutorial exercise must be completed using AutoCAD Release 11 with the Advanced Modeling Extension (AME).

Step #1

Begin this tutorial by loading the AME and constructing a solid box 4 units long by 2 units wide and 1 unit in height using the Solbox command. Begin this box at absolute coordinate 4.00,5.50. This slab will represent the base of the guide.

Command: **Solbox**
Corner of box: **4.00,5.50**
Cube/Length/<Other corner>: **L**
Length: **4.00**
Width: **2.00**
Height: **1.00**

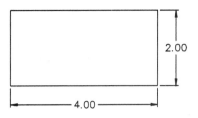

Step #2

Construct a solid box 1 unit long by 2 units wide and 1.5 units in height using the Solbox command. Begin this box at absolute coordinate 2.00,1.50. This slab will represent the vertical column of the guide.

Command: **Solbox**
Corner of box: **2.00,1.50**
Cube/Length/<Other corner>: **L**
Length: **1.00**
Width: **2.00**
Height: **1.50**

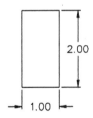

Step #3

Construct a solid box 2 units long by 1 unit wide and 1 unit in height using the Solbox command. Begin this box at absolute coordinate 5.50,1.50. This slab will represent the rectangular hole made into the slab that will be subtracted at a later time.

Command: **Solbox**
Corner of box: **5.50,1.50**
Cube/Length/<Other corner>: **L**
Length: **2.00**
Width: **1.00**
Height: **1.00**

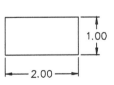

Step #4

Use the Solwedge command to draw a wedge 1.00 unit in length, 1.00 unit wide, and 1.00 unit in height. Begin this primitive at absolute coordinate 9.50,2.00. This wedge will be subtracted from the vertical column to form the inclined surface.

Command: **Solwedge**
First corner: **9.50,1.50**
Length/<Opposite corner>: **L**
Length: **1.00**
Width: **1.00**
Height: **1.00**

Step #5

Use the Vpoint command to view the four solid primitives in 3D. Use a new view point of 1,-1,0.75. Then use the Move command to move the vertical column at "A" to the top of the base at "B".

Command: **Vpoint**
Rotate/<View point> <0,0,0>: **1,-1,0.75**

Command: **Move**
Select objects: *(Select the solid box at "A")*
Select objects: *(Strike Enter to continue)*
Base point or displacement: **Endp**
of *(Select the endpoint of the solid at "A")*
Second point of displacement: **Endp**
of *(Select the endpoint of the base at "B")*

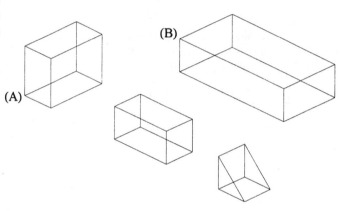

Step #6

Use the Solunion command to join the base and vertical column into one entity.

Command: **Solunion**
Select objects: *(Select the base at "A" and column at "B")*
Select objects: *(Strike Enter to perform the union)*
2 solids selected.

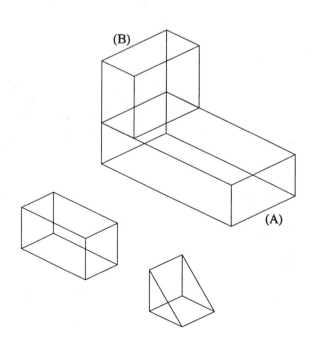

Step #7

Use the Move command to position the rectangle from its midpoint at "A" to the midpoint of the base at "B". In a moment, the small rectangle will be subtracted forming the rectangular hole in the base.

Command: **Move**
Select objects: *(Select the rectangle at "A")*
Select objects: *(Strike Enter to continue)*
Base point or displacement: **Mid**
of *(Select the midpoint of the rectangle at "A")*
Second point of displacement: **Mid**
of *(Select the midpoint of the base at "B")*

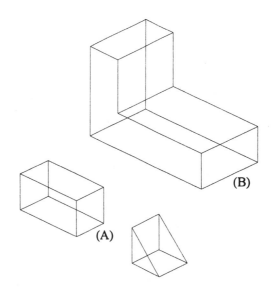

Step #8

Use the Solsub command to subtract the small rectangle from the base of the solid.

Command: **Solsub**
Source objects...
Select objects: *(Select the solid at "A")*
Select objects: *(Strike Enter to continue)*
1 solid selected.
Objects to subtract from them...
Select objects: *(Select the small rectangle at "B")*
Select objects: *(Strike Enter to perform the subtraction operation)*
1 solid selected.

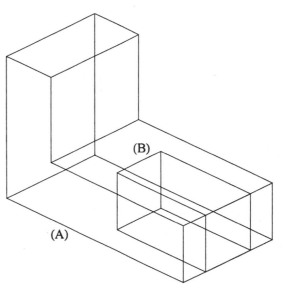

Step #9

Use the Rotate command to revolve the wedge at an angle of 90 degrees. This will begin preparing the wedge to be inserted onto the vertical column before subtracting.

Command: **Rotate**
Select objects: *(Select the wedge)*
Select objects: *(Strike Enter to continue)*
Base point: **Endp**
of *(Select the endpoint of the wedge at "A")*
<Rotation angle>/Reference: **90**

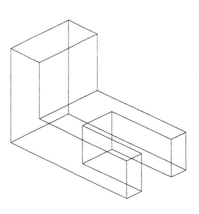

Step #10

Move the UCS icon to a new origin located on the wedge.

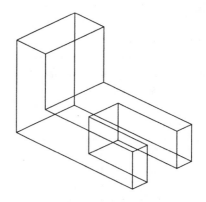

Command: **UCS**
Origin/ZAxis/3point/Entity/View/X/Y/Z/Prev/Restore/Save/
    Del/?/<World>: **O**
Origin point <0,0,0>: **Endp**
of *(Select the endpoint of the wedge at "A")*

Command: **Ucsicon**
ON/OFF/Noorigin/ORigin <ON>: **OR**

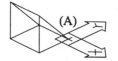

Step #11

The wedge needs to be rotated 90 degrees about the X axis
before being placed into position and subtracted from the
main solid. Use the Solmove command to accomplish this.
Once in the command, a new icon appears; it is the MCS or
Motion Coordinate System. This icon shows the orientation of the X, Y, and Z axes. For the motion description
type "RX-90" which will rotate the wedge 90 degrees in the
clockwise direction about the X axis.

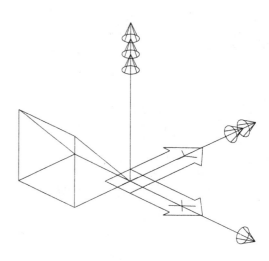

Command: **Solmove**
Select objects: *(Select the wedge)*
Select objects: *(Strike Enter to continue) Notice that a
    new icon appears*
<Motion description>/?: **RX-90**
<Motion description>/?: *(Strike Enter to exit this command)*

Step #12

Use the Move command to move the wedge from its endpoint at "A" to the endpoint of the vertical column at "B".

Command: **Move**
Select objects: *(Select the wedge)*
Select objects: *(Strike Enter to continue)*
Base point or displacement: **Endp**
of *(Select the endpoint of the wedge at "A")*
Second point of displacement: **Endp**
of *(Select the endpoint of the vertical column at "B")*

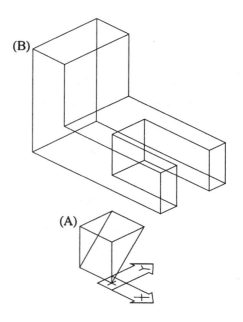

Step #13

Use the Solsub command to subtract the wedge from the main solid.

Command: **Solsub**
Source objects...
Select objects: *(Select the solid at "A")*
Select objects: *(Strike Enter to continue)*
1 solid selected.
Objects to subtract from them...
Select objects: *(Select the wedge at "B")*
Select objects: *(Strike Enter to perform the subtraction operation)*

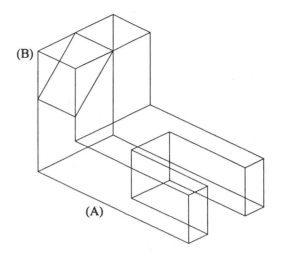

Step #14

An alternate method of creating the inclined surface is to use the Solchamfer command to chamfer the vertical column.

Command: **Solcham**
Select base surface: *(Select an edge along surface"A")*
<OK>/Next: *(If surface "A" highlights, Strike Enter to continue; If another surface highlights, Type "N" for next surface and step through the surfaces until surface "A" is highlighted)*
Select edges to be chamfered (Press Enter when done): *(Select the line at "B" and strike Enter)*
1 edge selected.
Enter distance along first surface <0.00>: **1.00**
Enter distance along second surface <0.00>: **1.00**

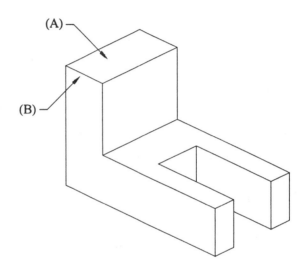

Step #15

The object is still currently displayed in wireframe mode. If a hidden line removal is to be performed, surfaces need to be added to the object. This can automatically be done using the Solmesh command.

Command: **Solmesh**
Select solids to be meshed. . .
Select objects: *(Select the solid model at "A")*
Select objects: *(Strike Enter to continue)*
1 solid selected.

The results of the Solmesh command may not be immediately apparent until a hidden line removal is performed using the Hide or Shade commands.

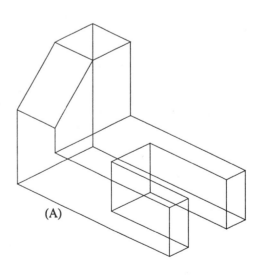

Step #16

Using the Hide command performs a hidden line removal on all surfaces of the object. The results are illustrated at the right.

Command: **Hide**
Regenerating drawing.
Removing hidden lines:

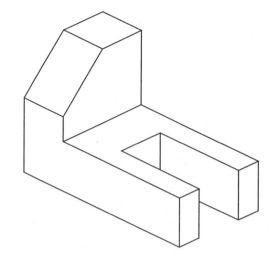

Step #17

To view the surfaces of the model, the system variable Splframe may expose all surfaces created by the Solmesh command. By default, this variable is Off. If set On, all surfaces needed to surface the model are displayed. Before showing surfaces, a screen regeneration must be made to update the drawing data base. The same is true when turning the variable Off.

Command: **Splframe**
New value for SPLFRAME <0>: **1**

Command: **Regen**

Command: **Splframe**
New value for SPLFRAME <1>: **0**

Command: **Regen**

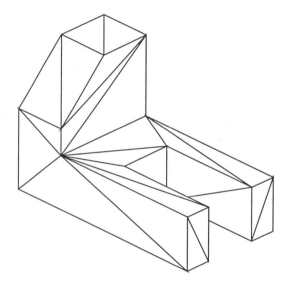

Step #18

Use the Shade command to view the solid model as a shaded image. Remember, the screen will momentarily go blank as the system calculates the hidden line removal along with the shading in the current color of the model. Use the Mslide command to document the image for later use.

Command: **Shade**
Regenerating drawing.
Shading xx% done.

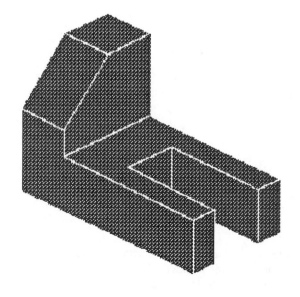

Tutorial Exercise #26
Collar.Dwg

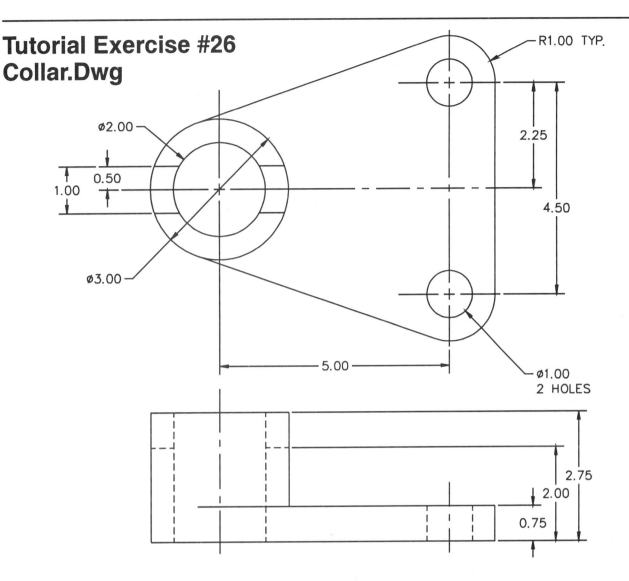

Ø2.00

0.50

1.00

Ø3.00

R1.00 TYP.

2.25

4.50

Ø1.00
2 HOLES

5.00

2.75

2.00

0.75

PURPOSE:
This tutorial is designed to construct a solid model of the Collar using the dimensions illustrated above.

SYSTEM SETTINGS:
Begin a new drawing called "Collar." Use the Units command to change the number of decimal places past the zero from 4 to 2. Keep the remaining default unit values. Using the Limits command, change the lower left corner from 0,0 to -3.00,-4.00 and change the upper right corner from 12,9 to 7.00,4.00. Use the Grid command and change the grid spacing from 1.00 to 0.50 units. Do not turn the snap or ortho On.

LAYERS:
Special layers do not have to be created for this tutorial exercise although an object layer may be created using yellow lines:

Name-Color-Linetyp
Object - Yellow - Continuous

SUGGESTED COMMANDS:
Begin this tutorial by laying out the Collar in plan view and drawing the basic shape outlined in the top view. Convert the entities into a polyline and extrude the entities to form a solid. Draw a cylinder and combine this entity with the base. Add another cylinder and then subtract it to form the large hole through the model. Add two small cylinders and subtract them from the base to form the smaller holes. Construct a solid box, use the Solmove command to move the box into position, and subtract it to form the cut across the large cylinder.

DIMENSIONING:
This tutorial does not require any special dimensioning.

PLOTTING:
This tutorial exercise may be plotted on "B"-size paper (11" x 17"). Use a plotting scale of 1=1 to produce a full size plot.

VERSION OF AUTOCAD:
This tutorial exercise must be completed using AutoCAD Release 11 with the Advanced Modeling Extension (AME).

Step #1

Begin the collar by drawing the three circles illustrated at the right using the Circle command. Place the center of the circle at "A" at 0,0.

Command: **Circle**
3P/2P/TTR/<Center point>: **0,0**
Diameter/<Radius>: **D**
Diameter: **3.00**

Command: **Circle**
3P/2P/TTR/<Center point>: **5.00,2.25**
Diameter/<Radius>: **1.00**

Command: **Circle**
3P/2P/TTR/<Center point>: **5.00,-2.25**
Diameter/<Radius>: **1.00**

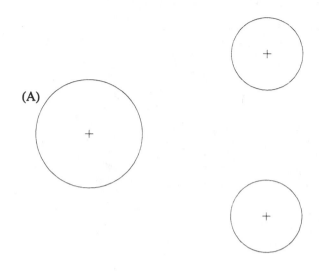

(A)

Step #2

Draw lines tangent to the three arcs using the Line command and the Osnap-Tangent option.

Command: **Line**
From point: **Tan**
to *(Select the circle at "A")*
To point: **Tan**
to *(Select the circle at "B")*
To point: *(Strike the Enter key to exit the command)*

Repeat the above procedure to draw lines from "C" to "D" and "E" to "F".

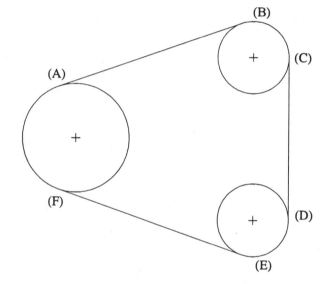

Step #3

Use the Trim command, select the three lines as cutting edges, and trim the circles.

Command: **Trim**
Select cutting edges...
Select objects: *(Select the three dashed lines at the right)*
Select objects: *(Strike Enter to continue)*
Select object to trim: *(Select the circle at "A")*
Select object to trim: *(Select the circle at "B")*
Select object to trim: *(Select the circle at "C")*
Select object to trim: *(Strike the Enter key to exit the command)*

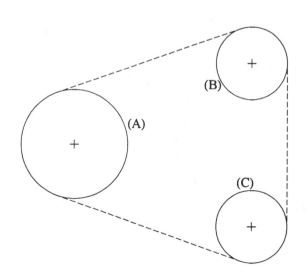

Step #4

If center markers were placed, use the Erase command to delete them now. The collar should appear similar to the illustration at the right.

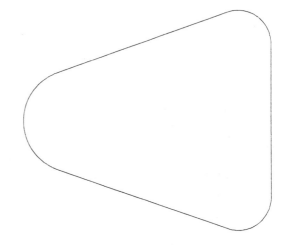

Step #5

Prepare to construct the collar by viewing the object in 3D using the Vpoint command and the coordinates 1,-1,1.

Command: **Vpoint**
Rotate/<View point><0.00,0.00,0.00>: **1,-1,1**

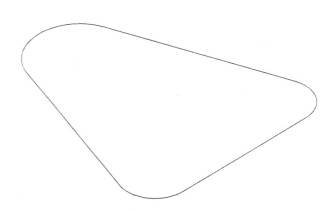

Step #6

Convert all entities into a polyline using the Join option of the Pedit command.

Command: **Pedit**
Select polyline: *(Select the arc at "A")*
Entity selected is not a polyline.
Do you want to turn it into one?<Y>: **Y**
Close/Join/Width/Edit Vertex/Fit curve/Spline curve/
　　　Decurve/Undo/eXit<X>: **J**
Select objects: **Window**
First corner: **3.00,-6.00**
Other corner: **1.50,8.00**
Select objects: *(Strike Enter to continue)*
5 segments added to polyline.
Close/Join/Width/Edit Vertex/Fit curve/Spline curve/
　　　Decurve/Undo/eXit<X>: **X**

(A)

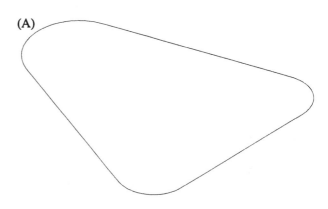

Step #7

Begin the conversion from 2D to solids by loading AME. This will take a few seconds depending on the speed of your machine. Next, use the Solext command to extrude the base to a thickness of 0.75 units.

Command: **Solext**
Initializing Advanced Modeling Extension.
Select polylines and circles for extrusion.
Select objects: *(Select the polyline)*
Select objects: *(Strike Enter to continue)*
Height of extrusion: **0.75**
Extrusion taper angle from Z <0>: **0**

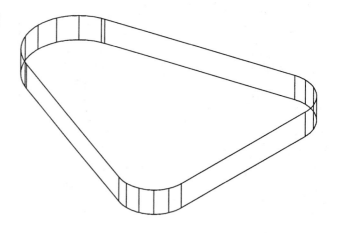

Step #8

Create a cylinder using the Solcyl command. Begin the center point of the cylinder at 0,0 and a diameter of 3.00 with a height of 2.75.

Command: **Solcyl**
Elliptical/<Center point>: **0,0**
Diameter/<Radius>: **D**
Diameter: **3.00**
Height of cylinder: **2.75**
Phase I - Boundary evaluation begins.
Phase II - Tessellation computation begins.
Updating the AME data base.

Notice that an initialization occurs each time an AME command is selected. The different phases deal with computation and construction calculations to create or modify the solid.

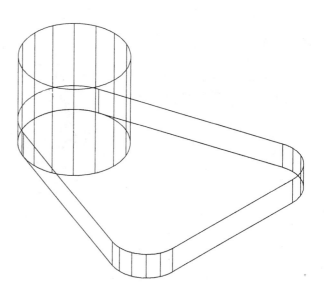

Step #9

Merge the cylinder just created with the extruded base to combine both entities into one using the Solunion command.

Command: **Solunion**
Select objects: *(Select the extruded base)*
Select objects: *(Select the cylinder)*
Select objects: *(Strike Enter to continue)*
2 solids selected.
Updating solid...

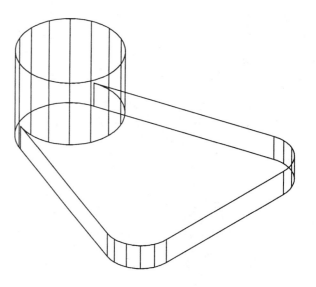

Step #10

Use the Solcyl command to create a 2.00-unit diameter cylinder representing a through hole. The height of the cylinder is 2.75 with the center point located at 0,0.

Command: **Solcyl**
Elliptical/<Center point>: **0,0**
Diameter/<Radius>: **D**
Diameter: **2.00**
Height of cylinder: **2.75**

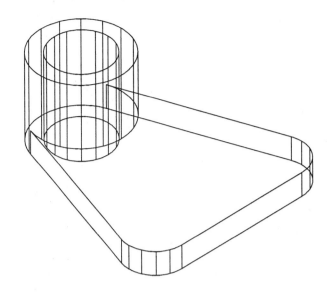

Step #11

To form a hole through the outer cylinder, the Solsub command will be used. Select the base as the source object and the inner cylinder as the object to subtract. Wait a moment while the computer performs its calculations to remove the cylinder from the model creating a hole.

Command: **Solsub**
Source objects...
Select objects: *(Select the base)*
Select objects: *(Strike Enter to continue)*
1 solid selected.
Objects to subtract from them...
Select objects: *(Select the 2.00-diameter cylinder just drawn)*
Select objects: *(Strike Enter to continue)*
1 solid selected.

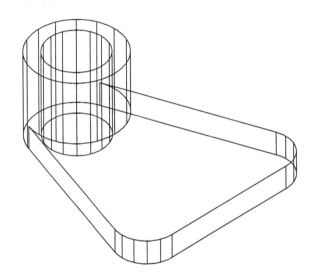

Step #12

Begin placing the two small drill holes in the base by using the Solcyl command. The absolute coordinate value of 5.00,2.25 will be used as the center of the cylinder.

Command: **Solcyl**
Elliptical/<Center point>: **5.00,2.25**
Diameter/<Radius>: **D**
Diameter: **1.00**
Height of cylinder: **0.75**

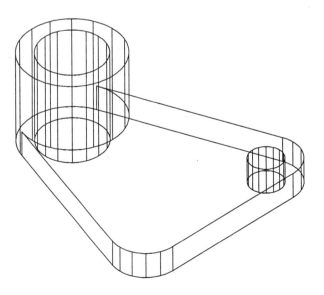

Step #13

Rather than create another cylinder (with center point at 5.00,-2.25) use the Copy command to merge existing AutoCAD commands into the creation of this model of the collar. Use the Osnap-Center option to assist the Copy command.

Command: **Copy**
Select objects: *(Select the 1.00-diameter cylinder just drawn)*
Select objects: *(Strike Enter to continue)*
<Base point or displacement>/Multiple: **Cen**
of *(Select the arc at "A")*
Second point of displacement: **Cen**
of *(Select the arc at "B")*

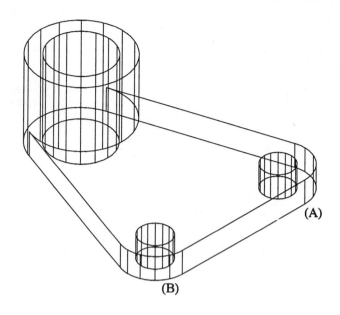

(A)

(B)

Step #14

Subtract both 1.00-diameter cylinders from the base of the model using the Solsub command.

Command: **Solsub**
Source objects...
Select objects: *(Select the base)*
Select objects: *(Strike Enter to continue)*
1 solid selected.
Objects to subtract from them...
Select objects: *(Select the 1.00-diameter cylinder "A")*
Select objects: *(Select the 1.00-diameter cylinder "B")*
Select objects: *(Strike Enter to continue)*
Updating solid...

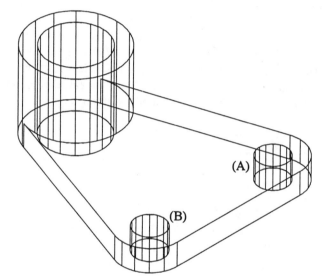

(A)

(B)

Step #15

Begin constructing the rectangular slot which will pass through the two cylinders. Use the Solbox command to accomplish this. Construct the box locating its bottom using the current elevation of 0. The size of the box is 4 units long by 1 unit wide and 0.75 units deep. Use the origin point 0,0 as reference for drawing the box.

Command: **Solbox**
Corner of box: **-2.00,0.50**
Cube/Length/<Other corner>: **2.00,-0.50**
Height: **0.75**

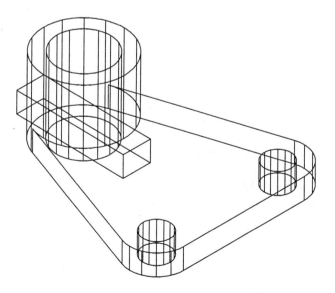

Step #16

Geometric primitives are easily constructed in any location and moved into correct positions using the Solmove command. This command affects only solid model entities. When issuing the Solmove command, the familiar prompt "Select objects" appears prompting the operator to select the solid object to move. Once an object has been selected, the MCS appears. This icon alerts the operator of the X, Y, and Z axes locations in relation to the object selected. The X axis is symbolized with one cone at its axis endpoint; the Y axis has two cones; the Z axis has three cones. Along with directions, the operator has numerous key-ins to perform rotations and movement (called translations) of the solid object selected. Follow the next step to move the solid box up along the Z axis the distance of 2.00 units. When the command is exited, the MCS icon will disappear.

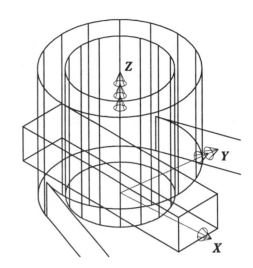

Step #17

Use the Solmove command to position the box in its correction position. The icon that appears on the screen is used to guide you in the correct direction for the move. In addition, the move will occur by entering a value in the specified direction prefaced by the letter **T** which stands for translate or move.

Command: **Solmove**
Select objects: *(Select the box)*
Select objects: *(Strike Enter to continue)*
1 solid selected.
<Motion description>/?: **TZ2.00**
<Motion description>/?: *(Strike the Enter key to exit)*

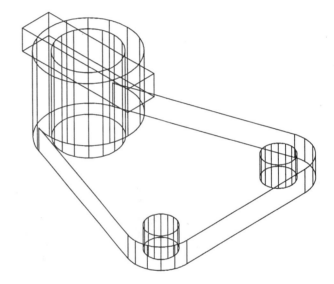

Step #18

Use the Solsub command to subtract the rectangular box from the model. Your model should appear similar to the illustration at the right.

Command: **Solsub**
Source objects...
Select objects: *(Select the base)*
Select objects: *(Strike Enter to continue)*
1 solid selected.
Objects to subtract from them...
Select objects: *(Select the rectangular box)*
Select objects: *(Strike Enter to continue)*

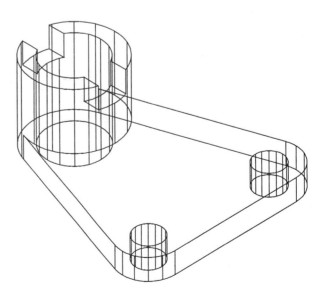

Step #19

Although a solid image of the collar has been created, it has been produced in wireframe mode. If the Hide command were to be used, the results would not have any hidden surfaces removed. The wireframe needs to be converted to a surface mesh using the Solmesh command. The Hide command will produce an image similar to the illustration at the right.

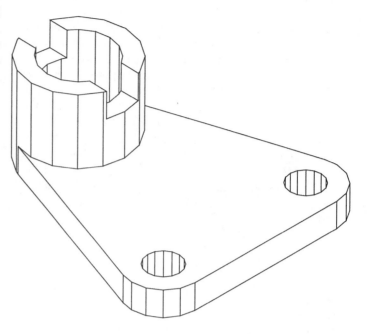

Command: **Solmesh**
Select solids to be meshed. . .
Select objects: *(Strike the Enter key to continue)*
1 solid selected.

Command: **Hide**
Regenerating drawing.
Removing hidden lines:

Step #20

Before continuing on, let's view the model with all surfaces being visible. Set the Splframe system variable from 0 to a value of 1 or On. Then use the Regen command to force a regeneration of the display screen. The results may appear similar to the illustration at the right.

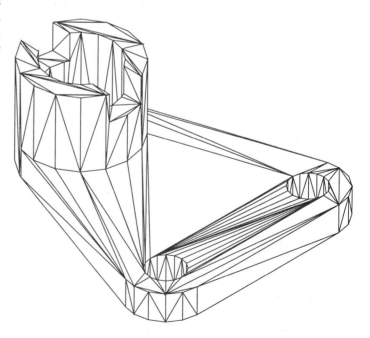

Command: **Splframe**
New value for SPLFRAME <0>: **1**

Command: **Regen**
Regenerating drawing.

To return the model back to hidden surfaces, reset the Splframe system variable back to 0 and force a screen regeneration using the Regen command.

Command: **Splframe**
New value for SPLFRAME <1>: **0**

Command: **Regen**
Regenerating drawing.

Step #21

For a model that has been surfaced using the Solmesh command, use the Shade command to perform a colorization of the model in addition to removing all hidden lines. The shading will occur in the current color of the solid model.

Command: **Shade**
Regenerating Drawing.
Shading xx% done.

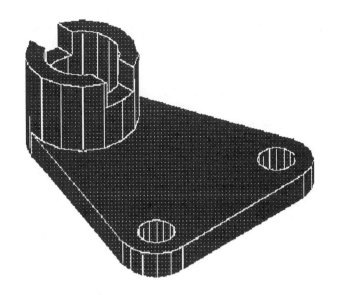

To convert the mesh generated solid back to a wireframe model, used the Solwire command.

Command: **Solwire**
Select solids to be wired...
Select objects: *(Select the base of the solid model)*
Select objects: *(Strike the Enter key to continue)*
1 solid selected.

Step #22

Important information may be extracted from the solid model to be used for design and analysis purposes. This information in the form of calculations is illustrated at the right and is obtained then using the Solmassp command. The following properties are calculated by this command:

    Mass
    Volume
    Bounding Box
    Centroid
    Moments of Inertia
    Products of Inertia
    Radii of Gyration
    Principle Moments about Centroid

All of the above values are calculated based on the current material type.

Command: **Solmassp**
Select objects: *(Select the base of the collar)*
Select objects: *(Strike Enter to continue)*
1 solid selected.
Calculating mass properties.

Ray projection along X axis, level of subdivision: 3.
Mass: 223.2 gm
Volume: 28.39 cu cm (Err: 3.397)

Bounding box:	X: -1.501 — 6.001 cm
	Y: -0.0009944 — 2.75 cm
	Z: -3.251 — 3.251 cm

Centroid:	X: 2.36 cm (Err: 0.4343)
	Y: 0.7067 cm (Err: 0.0792)
	Z: 0.07384 cm (Err: 0.2559)

Moments of inertia:	X: 612.2 gm sq cm (Err: 76.16)
	Y: 2667 gm sq cm (Err: 442.6)
	Z: 2444 gm sq cm (Err: 383.2)
Products of inertia:	XY: 223.3 gm sq cm (Err: 52.52)
	YZ: -4.466 gm sq cm (Err: 35.63)
	ZX: 82.77 gm sq cm (Err: 188.2)

Radii of gyration:	X: 1.656 cm
	Y: 3.457 cm
	Z: 3.309 cm

Principal moments (gm sq cm) and X-Y-Z directions about centroid:

        I: 472.7 along [0.985 -0.1555 0.07419]
        J: 1447 along [0.154 0.9877 0.02561]
        K: 1092 along [-0.07726 -0.0138 0.9969]

Write to file <N>?

Step #23 — Extracting a Feature

Features of solid models may be extracted for design check-ing using the Solfeat command. This command does not produce a primary view of the entire object; rather, it creates the profile of a single feature. The command high-lights the feature selected in the model and the user either accepts the feature or continues on using <Next> until the desired feature is selected for extraction.

Command: **Solfeat**
Edge/<Face>: *(Strike the Enter key to accept "Face")*
Select a face: *(Select the bottom of the base at "A")*
<OK>/Next: *(If the base is not highlighted, select "Next")*
<OK>/Next: *(When the base highlights, strike the Enter key)*

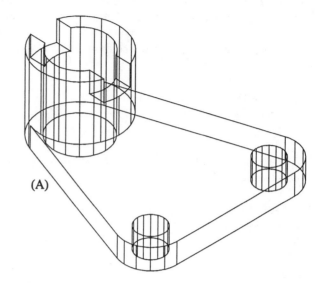

(A)

Step #24

Since the extracted feature lies directly on top of the model, use the Move command to change its position. When finished use Zoom-All to view the model and feature.

Command: **Move**
Select objects: **L**
Select objects: *(Strike Enter to continue)*
Base point or displacement: **0,10,0**
Second point of displacement: *(Strike Enter to continue)*

Command: **Zoom**
All/Center/Dynamic/Extents/Left/Previous/Vmax/Window/
    <Scale(X/XP)>: **All**
Regenerating drawing.

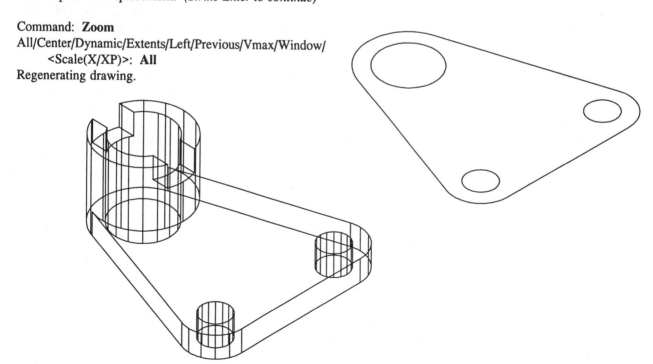

Step #25 — Extracting a Section

As with a feature, it is also possible to extract a section from the solid model using the Solsect command and have that section cross-hatched automatically using the Solhpat command. The section is performed based on the position of the current UCS. The UCS acts as a cutting edge where the model is sliced. The result is a section placed at the current UCS position. The Move command is used here to move the section to a more convenient location. Begin by positioning the UCS icon illustrated at the right.

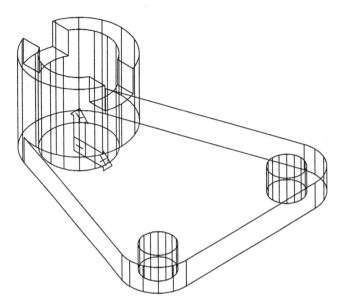

Command: **Ucsicon**
ON/OFF/All/Noorgin/ORgin <ON>: **ORigin**

Command: **Ucs**
Origin/ZAxis/3point/Entity/View/X/Y/Z/Prev/Restore/Save/
    Del/?/<World>: **X**
Rotation angle about X axis <0>: **90**

Step #26

Set the Solhpat command to Ansi31, use the Solsect command to extract the section, and move the section in the positive Z direction illustrated at the right. Perform a Zoom-Extents when completed.

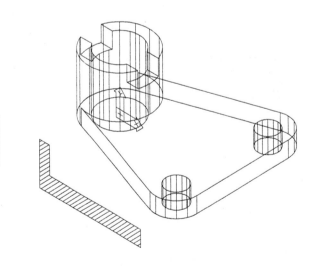

Command: **Solhpat**
Hatch pattern <None>: **Ansi31**

Command: **Solsect**
Select objects: *(Select the solid model of the collar)*
Select objects: *(Strike the Enter key to continue)*
Updating solid...
Done.

Command: **Move**
Select objects: *(Select the hatch pattern and profile)*
Select objects: *(Strike Enter to continue)*
Base point or displacement: **0,0,5**
Second point of displacement: *(Strike the Enter key to
    continue)*

Command: **Zoom**
All/Center/Dynamic/Extents/Left/Previous/Vmax/Window/
    <Scale(X/XP)>: **All**
Regenerating drawing.

Step #27 — Extracting a Profile

One big advantage of creating a solid model of the Collar is the ability to extract profiles based on the current viewing angle. During this creation process, two new layers will automatically be created; one consisting of hidden lines and the other consisting of object lines used to describe the Collar. Continue with the next series of steps used to create a solid profile using the Solprof command. Reposition the User Coordinate System by rotating the icon along the X axis in the −90 degree direction. Erase all entities created with the Solfeat and Solsect commands.

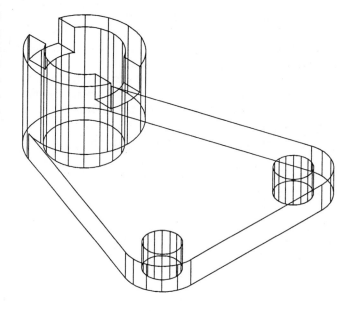

Command: **Ucs**
Origin/ZAxis/3point/Entity/View/X/Y/Z/Prev/Restore/Save/
    Del/?/<World>: **X**
Rotation angle about X axis <0>: **−90**

Step #28

The only way to successfully use the Solprof command is to first turn the Tilemode system variable off; in other words, set this variable to a value of "0". The screen will immediately go blank and will display the paper space icon in the lower left corner of the display screen. Use the following commands to open one viewport in paper space using the Mview command.

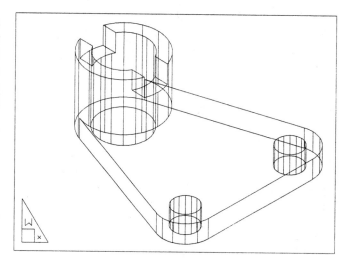

Command: **Tilemode**
New value for TILEMODE <1>: **0**

Command **Mview**
ON/OFF/Hideplot/Fit/2/3/4/Restore/<First point>: **Restore**
?/Name of window configuration to insert<*ACTIVE*>:
    (Enter)
Fit/<First point>: **Fit**

Step #29

Use the Linetype command to load the hidden linetype. Switch back to Model space using the Mspace command and execute the Solprof command from the keyboard, pull down menu, screen menu, or digitizing tablet menu. Select the object at the right and answer "Yes" to the prompt to display hidden lines on a separate layer.

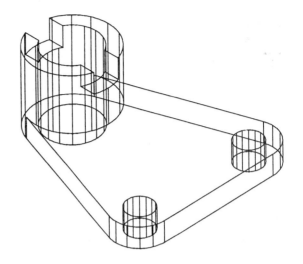

Command: **Linetype**
?/Create/Load/Set: **Load**
Linetype(s) to load: **Hidden**
File to search <acad>: *(Strike Enter to continue)*
Linetype HIDDEN loaded.
?/Create/Load/Set: *(Strike Enter to exit this command)*

Command: **MS** *(for Model Space)*

Command: **Solprof**
Select objects: *(Select the solid model at the right)*
Select objects: *(Strike Enter to continue)*
Display hidden profile lines on a separate layer? <N>: **Yes**
1 solid selected.

Step #30

The following layers are automatically created as a result of using Solprof: *(The exact names of the hidden and visible vary for every Solprof session.)*

      0-PH-2 (Holds hidden line information)
      0-PV-2 (Holds visible profile line information)

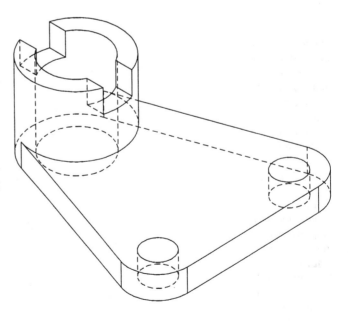

Use the layer command, turn off the layer the object was originally drawn in to show the solid model appearing similar to the illustration at the right.

Command: **Layer**
?/Make/Set/New/ON/OFF/Color/Ltype/Freeze/Thaw: **Off**
Layer name(s) to turn off: **0**
Really want layer 0 (the CURRENT layer off?)<N>: **Yes**
?/Make/Set/New/ON/OFF/Color/Ltype/Freeze/Thaw:
      (Strike Enter to exit this command)

Problems for Unit 10

Directions for Problems 10–1 through 10–25
Use Advanced Modeling Extension commands to construct a solid model of each object.

Problem 10–1

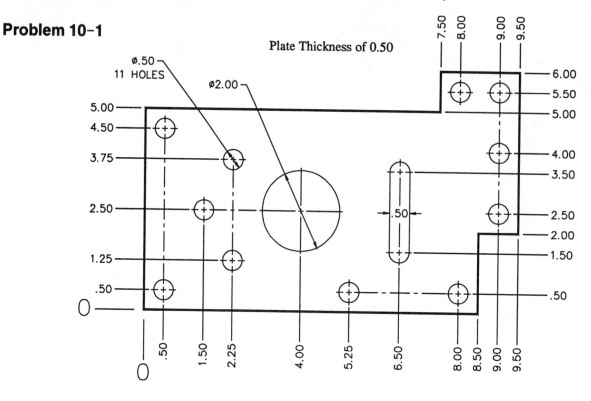

Plate Thickness of 0.50

Ø.50
11 HOLES

Ø2.00

Problem 10–2

Plate Thickness of 0.75

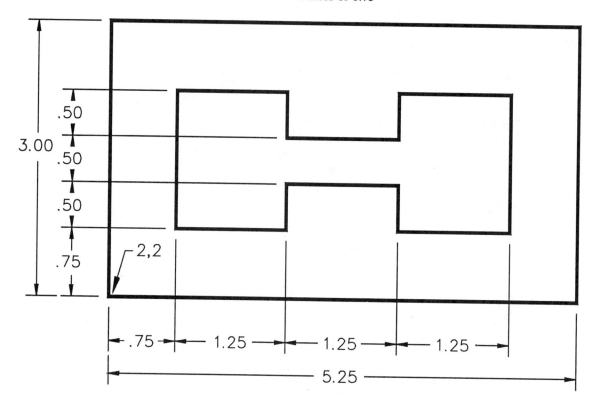

Problem 10-3

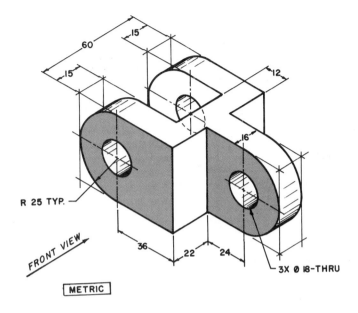

R 25 TYP.

FRONT VIEW

METRIC

60

15

15

12

16

36

22

24

3X Ø 18-THRU

Problem 10-4

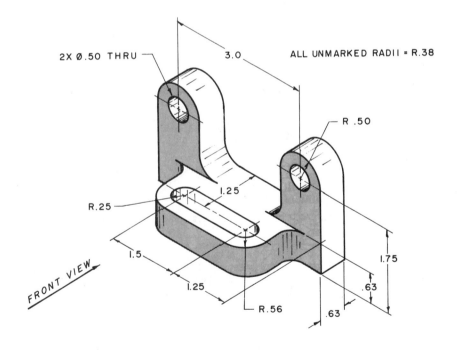

2X Ø.50 THRU

3.0

ALL UNMARKED RADII = R.38

R .50

R.25

1.25

1.5

1.25

R.56

.63

1.75

.63

FRONT VIEW

Problem 10-5

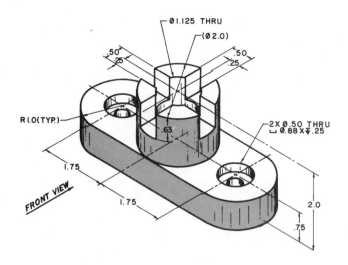

Ø1.125 THRU
(Ø2.0)
.50
.25
.50
.25
R 1.0 (TYP.)
.63
2X Ø.50 THRU
⌴ Ø.88 X ↧ .25
1.75
1.75
2.0
.75
FRONT VIEW

Problem 10-6

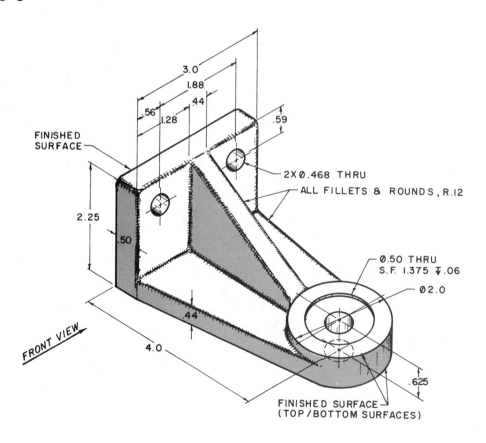

3.0
1.88
.44
.56
1.28
.59
FINISHED
SURFACE
2X Ø.468 THRU
ALL FILLETS & ROUNDS, R.12
2.25
.50
Ø.50 THRU
S.F. 1.375 ↧ .06
Ø2.0
.44
4.0
.625
FINISHED SURFACE
(TOP/BOTTOM SURFACES)
FRONT VIEW

Problem 10-7

Problem 10-8

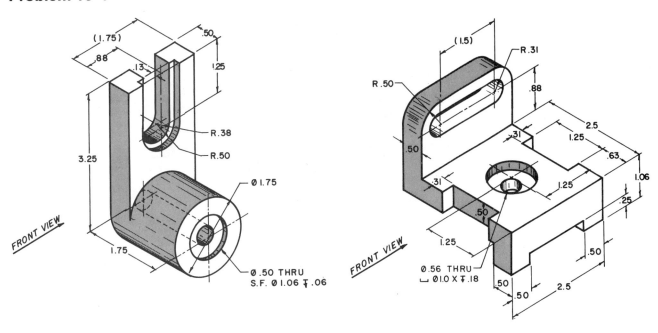

Problem 10-9

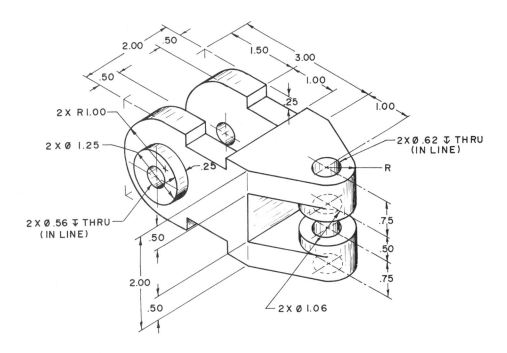

Problem 10-10

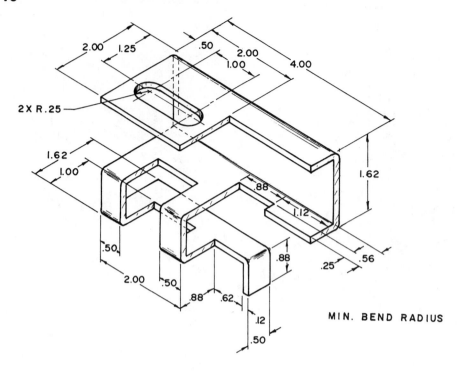

MIN. BEND RADIUS

Problem 10-11

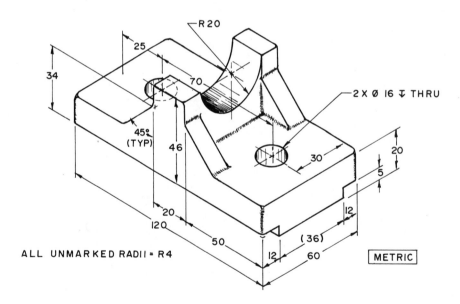

ALL UNMARKED RADII = R4

METRIC

Problem 10-12

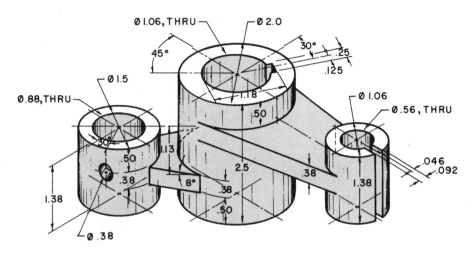

Problem 10-13

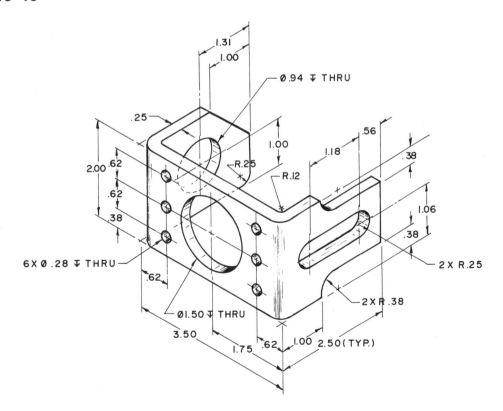

Problem 10-14

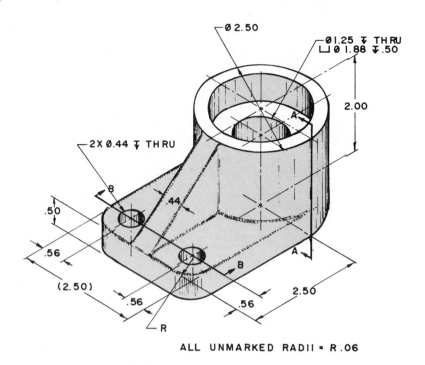

Ø 2.50

Ø1.25 ⊤ THRU
⊔ Ø 1.88 ⊽.50

2.00

2X Ø.44 ⊤ THRU

A

.50

.56

.44

B

.56

(2.50)

B

A

.56

2.50

R

ALL UNMARKED RADII ▪ R.06

Problem 10-15

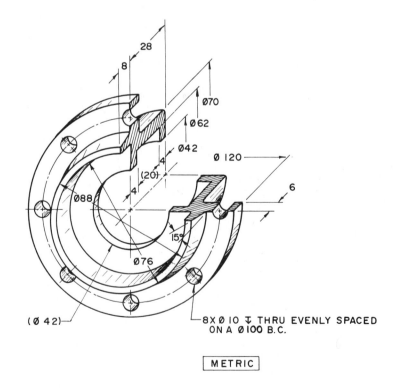

28

8

Ø70

Ø62

Ø42

Ø 120

6

(20)

4

4

Ø88

Ø76

15°

(Ø 42)

8X Ø 10 ⊤ THRU EVENLY SPACED
ON A Ø100 B.C.

METRIC

Problem 10-16

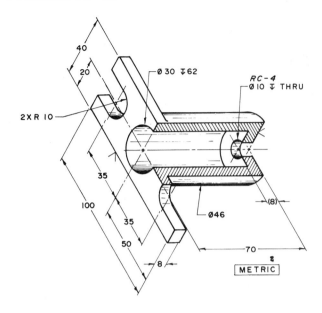

Problem 10-18

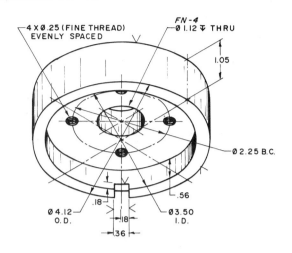

Problem 10-19

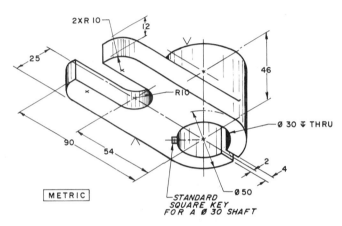

Problem 10-17

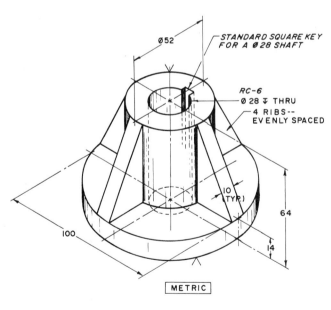

Problem 10-20

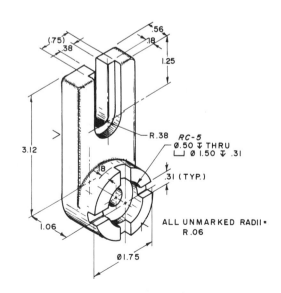

Problem 10-21

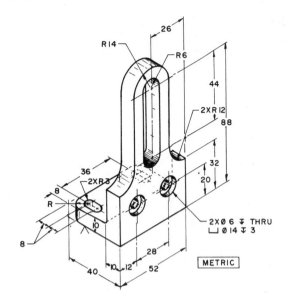

R14
R6
26
44
2X R12
88
36
2X R3
32
20
8
R
10
8
2X Ø 6 ↧ THRU
⊔ Ø 14 ↧ 3
28
40   10 12   52
METRIC

Problem 10-23

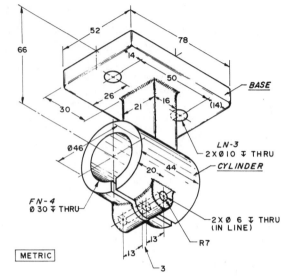

52
66
78
14
50
26
30   21   16   (14)
BASE
Ø46
20   44
LN-3
2X Ø 10 ↧ THRU
CYLINDER
FN-4
Ø 30 ↧ THRU
2X Ø 6 ↧ THRU
(IN LINE)
13   13   R7
3
METRIC

<u>NOTE</u>: CYLINDER IS CENTERED ON THE BASE

Problem 10-22

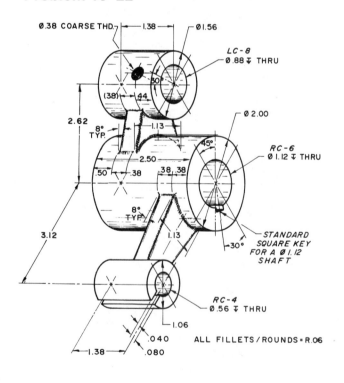

Ø 0.38 COARSE THD.
1.38
Ø 1.56
LC-8
Ø 0.88 ↧ THRU
(38)   44
30°
2.62
8°
TYP.
1.13
Ø 2.00
2.50
45°
RC-6
Ø 1.12 ↧ THRU
.50   .38   .38 .38
3.12
8°
TYP.
1.13
-30°
STANDARD
SQUARE KEY
FOR A Ø 1.12
SHAFT
RC-4
Ø 0.56 ↧ THRU
1.06
.040
1.38   .080
ALL FILLETS / ROUNDS = R.06

Problem 10-24

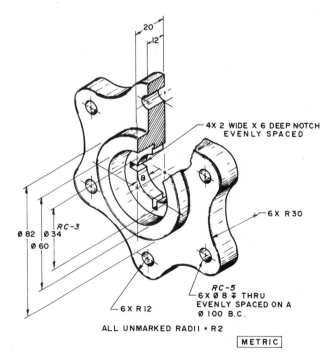

20
12
4X 2 WIDE X 6 DEEP NOTCH
EVENLY SPACED
8
6X R30
Ø 82   Ø 34
Ø 60
RC-3
RC-5
6X Ø 8 ↧ THRU
EVENLY SPACED ON A
Ø 100 B.C.
6X R12
ALL UNMARKED RADII = R2
METRIC

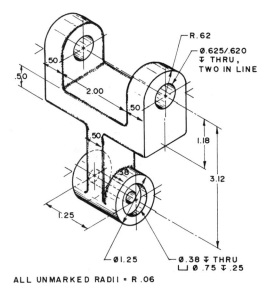

R.62

Ø.625/.620
⊤ THRU,
TWO IN LINE

.50

.50

2.00

.50

.50

1.18

3.12

.38

1.25

Ø1.25

Ø.38 ⊤ THRU
⌴ Ø.75 ⊤.25

ALL UNMARKED RADII = R.06

Engineering Graphics Tutorial/Problem Disk Boot-up Information

The enclosed diskette contains the following subdirectories and files:

UNIT_4
   DIMEX.DWG
   TBLK-ISO.DWG
   BAS-PLAT.DWG

UNIT_5
   DFLANGE.DWG
   COUPLER.DWG

UNIT_6
   BRACKET.DWG

UNIT_9
   INCLINE.DWG
   9-1.DWG through 9-20.DWG

Also included on the diskette are four batch files that will copy all drawing files into the current hard disk subdirectory:

4.BAT
5.BAT
6.BAT
9.BAT

Units 4, 5, 6, and 9 require the use of files already created. All files have been created using AutoCAD Release 10, with the exception of TBLK-ISO.DWG and BAS-PLAT.DWG. These two files require the use of AutoCAD Release 11. To load these files into your computer, follow these steps:

1. Change subdirectories to the area where AutoCAD is found. For example, if AutoCAD is located on the "C" drive and in a subdirectory called "\ACAD", enter the following at the DOS prompt:

C:\> CD\ACAD

This will change subdirectories from the root to the AutoCAD subdirectory.

2. Enter the appropriate batch file to load the tutorial drawings, depending on the unit desired. For example, if you want to load all the drawings from the UNIT_9 subdirectory, enter the following at the DOS prompt:

C:\ACAD> A:9

This will run a batch file called 9.BAT from the "A" drive and automatically copy all drawing files into the AutoCAD subdirectory. Follow the same procedure for copying the drawings required to perform the tutorial exercises found in UNIT_4, UNIT_5, and UNIT_6.

Index